Polycystic Kidney Disease

Methods in Signal Transduction

Series Editors: Joseph Eichberg Jr. and Michael X. Zhu

The overall theme of this series continues to be the presentation of the wealth of up-to-date research methods applied to the many facets of signal transduction. Each volume is assembled by one or more editors who are preeminent in their specialty. In turn, the guiding principle for editors is to recruit chapter authors who will describe procedures and protocols with which they are intimately familiar in a reader-friendly format. The intent is to assure that each volume will be of maximum practical value to a broad audience, including students and researchers just entering an area, as well as seasoned investigators.

Lipid Second Messengers
Suzanne G. Laychock and Ronald P. Rubin

G Proteins: Techniques of Analysis
David R. Manning

Signaling through Cell Adhesion Molecules
Jun-Lin Guan

G Protein-Coupled Receptors
Tatsuya Haga and Gabriel Berstein

G Protein-Coupled Receptors: Structure, Function, and Ligand Screening
Tatsuya Haga and Shigeki Takeda

Calcium Signaling, Second Edition
James W. Putney, Jr.

Analysis of Growth Factor Signaling in Embryos
Malcolm Whitman and Amy K. Sater

Signal Transduction in the Retina
Steven J. Fliesler and Oleg G. Kisselev

Signaling by Toll-Like Receptors
Gregory W. Konat

Lipid-Mediated Signaling
Eric J. Murphy and Thad A. Rosenberger

TRP Channels
Michael Xi Zhu

Cyclic Nucleotide Signaling
Xiaodong Cheng

Gap Junction Channels and Hemichannels
Donglin Bai and Juan C. Sáez

Signaling Mechanisms Regulating T Cell Diversity and Function
Jonathan Soboloff and Dietmar J. Kappes

Lipid-Mediated Signaling Transduction, Second Edition
Eric Murphy, Thad Rosenberger, and Mikhail Golovko

Calcium Entry Channels in Non-Excitable Cells
Juliusz Ashot Kozak and James W. Putney, Jr.

Autophagy and Signaling
Esther Wong

Signal Transduction and Smooth Muscle
Mohamed Trebak and Scott Earley

Polycystic Kidney Disease
Jinghua Hu and Yong Yu

For more information about this series, please visit: https://www.crcpress.com/Methods-in-Signal-Transduction-Series/book-series/CRCMETSIGTRA?page=&order=pubdate&size=12&view=list&status=published,forthcoming

Polycystic Kidney Disease

Edited by
Jinghua Hu
Department of Biochemistry and Molecular Biology
Mayo Clinic
Rochester, Minnesota

Yong Yu
Department of Biological Sciences
St. John's University
Queens, New York

CRC Press
Taylor & Francis Group
Boca Raton London New York

CRC Press is an imprint of the
Taylor & Francis Group, an **informa** business

Cover image from Kevin Yu, a sophomore at Hunter College High School, New York. From the amateur but insightful perspective of a young and curious mind, Kevin Yu incorporates many components in the context of polycystic kidney disease to vividly reflect the endeavor devoted to curing the disease. In his drawing, we, the scientists, are hard-working birds who are trying their best to remove cysts from the diseased kidney. We are fortunate to have this image, which projects our belief that the communication between scientists and laymen, also true for the multidiscipline interactions among scientists, does not have to be intimidating.

CRC Press
Taylor & Francis Group
6000 Broken Sound Parkway NW, Suite 300
Boca Raton, FL 33487-2742

Printed on acid-free paper

International Standard Book Number-13: 978-1-138-60389-9 (Hardback)

Visit the Taylor & Francis Web site at
http://www.taylorandfrancis.com

and the CRC Press Web site at
http://www.crcpress.com

Contents

Preface

Autosomal dominant polycystic kidney disease (ADPKD) is one of the most common, life-threatening genetic diseases. Twenty-five years after the first cloning of *PKD1*, one the two major causal loci for ADPKD, and the subsequent intensive investigation, the pace of progress in understanding the pathogenesis of ADPKD, the function of polycystins, and the development of therapeutics has been inadequate due to enormous technical challenges and availability of appropriate experimental models. The approval of the very first pharmacological treatment, tolvaptan, by the U.S. Food and Drug Administration (FDA) in 2018 has once again inspired tremendous interest in ADPKD research in the field. In the meantime, the fact that tolvaptan is moderately effective with significant side effects encourages more research to address the unmet clinical needs of ADPKD patients. We will thus find an ever-greater need from new biologists and clinicians seeking proper methodology to study ADPKD and polycystins.

This book focuses on methodology and concepts including research approaches, practical protocols, and molecular diagnosis used in the study of polycystins and ADPKD. We envision this book to serve as a reference source with sufficient information and practical descriptions for those interested in ADPKD and polycystins, ranging from biologists and clinicians in the field to interested students. However, this book is not a comprehensive survey of all experimental approaches used in studying ADPKD, but rather focuses on the protocols that specifically analyze polycystins' structure and function with biochemical, epigenetic, electrophysiological, and structural methods, as well as the best-available functional models, ranging from state-of-the-art *in vitro* "kidney-in-a-dish" organoids to *in vivo* model organisms including simple nematode, vertebrate zebrafish, and preclinical rodent models. Due to the emerging evidence suggesting that cardiovascular anomalies, a major cause of morbidity and mortality in ADPKD patients, are likely primary manifestations independent of hypertension in ADPKD, as well as the discoveries supporting that cysts in PKD exhibit an altered energy metabolism similar to the Warburg effect in many cancer cells, the book extends two chapters specifically on the study of cardiovascular disorders and metabolism in ADPKD. The book assumes a rudimentary knowledge of cell biology, biochemistry, and electrophysiology. The references are by no means exhaustive, and we apologize to all our colleagues whose work has not been included or cited.

We could not have completed this book without the help of many people. We wish to express our gratitude to all contributing authors who invested their priceless time to expedite this project to fruition. We would like to thank Dr. Michael Zhu for inviting us to start this project and thank the publishing team at CRC Press/Taylor & Francis for keeping the editing process on track and being extremely patient with repeated delays due to unexpected but reasonable scenarios that occurred frequently in the overloaded lives of scientists. Finally, we want to thank our families for putting up with us and recognizing the value of our task.

Jinghua Hu
Yong Yu

Editors

Jinghua Hu, PhD, is an associate professor of biochemistry and molecular biology in the Department of Biochemistry and Molecular Biology and an associate professor of medicine in the Division of Nephrology and Hypertension at the Mayo Clinic. Dr. Hu received a PhD from the Chinese Academy of Science in 2001. He completed his postdoctoral training with Dr. Maureen M. Barr at the University of Wisconsin–Madison and was recruited as an assistant professor of medicine and BMB in 2008 and promoted to associate professor in 2014. Dr. Hu's research systematically explores cilia-related human diseases, collectively termed ciliopathies, by using various disease models from the inexpensive and efficient nematode *C. elegans* in cultured mammalian cells and genetically engineered rodent models. His lab has developed numerous models of ciliopathies, with the goals of understanding the pathogenesis and testing the potential for diagnostic/therapeutic purposes. Dr. Hu has administered numerous extramural funding sources and presently holds two R01 grants from NIDDK and is the co-principal investigator on another two R01s with his Mayo colleagues. Dr. Hu serves as a director of the Model Organism Core in the NIH-funded Mayo Translational PKD Center and is a member of the Scientific Advisory Committee of the PKD foundation.

Yong Yu, PhD, is an associate professor and the director of graduate studies in the Department of Biological Sciences at St. John's University, Queens, New York. He received a BS from the Ocean University of China, Qingdao, China, in 1996, and a PhD from the Shanghai Institute of Plant Physiology and Ecology, Chinese Academy of Sciences, in 2001. He was trained as a postdoctoral research scientist first in the Center for Molecular Recognition with Dr. Arthur Karin, then in the Department of Biological Sciences with Dr. Jian Yang at Columbia University, New York, from 2001 to 2006. He was an associate research scientist at Columbia University from 2006 to 2012. In 2012, he became an assistant professor in the Department of Biological Sciences at St. John's University and was promoted to associate professor in 2016. Dr. Yu's research interests include the structure and function of receptor and ion channels, especially transient receptor potential (TRP) channels and polycystins. Dr. Yu's main contributions in PKD research include molecular determination of the subunit stoichiometry of the polycystin-1/polycsytin-2 complex, generation of the first gain-of-function mutants of the polycystin-2 channel and the polycystin-1/polycystin-2 complex channel, and characterization of the ion channel function of polycystin-1 in the polycystin-1/polycystin-2 complex and of PKD1L3 in the PKD1L3/polycystin-L complex.

Contributors

Ewud Agborbesong
Division of Nephrology and Hypertension
Department of Internal Medicine
Department of Biochemistry and Molecular Biology
Mayo Clinic
Rochester, Minnesota

Maureen Barr
Department of Genetics
Rutgers University
Piscataway, New Jersey

James P. Calvet
Departments of Biochemistry and Molecular Biology
Jared Grantham Kidney Institute
University of Kansas Medical Center
Kansas City, Kansas

Erhu Cao
Department of Biochemistry
University of Utah Health
Salt Lake City, Utah

Ying Cao
Shanghai Tenth People's Hospital
School of Life Sciences and Technology
Tongji University
Shanghai, China

Xing-Zhen Chen
Membrane Protein Disease Research Group
Department of Physiology
Faculty of Medicine and Dentistry
University of Alberta
Edmonton, Alberta, Canada

Nelly M. Cruz
Division of Nephrology
Kidney Research Institute, and Institute for Stem Cell and Regenerative Medicine
Department of Medicine
Department of Pathology
University of Washington School of Medicine
Seattle, Washington

Paul G. DeCaen
Department of Pharmacology
Northwestern University
Feinberg School of Medicine
Chicago, Illinois

Benjamin S. Freedman
Division of Nephrology
Kidney Research Institute, and Institute for Stem Cell and Regenerative Medicine
Department of Medicine
Department of Pathology
University of Washington School of Medicine
Seattle, Washington

Amirreza Haghighi
Division of Nephrology
University Health Network and University of Toronto
Toronto, Ontario, Canada

Sara J. Holditch
Department of Medicine
Division of Renal Diseases and Hypertension
Anschutz Medical Campus
University of Colorado–Denver
Aurora, Colorado

Katharina Hopp
Department of Medicine
Division of Renal Diseases and Hypertension
Anschutz Medical Campus
University of Colorado Denver
Aurora, Colorado

Qiong Huang
Department of Medicine
Division of Nephrology
University of Maryland School of Medicine
Baltimore, Maryland

Sonu Kashyap
Department of Anesthesiology and Kogod Center on Aging
Mayo Clinic
Rochester, Minnesota

Matthew Lanktree
Division of Nephrology
McMaster University
Hamilton, Ontario, Canada

Bin Li
Department of Biological Sciences
St. John's University
Queens, New York

Jingyu Li
Shanghai Tenth People's Hospital
School of Life Sciences and Technology
Tongji University
Shanghai, China

Xiaogang Li
Division of Nephrology and Hypertension
Department of Internal Medicine
Department of Biochemistry and Molecular Biology
Mayo Clinic
Rochester, Minnesota

Brenda S. Magenheimer
Departments of Biochemistry and Molecular Biology
Jared Grantham Kidney Institute
University of Kansas Medical Center
Kansas City, Kansas

Robin L. Maser
Departments of Biochemistry and Molecular Biology
Department of Clinical Laboratory Sciences
Jared Grantham Kidney Institute
University of Kansas Medical Center
Kansas City, Kansas

Amitabha Mukhopadhyay
Department of Pharmacology
Feinberg School of Medicine
Northwestern University
Chicago, Illinois

Raphael A. Nemenoff
Department of Medicine
Division of Renal Diseases and Hypertension
Anschutz Medical Campus
University of Colorado Denver
Aurora, Colorado

Courtney Ng
Department of Biological Sciences
St. John's University
Queens, New York

Leo C.T. Ng
Department of Pharmacology
Feinberg School of Medicine
Northwestern University
Chicago, Illinois

Eduardo Nunes Chini
Department of Anesthesiology and Kogod Center on Aging
Mayo Clinic
Rochester, Minnesota

Patricia Outeda
Department of Medicine
Division of Nephrology
University of Maryland School of Medicine
Baltimore, Maryland

Stephen C. Parnell
Departments of Biochemistry and Molecular Biology
Jared Grantham Kidney Institute
University of Kansas Medical Center
Kansas City, Kansas

York Pei
Division of Nephrology
University Health Network and University of Toronto
Toronto, Ontario, Canada

Ji-Bin Peng
Division of Nephrology
Department of Medicine
Nephrology Research and Training Center
University of Alabama at Birmingham
Birmingham, Alabama

Feng Qian
Department of Medicine
Division of Nephrology
University of Maryland School of Medicine
Baltimore, Maryland

Xueweng Song
Division of Nephrology
University Health Network and University of Toronto
Toronto, Ontario, Canada

Jingfeng Tang
National "111" Center for Cellular Regulation and Molecular Pharmaceutics
Hubei University of Technology
Wuhan, Hubei, China

Thuy N. Vien
Department of Pharmacology
Northwestern University
Feinberg School of Medicine
Chicago, Illinois

Rebecca Walker
Department of Medicine
Division of Nephrology
University of Maryland School of Medicine
Baltimore, Maryland

Juan Wang
Department of Genetics
Rutgers University
Piscataway, New Jersey

Lingyun Wang
Division of Nephrology
Department of Medicine
Nephrology Research and Training Center
University of Alabama at Birmingham
Birmingham, Alabama

Qinzhe Wang
Department of Biochemistry
University of Utah Health
Salt Lake City, Utah

Zhifei Wang
Department of Biological Sciences
St. John's University
Queens, New York

Terry Watnick
Department of Medicine
Division of Nephrology
University of Maryland School of Medicine
Baltimore, Maryland

Hangxue Xu
Department of Medicine
Division of Nephrology
University of Maryland School of Medicine
Baltimore, Maryland

Yong Yu
Department of Biological Sciences
St. John's University
Queens, New York

Wang Zheng
Membrane Protein Disease Research Group
Department of Physiology
Faculty of Medicine and Dentistry
University of Alberta
Edmonton, Alberta, Canada

1 Biochemical Analysis of the Polycystin-1 Complexity Generated by Proteolytic Cleavage at the G Protein-Coupled Receptor Proteolysis Site

Rebecca Walker, Hangxue Xu, Qiong Huang, and Feng Qian

CONTENTS

1.1 INTRODUCTION

Polycystin-1 (PC1) and Polycystin-2 (PC2) play an essential role in renal tubular morphogenesis.[1–4] They are encoded by *PKD1* and *PKD2*, respectively, genes which are mutated in 85%–90% and 10%–15%, respectively, of probands with autosomal dominant polycystic kidney disease (ADPKD).[5] ADPKD is characterized by cyst formation that can initiate during fetal development and continue throughout the lifetime, leading to kidney failure, usually after the sixth decade of life.[6] Cysts are formed when the activities of the polycystins fall below a threshold level.[7,8] PC1 is a large receptor-like protein with 11 transmembrane domains which undergoes cis-autoproteolytic cleavage at the juxtamembrane G protein-coupled receptor (GPCR) proteolysis site (GPS) motif.[9–13] GPS cleavage is a key posttranslational modification of PC1 that is essential for its full biological function,[11,12–16] and disruption of cleavage plays a critical role in cystogenesis of ADPKD.[17–19] PC2, also called TRPP2, is a calcium-permeable, nonselective cation channel of the transient receptor potential (TRP) channel superfamily[20,21] and modulates calcium release from the endoplasmic reticulum (ER).[22] PC1 and PC2 interact[23–25] to form a receptor-ion channel complex in primary cilia and the plasma membrane of renal epithelial cells.[26,27]

The ciliary PC1/2 complex, or polycystin complex, was reported to be required to mediate calcium influx in response to mechanical signals, and impaired mechanosensation through primary cilia has been proposed as a pathogenic mechanism in ADPKD.[27,28] However, the function of the polycystin complex in cilia and its involvement in cilium-associated calcium fluxes are intensely debated.[29] PC1 and PC2 also localize to other subcellular locations and may each have independent functions. For example, PC1 is also associated with mitochondria-associated ER membranes or mitochondria matrix, thereby affecting mitochondrial function and energy metabolism.[30,31]

There are many obstacles that make the analysis of the endogenous polycystin complex challenging. First, PC1 is intrinsically difficult to detect because of its low abundance, large size, and many transmembrane domains. Detection of endogenous PC1 molecules in tissues and cells requires high-quality anti-PC1 antibodies that are rigorously validated with genetically defined positive and negative controls. Second, PC1 is biochemically complex due to cis-autoproteolytic cleavage at the juxtamembrane GPS motif, generating multiple PC1 forms. Analysis of the full complement of PC1 products requires a combination of specific antibodies directed to different regions of PC1 under defined experimental conditions that can discriminate these forms. Third, each PC1 form undergoes differential posttranslational *N*-glycosylation, giving rise to distinct species differing in the type of *N*-glycans attached. *N*-glycosylation analysis is usually required to further distinguish between the PC1 products. PC1 is additionally proteolytically cleaved at other undefined sites outside of the GPS, generating smaller products such as P100[32] or multiple C-terminal tail fragments.[33–35] These cleavage events are far more limited in extent, and their *in vivo* significance remains largely unclear. Currently available reagents are unable to directly assess, by immunofluorescence-based methods, the unique complexity of PC1 generated through modifications by GPS cleavage and *N*-glycosylation. However, these properties can be utilized to distinguish the various PC1 species by biochemical methods and therefore reveal information regarding which PC1 species resides in the tissues and cells. This chapter will describe biochemical complexity of PC1 due to GPS cleavage and provide detailed protocols for analyzing the full complement of PC1 products and the polycystin complex *in vivo*.

1.2 POLYCYSTIN-1 CLEAVAGE AT THE GPS MOTIF

PC1 is a 4302-amino acid (aa), 11-transmembrane glycoprotein with a large N-terminal extracellular ectodomain of 3072 aa and a short cytoplasmic C-terminal tail (CTT) of ~200 aa[36] (Figure 1.1a). The ectodomain contains a set of domains involved in protein–protein interactions and a ~1000-aa receptor for egg jelly (REJ) module[37] harboring four FnIII domains.[38,39] The CTT is responsible for regulating a number of intracellular signaling pathways[15,40] including calcium,[41,42] Wnt,[43] and mTOR.[44] The CTT fragment can bind heterotrimeric G proteins *in vitro*[45] and mediates AP-1 transcription factor activation via heterotrimeric G proteins, suggesting that PC1 could be an atypical GPCR.[46–49]

Situated at the base of the ectodomain is the 50-aa GPS motif[16,17,50] (Figure 1.1a). The GPS motif was first identified in a neuronal GPCR, CIRL/latrophilin,[51] and has recently been recognized as a part of the larger GPCR autoproteolysis-inducing (GAIN) domain that is also present in PC1.[19] The GAIN domain is a defining feature of the adhesion GPCRs (aGPCRs),[19,52] the second largest subgroup of GPCRs in the human genome.[53,54] The GPS/GAIN domain is the primary site for PC1 processing, allowing a much greater complexity of the PC1 pool to be generated and allowing for modulation of PC1 function. PC1 is cleaved between leucine (L) and threonine (T) at the highly conserved HL↓T tripeptide sequence within the GPS[10] (see Figure 1.1a), resulting in two cleavage products, $PC1_{NTF}$ and $PC1_{CTF}$. The reaction takes place in the ER shortly after PC1 synthesis. GPS cleavage of PC1 occurs via a cis-autoproteolytic mechanism[10] (Figure 1.1b). This mechanism was found in an aGPCR, EMR2,[56] and is similar to those found in other autoproteolytic proteins such as the Ntn hydrolases[57,58] and hedgehog proteins.[59] Cis-autoproteolysis is a self-catalyzed chemical rearrangement based on the ability of the nucleophilic Thr residue of the tripeptide HL↓T to initiate a proximal N–O acyl rearrangement. This converts the peptide (amide) bond to a more reactive ester intermediate.[60] A second nucleophile, such as a water molecule, attacks the ester bond, leading to the irreversible cleavage of the scissile bond.[10]

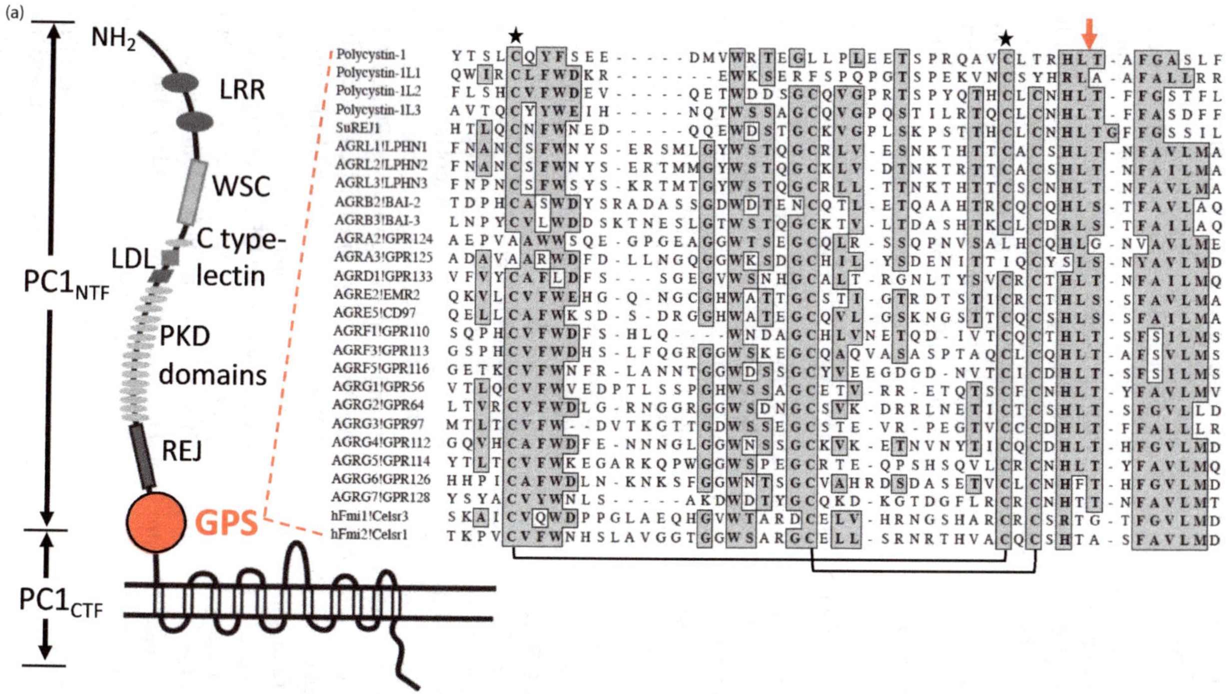

FIGURE 1.1 Polycystin-1 cis-autoproteolytic cleavage at juxtamembrane GPS and biochemical complexity. (a) Schematic diagram of the domain organization of polycystin-1. LRR, leucine-rich repeat; WSC, cell-wall and stress-response component (a putative carbohydrate binding domain); C-type lectin, a calcium-dependent carbohydrate binding domain; LDL, low density lipoprotein class A module; REJ, receptor for egg jelly module; GPS: G protein-coupled receptor proteolysis site. Cleavage occurs at the HL↓T tripeptide of the GPS and produces the $PC1_{NTF}$ and 11-transmembrane containing $PC1_{CTF}$. A multiple sequence alignment of the GPS sequences of polycystin-1 family proteins and adhesion GPCRs is shown to the right of the diagram, with the red arrow depicting the cleavage site. The two pairs of cysteine residues are bracketed at bottom. Note that polycystin-1 and polycystin-1L1 contain only one pair of cysteine as indicated by stars on the top.

(Continued)

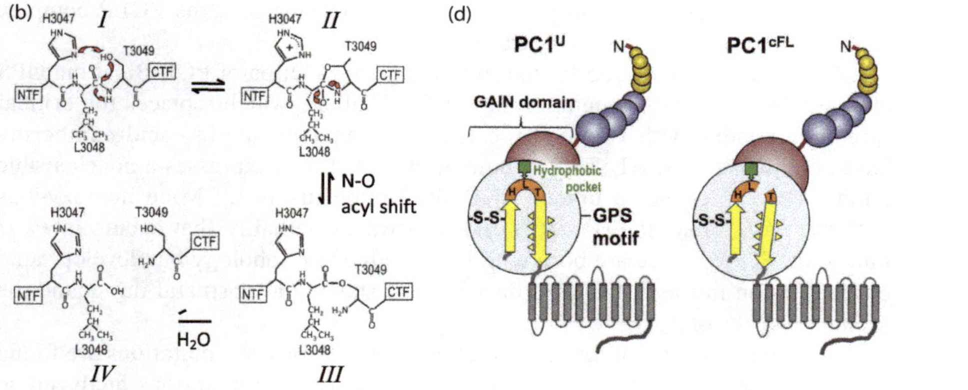

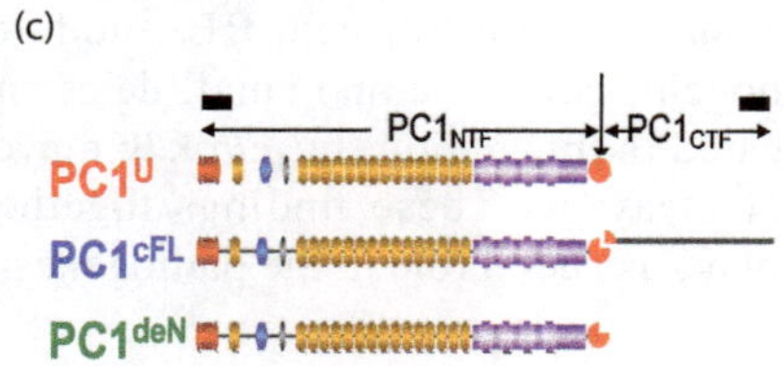

Product	Structure	Properties
$PC1^{U}$	Uncleaved Full Length	Predominant in early developing kidneys
$PC1^{cFL}$	cleaved Full Length. $PC1_{NTF}$ and $PC1_{CTF}$ are non-covalently associated	Predominant post-natally. Detected at cell surface and in cilia.
$PC1^{deN}$	detached $PC1_{NTF}$. Free from $PC1_{CTF}$	Most abundant form post-natally

FIGURE 1.1 (Continued) Polycystin-1 cis-autoproteolytic cleavage at juxtamembrane GPS and biochemical complexity. (b) Proposed cis-autoproteolytic mechanism for PC1 cleavage at GPS. A tight strain at HLT3049 is generated during the folding of the newly synthesized PC1 polypeptide in the ER, which confers the ability of Thr3049 to initiate cis-autoproteolysis. The active hydroxyl of Thr3049 launches a nucleophile attack on the carbonyl of Leu3048 (step I), resulting in the formation of a transitional tetrahedral intermediate (step II) and then an ester intermediate via an N–O acyl shift (step III). The hydrolysis of the ester leads to cleavage of the peptide bond between Leu3048 and Thr3049 (step IV). Polycystin-1 Cis-autoproteolytic Cleavage at Juxtamembrane GPS and Biochemical Complexity. (c) Biochemical complexity of polycystin-1 species. Three PC1 species can typically be detected within cells: uncleaved $PC1^{U}$, heterodimeric $PC1^{cFL}$, and detached $PC1^{deN}$. The table on the right summarizes properties of the PC1 species. (d) A model of GPS structure and cleavage of polycystin-1. The GPS motif is a part of the larger GPCR-autoproteolysis inducing (GAIN) domain. The final two β-strands of the GPS motif are held in a tight conformation by the only S–S bond and hydrophobic interactions between side chains (yellow triangles). A tight kink is held between these two β-strands with a critical HL↓T at the head. The L residue interacts with a hydrophobic pocket in the GPS, contributing to proper folding. After cleavage (*right image*), the tight kink is relaxed, and the final β-strand tilts slightly but remains tightly bound with the remainder of the GPS.

1.3 CRITICAL ROLE OF POLYCYSTIN-1 GPS CLEAVAGE FOR ITS NORMAL FUNCTION AND IN ADPKD

GPS cleavage plays a critical role in regulating PC1 trafficking to primary cilia.[12,61] Indeed, PC1 requires physical interaction with PC2 to exit from the ER in the form of a PC1/2 complex.[12,61,62] We have shown that PC1 binds adaptor Rabep1 and, once arriving at the trans-Golgi, the PC1/2-bound Rabep1 recruits GGA1[63] and the small GTPase Arl3[64] to enable subsequent ciliary targeting.[12] GPS cleavage is not required for PC1/2 complex formation yet in the absence of cleavage, the PC1/2 complex cannot reach the Golgi for ciliary trafficking.

GPS cleavage is required for the full biological function of PC1. Blocking GPS cleavage by the T3041V mutation in the *Pkd1^V* allele, which replaces the critical threonine residue with valine at the HL↓T[3041] sequence (amino acid numbering based on mouse PC1; HL↓T[3049] in human PC1) and thus expresses a noncleavable mutant ($PC1^V$), causes a unique PKD phenotype in mice.[11] Mice homozygous with the *Pkd1^V* mutation (*Pkd1^{V/V}*) escape perinatal lethality that occurs in *Pkd1* knockout. *Pkd1^{V/V}* mice are born with intact kidney morphology but develop rapid cystic dilation in the distal nephron segments starting at postnatal day 3 and die around 3 weeks of age.[11]

Increasing numbers of disease-associated *PKD1* missense mutations are found to disrupt GPS cleavage.[12,18,19,61] The cleavage-disrupting mutations analyzed so far are mostly located in the GAIN/GPS domain and the adjacent REJ module. Importantly, approximately 30% of all pathogenic missense and small deletion/insertion mutations identified to date are located in this region (http://pkdb.mayo.edu/) and have the potential to affect PC1 cleavage. These findings together suggest that defective GPS cleavage of PC1 plays a crucial role in the pathogenesis of ADPKD.

1.4 POLYCYSTIN-1 BIOCHEMICAL COMPLEXITY GENERATED BY GPS CLEAVAGE

GPS cleavage results in significant biochemical complexity of the PC1 pool and has functional implications essential to the maintenance of cellular operations. At least three distinct PC1 forms exit in the tissues and cells[9]: uncleaved ($PC1^U$), cleaved full-length ($PC1^{cFL}$), and the detached $PC1_{NTF}$ form ($PC1^{deN}$) (Figure 1.1c). The ratio of cleavage products to full-length PC1 appears to change in the kidney during the development and may have functional consequences for kidney development.[66] It is possible that this ratio must remain correctly balanced throughout the nephron since cleavage appears to have a specific function in distinct nephron segments of the kidney.

$PC1^U$: Cleavage is developmentally regulated in the kidney, resulting in changing the proportion of the $PC1^U$ form during successive embryonic stages.[66] The proportion of $PC1^U$ is maximal at E13.5 (>50%) and decreases gradually to approximately 20% by E15.5. Levels continue to decline until almost undetectable amounts remain at birth. However, in proximal tubule (PT) cells, a significant portion of PC1 remains uncleaved (~50%) even after birth.[11] Analyses of the PC1 mutant expressed in *Pkd1^{V/V}* mouse model showed that the noncleavable $PC1^V$ mutant protein is expressed at a comparable level to wildtype (WT) $PC1^U$. $PC1^V$ allows postnatal survival and is thus fully sufficient to rescue the embryonic cystogenesis and lethality. These data raised the possibility that uncleaved PC1 may play a key role in embryonic kidney development for proper renal epithelial tubular differentiation and maturation.[16]

$PC1^{cFL}$: GPS cleavage results in the N-terminal fragment ($PC1_{NTF}$) and 11-transmembrane C-terminal fragment ($PC1_{CTF}$).[17] A unique outcome of the reaction is that the two fragments remain tightly and noncovalently associated to form a

heterodimeric molecule termed $PC1^{cFL}$.[9] During embryonic kidney development, the cleaved PC1 form increases gradually in proportion and becomes the predominant form of PC1 in distal nephrons (>90%)[66] at birth as in most adult tissues.[9] These considerations support the notion that $PC1^{cFL}$ may have a key function for maintaining intact distal nephron morphology during postnatal maturation and throughout adulthood.[11]

$PC1^{deN}$: GPS cleavage also gives rise to a distinct pool of $PC1_{NTF}$. This fragment, termed $PC1^{deN}$, is derived from dissociation of the two halves of $PC1^{cFL}$. $PC1^{deN}$ is a major endogenous PC1 form in many tissues, is predominantly Endo-H resistant and is associated with the plasma membrane of renal epithelial cells.[9] In analogous models of adhesion GPCRs, detachment of the N-terminal fragment has been suggested to represent an activation mechanism. Indeed, constitutive activation of an aGPCR (GPR56) has been achieved by deletion of the N-terminal sequence, mimicking cleavage.[67] The function of this form of PC1 currently remains unclear.

1.5 STRUCTURAL BASIS OF POLYCYSTIN-1 GPS CLEAVAGE AND THE HETERODIMERIC ASSOCIATION

The function of the cleaved form of the PC1 molecule may be dependent on its heterodimeric composition.[9,17] One interesting clue comes from the structural studies of aGPCRs, which contain a GPS motif at the same juxtamembrane position and a large number of extracellular domains as PC1, although their actual domains and folds differ from the latter.[53,54] X-ray crystallographic analyses of aGPCRs show that the GPS motif forms five β-strands and is integrated into the ~320 aa GAIN domain that is also present in PC1 (Figure 1.1d).[19] In the structure of the uncleaved GAIN domain of the aGPCR, the HL↓T consensus sequence is positioned at a sharply kinked loop between the last two β-strands (β1 and β2). This distorted and strained geometry provides the driving force for an N–O acyl shift of its Thr (T) residue, and thus facilitates ester formation and cleavage. Work from the aGPCR structure identified three structural elements that are important for keeping the sharp β-turn in place: (1) two disulfide (S–S) bonds between neighboring β-strands, (2) an extensive network of hydrophobic interactions between the β1-strand and adjacent residues within the GPS, and (3) the trapping of the Leu (L) of the HL↓T within a hydrophobic pocket. Cleavage relaxes the sharp kink of the β-strands, resulting in a small tilt in the last β-strand (β1), which remains tightly bound to the remainder of the GPS through the extensive network of hydrophobic interactions between side chains. This conformational change in the β1 is thought to be functionally important and may be involved in activating the molecule.[19]

The analogous structural elements with the final two β-strands are depicted for PC1 in Figure 1.1d. Interestingly, PC1 only has one S–S bond, C3015–C3043 between neighboring β-strands. Mutation of the S–S bond prevents cleavage.[17] The presence of only one S–S bond in PC1, in comparison to aGPCRs, may weaken the association of the β-strands, leading to less energetically costly cleavage or a reduction of the strength of the subunit association following cleavage. The hydrophobic pocket is also essential for cleavage. Typically, the leucine residue within the HL↓T is expected to interact with the hydrophobic pocket (Figure 1.1d, green) as in aGPCRs.[19] Mutation of this residue to a charged residue would likely affect the interaction with the hydrophobic pocket and thereby alter the folding of the β-strands to render the GPS incompatible with cleavage.[10,17,61]

Cleavage, allowing conformational changes in the PC1 molecule, as well as releasing fragments of the protein, appears to be essential for proper function of PC1. Within the aGPCR field, the structural composition of heterodimers is proposed to facilitate cell guidance.[19,67,68] Similarly, within the kidney, the heterodimeric structure of $PC1^{cFL}$ is likely to be important in enabling specific biological functions.

1.6 PRINCIPLE OF ANALYZING THE FULL COMPLEMENT OF ENDOGENOUS POLYCYSTIN-1 PRODUCTS IN TISSUES AND CELLS

As previously stated, three distinct PC1 forms compose the PC1 pool present and distinguishable in cell and tissue preparations: uncleaved ($PC1^{U}$), heterodimeric ($PC1^{cFL}$), and the detached $PC1_{NTF}$ form ($PC1^{deN}$)[9] (Figure 1.1c). Current available immunoreagents cannot directly discriminate between all the PC1 products. Antibodies that are directed against one cleavage fragment of PC1 cannot detect the other cleavage fragment by SDS-PAGE. Therefore, analysis of the full complement of PC1 products requires the use of a combination of specific PC1 antibodies directed to $PC1_{NTF}$ and $PC1_{CTF}$ underdefined experimental conditions that distinguish between the various forms as illustrated in Figure 1.2.

The $PC1^{U}$ form can be easily distinguished from the cleaved forms of PC1 by slower migration on Western blot. While $PC1^{cFL}$ presents as a single molecule in the tissues, it undergoes subunit dissociation by the denaturing condition of SDS-PAGE. The dissociated $PC1_{NTF}$ and $PC1_{CTF}$ subunits migrate according to their respective sizes, and these are different from that of $PC1^{U}$ (Figure 1.2). To distinguish $PC1^{cFL}$ from the $PC1^{deN}$, $PC1^{cFL}$ can be specifically captured by immunoprecipitation with an antibody against $PC1_{CTF}$, which does not recognize $PC1_{NTF}$ and $PC1^{deN}$. The immunoprecipitated $PC1^{cFL}$ can subsequently be analyzed with the $PC1_{NTF}$- and $PC1_{CTF}$-specific antibodies for the subunit constituents (see Protocol I). This approach will simultaneously capture $PC1^{U}$, allowing a direct comparison of the two forms in a given sample in the same lane of Western blots. The $PC1^{deN}$ form is present in the flow through lysate (L^{Δ}) that is depleted from $PC1^{U}$ and $PC1^{cFL}$ (see Protocol II) and can be analyzed separately (Figure 1.2).

One important consideration is the heavy *N*-glycosylation of $PC1^{cFL}$ and $PC1^{deN}$ in tissues, which significantly increases their molecular weight and gives rise to their doublet bands on Western blot. In the case of $PC1^{cFL}$, this causes the upper band of the $PC1_{NTF}$ doublet to migrate to a position that partially overlaps with that of the $PC1^{U}$, obscuring the distinction between the two PC1 forms even under optimized polyacrylamide gel electrophoresis conditions. This issue can be resolved by *N*-glycosylation analysis using the *N*-deglycosylases PNGase-F (peptide *N*-glycosidase F) and Endo-H (endoglycosidase H)[9,12,69,70] (see Protocol III). PNGase-F removes all types of *N*-linked glycans (high mannose, hybrid, and complex

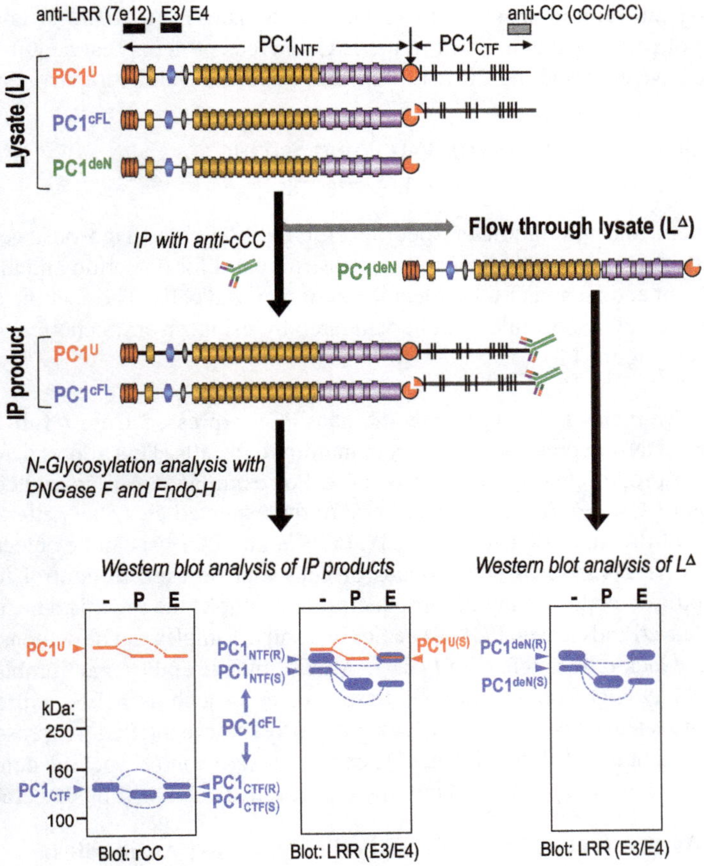

FIGURE 1.2 Principle of biochemical analysis of full complement of polycystin-1 products. Lysates (L) of tissues and cells typically contain three species of PC1: $PC1^{U}$, $PC1^{cFL}$, and $PC1^{deN}$. The cCC antibody binds to the C-terminus of PC1 (region indicated by gray box). Immunoprecipitation with cCC will only capture $PC1^{U}$ and $PC1^{cFL}$, while $PC1^{deN}$ washes out into the flow through fraction (L^{Δ}). After SDS-PAGE and Western blot of the IP product, $PC1^{U}$ will appear as the highest band (red), that is, uncleaved full length. $PC1^{cFL}$ de-tethers and produces two bands: $PC1_{NTF}$ (blue, LRR blot) and $PC1_{CTF}$ (blue, rCC blot). The expected mobility shift pattern for all three PC1 species upon PNGase-F (P) and Endo-H (E) digestion is schematically shown at the bottom. Endo-H resistant or sensitive form is appended by the suffix "R" or "S," respectively (e.g., $PC1_{NTF(R)}$ or $PC1_{NTF(S)}$).

N-glycans) from glycoproteins, while Endo-H removes only high mannose and some hybrid types of *N*-linked carbohydrates. The use of the two *N*-deglycosylases not only helps to distinguish between $PC1^{U}$ from of $PC1^{cFL}$ and $PC1^{deN}$ in tissues and cells, but also serves to monitor their trafficking along the secretory pathway.[69,70] The general rationale is that *N*-glycans of glycoproteins in the ER are all high mannose and are susceptible to removal by cleavage using PNGase-F and Endo-H, whereas complex *N*-glycans acquired in the medial/trans-Golgi compartment are resistant to removal by Endo-H but remain sensitive to PNGase-F. Sensitivity to Endo-H is therefore indicative of proteins that are still in the ER, whereas proteins that acquire Endo-H resistance have egressed the ER and passed through the Golgi compartment.

1.6.1 Quality Controls for Validating Specific Detection of Polycystin-1 Antibodies

Development and validation of specific PC1 antibodies by rigorous, genetically defined positive and negative controls are instrumental for detection and analysis of low-abundant endogenous PC1 molecules in tissues and cells. The specific detection of endogenous PC1 molecules can be validated by using several genetically defined controls (see Figure 1.3).

1. *Positive control*: A tagged recombinant PC1 expressed from a full-length *Pkd1* cDNA expression construct in mammalian cells. This allows detection of the proper cleavage pattern of PC1. For example, the Flag-tagged PC1 (Figure 1.3a, *Pkd1* cDNA) gives rise to the expected cleavage pattern, and each of the three PC1 forms ($PC1^{U}$, $PC1^{cFL}$, and $PC1^{deN}$) can be detected.
2. *Negative control*: $Pkd1^{-/-}$ tissues or cells can be used to control for the specific detection of any PC1 products. For example, the specific detection of uncleaved and cleaved PC1 molecules in positive samples can be demonstrated by the lack of the signals in $Pkd1^{-/-}$ MEFs (murine embryonic fibroblasts).
3. *GPS cleavage control*: $Pkd1^{V/V}$ tissues or cells such as MEFs expressing the noncleavable $PC1^{V}$ mutant, or tagged recombinant $PC1^{V}$ expressed in mammalian cells from the cDNA, can be used to control specific detection of PC1 cleavage products. Only the uncleaved species will be detectable.

1.6.1.1 Assessment of Earlier Published Polycystin-1 Antibodies

The literature describing endogenous PC1 expression patterns is replete with examples of poorly controlled antibody validation. In many cases, early descriptions of PC1 expression patterns by Western blot or immunolocalization using both homemade and commercial PC1 antibodies produced inconsistent and conflicting results. Indeed, the diverse banding pattern, complex posttranscriptional modifications, and cellular localization of PC1 has led to much debate over the true banding pattern and localization of the protein (Table 1.1). While many of the published studies might have accurately described some aspect of the true pattern and localization of PC1, an unknown variable is the specificity of the antibody used.

The quality of most published and commercial antibodies was determined by their ability to (1) detect the recombinant protein that was used for immunization or (2) detect a high molecular weight band of >400 kDa (approximating the predicted MW of uncleaved PC1) in normal cells and tissues on Western blot. However, many of them were not subject to more rigorous quality controls using irrefutable genetically negative controls (Table 1.1). The importance of adequate controls was dramatically illustrated in a study by Nauta et al.[71], who generated 14 polyclonal PC1 antibodies but concluded that most, if not all, of them detected signals that were not related to PC1 but due to off-target cross-reactivity.

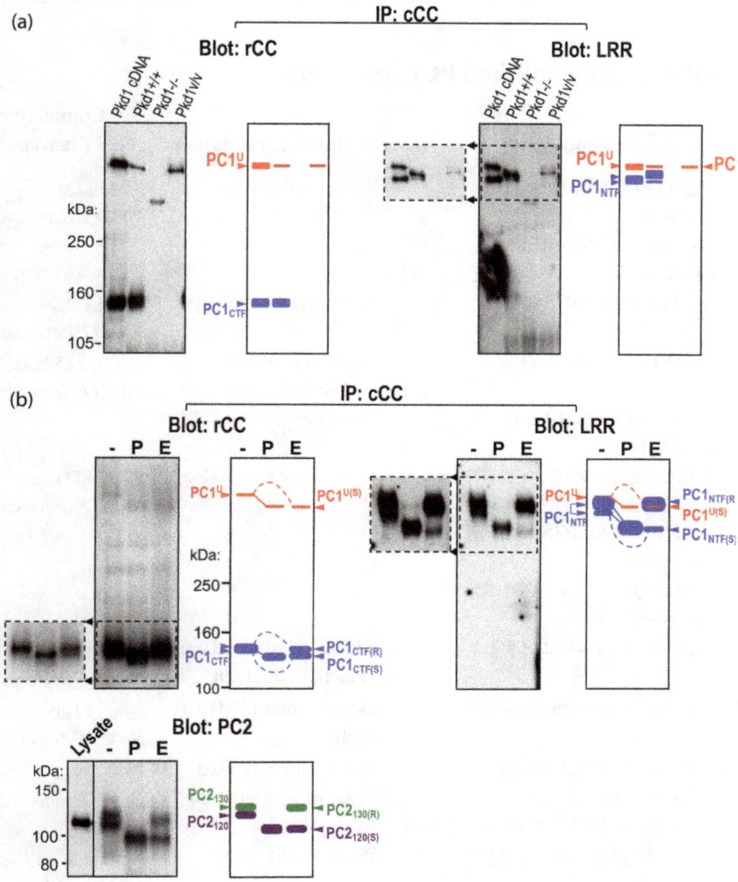

FIGURE 1.3 Expected results and interpretation of polycystin-1 cleavage pattern. (a) Representative results of PC1 banding patterns for MEFs with various genotypes as indicated. After performing a cCC IP, $PC1^U$ and $PC1^{cFL}$ species can be distinguished by their characteristic banding patterns with rCC ($PC1_{CTF}$-specific) and LRR ($PC1_{NTF}$-specific) antibodies, as illustrated by the color-coded schematic diagram adjacent to the blot. Exogenously expressed PC1 (*Pkd1* cDNA) serves as a positive control. (b) N-glycosylation analysis of the polycystin complex captured from WT MEFs lysate by cCC antibody. The PC1 species can be further differentiated by digestion with *N*-glycosylation enzymes. Probing with rCC antibody (*left blot*) detects $PC1^U$ (red) and $PC1_{CTF}$ of $PC1^{cFL}$ (blue). PNGase-F (P) and Endo-H (E) sensitivity is shown in the rCC blot, and $PC1^U$ and $PC1_{NTF}$ sensitivity is shown in the LRR blot (*right*), aided by the same color-coded schematic diagram adjacent to the blot. Endo-H resistant or sensitive form is appended by the suffix "R" or "S," respectively (for example $PC1_{NTF(R)}$ or $PC1_{NTF(S)}$). Probing with PC2 antibody shows the characteristic *N*-glycosylation pattern of PC2 within the polycystin complex (*bottom panels*). The total lysate contains a single PC2 band of 120 kDa. Two PC2 species, $PC2_{120}$ (green) and $PC2_{130}$ (purple), are detected within the polycystin complex after IP. Both of these PC2 species within the polycystin complex are sensitive to PNGase-F (P) digestion, whereas only the lower band is sensitive to Endo-H (E) digestion ($PC2_{120\,(s)}$).

TABLE 1.1
Assessment of Some Published PC1 Antibodies

Ref	Immunogens[a,b]	Cellular Localization	Comment[c,d] on GPS Cleavage Pattern
72	P1: GEEIVAQGKRSDPRS (aa 2778–92, h) P2: **LSKVKEFRHKVRFEG** (**aa 4142–56**, h)	Basolateral/apical, cytoplasmic	~642 kDa *No CTF detected*
73	Fusion protein: **aa 4075–4293**, m	Multi-organ expression	~400 kDa, *No CTF detected*
74,75	P1: **cKRLHPDEDDTLVE** (**aa 3619–31**, h) P2: **cELGLSLEESRDRLR** (**aa 3821–34**, h)	Plasma membrane, intracellular, adhesion junctions	~400, 245 kDa, *No CTF detected*
76	P1: **SRSGHRHLDGDRAFHRN** (**aa 3147–63**, h) P2: **QGRRSSRAPAGSSRGPSPG** (**aa 4246–64**, h) P3: **VHGNQSSPELGPPRLRQVR** (**aa 3734–52**, h)	Wide tissue expression	~200 kDa, *No CTF detected*
77–79	Fusion protein: **aa 4070–4302**, h	Kidney tubules, adherens junction, lipid raft	≥400 kDa *No CTF detected*
80	Exon 46 fusion protein (**aa 4148–4302**, h)	Desmosomes in MDCK cells	~400 kDa *No CTF detected*
81	MR3: EPYLAVYLHSEPRPNEHN (aa 2939–56, h) PKCR: **murine PC1 C-terminal 213 aa**	PKCR antibody used only on Western blot	~400 kDa *No CTF detected*
27	p96521: DACPPEVDFLKQDCTEE (aa 866-882, m)	Primary cilia	No signal on $Pkd1^{-/-}$ cilia (immunofluorescence)
82,83	7e12: LRR (281–751 nt, h)	Used mostly for Western blot	No signal in $Pkd1^{-/-}$ tissues (Western blot)

[a] Underline indicates immunogen amino acids located in $PC1_{NTF}$; h, human; m, mouse.
[b] Bold indicates immunogen amino acids located in $PC1_{CTF}$.
[c] Underline indicates the size of the reported PC1 bands detected.
[d] Italic indicates issue with Western blot detection.

The fundamental feature of PC1 cleavage pattern at GPS[9–13] can be exploited to retrospectively assess the specificity of some previously published, but poorly validated, PC1 antibodies. This is particularly poignant for $PC1_{CTF}$-directed antibodies, which can be diagnosed by the ability to detect ~150 kDa $PC1_{CTF}$ as the most prominent band, in addition to the weaker uncleaved $PC1^U$ band. However, many $PC1_{CTF}$ antibodies detected the >400 kDa band, but failed to detect the far more abundant $PC1_{CTF}$ on the same blot (Table 1.1, comment on GPS cleavage pattern). The specificity of these antibodies is therefore highly questionable and the presumed "PC1 bands" detected were likely due to unknown nonspecific cross-reactivity.

1.6.1.2 Validated $PC1_{CTF}$-Specific Antibodies

The first rigorously validated $PC1_{CTF}$-specific antibodies that explicitly detected endogenous PC1 products were rabbit and chicken polyclonal antibodies directed to the C-terminal tail of mouse PC1 (amino acid residues 4123–4291).[11,9] The rabbit antibody (rCC) was affinity purified with the PC1 C-terminal tail antigen immobilized to a resin column (ThermoFisher Scientific, AminoLink Plus Coupling Resin), while the chicken IgY antibody (cCC) was purified by a commercially available kit (such as the EGGstract IgY Purification System, Promega). A combination of the two antibodies enables detection of endogenous $PC1^{U}$ and $PC1^{cFL}$ (through $PC1_{CTF}$), whereby PC1 proteins are immunoprecipitated with cCC and subsequently detected with rCC by Western blotting. The antibodies can also be used to immunodeplete $PC1^{U}$ and $PC1^{cFL}$, allowing $PC1^{deN}$ to be analyzed with anti-$PC1_{NTF}$-specific antibodies following immunodepletion.

Another validated $PC1_{CTF}$-specific antibody is E8, a rat monoclonal antibody directed to the third extracellular loop of mouse PC1 (aa 3682–3882).[47,84] Epitope mapping shows that it reacts with a small region between aa 3736 and 3746 (human PC1). A remarkable property of this antibody is the ability to detect endogenous $PC1^{U}$ and $PC1_{CTF}$ products directly on Western blot without prior immunoprecipitation. The above-mentioned validated $PC1_{CTF}$-specific antibodies are also described on the Baltimore Polycystic Kidney Disease Research and Clinical Core Center's website (http://www.baltimorepkdcenter.org/antibody/index.shtml).

1.6.1.3 Validated $PC1_{NTF}$-Specific Antibodies

$PC1_{NTF}$-specific antibodies can be used to detect $PC1^{cFL}$ form in combination with a $PC1_{CTF}$-specific antibody such as cCC, as illustrated in Figure 1.3. There are a few validated $PC1_{NTF}$-specific antibodies. The commercially available mouse monoclonal antibody 7e12 (originally developed by Dr. Christopher Ward)[83] directed against the N-terminal leucine-rich repeat (LRR) domain of human PC1 is widely used for detecting endogenous PC1 by Western blot analysis. The 7e12 antibody detects bands corresponding to $PC1^{U}$ and $PC1^{deN}$.[9,62,85] Additional monoclonal antibodies directed to the C-lectin domain (E3 and E4) were later developed to detect $PC1_{NTF}$[12] and are further described on the Baltimore Polycystic Kidney Disease Research and Clinical Core Center's website (http://www.baltimorepkdcenter.org/antibody/index.shtml).

1.6.2 Protocol I: Immunoprecipitation and Western Blot Analysis of $PC1^{U}$ and $PC1^{cFL}$ Molecules

Sample preparation

1. Homogenize mouse tissues in lysis buffer (25 mM sodium phosphate pH 7.2, 150 mM NaCl, 10% glycerol, 1 mM EDTA, 1% Triton, protease inhibitor, Roche: 11873580001) by tissue homogenizer (Polytron-MP 2100, Kinematica AG). For cells, homogenization by pipetting up and down is usually sufficient.

   Notes:

   i. 10% glycerol and 1% Triton allow thorough homogenization of tissues or cells while preserving $PC1_{NTF}$ and $PC1_{CTF}$ association. A critical parameter for PC1 proteins for immunoprecipitating intact $PC1^{cFL}$ is the use of mild detergent, i.e., Triton X100 (0.5%–1%). Commonly used RIPA buffer, which usually contains 0.1% SDS, disrupts the heterodimeric association, revealing the noncovalent nature of $PC1_{NTF}$ and $PC1_{CTF}$ association.[9] Therefore, RIPA buffer (0.1% SDS) is too harsh for working with $PC1^{cFL}$.

ii. Use 1 mL lysis buffer for 0.1 g of tissue or confluent cultured cells collected from one 150 mm tissue culture plate.

2. Rotate the sample for 1 h at 4°C.
3. Centrifuge at 16,000 × g in a tabletop centrifuge for 10 min to pellet cell debris and nuclei. Collect clear supernatant and measure the protein concentration.

Immunoprecipitation (IP)

4. For a typical IP reaction, use an anti-$PC1_{CTF}$ antibody. Add 1–2 μL of chicken IgY cCC to the lysate containing ~2 mg protein in a volume of 1 mL.
5. Rotate the sample for 2 h at 4°C.
6. Add 30 uL goat anti-chicken IgY-agarose (Aves Labs #P-1010, wash 3 times with lysis buffer before use).
7. Rotate overnight at 4°C.
8. Spin sample at 400 × g for 2 min; discard the supernatant; wash the agarose beads with lysis buffer 3 times.
9. Add 50 uL 1 × SDS-sample buffer (50 mM Tris-HCl pH 6.8, 2% SDS, 10% glycerol, 1% β-mercaptoethanol, 12.5 mM EDTA, 0.02% bromophenol blue) and incubate at 95°C for 3–5 min to elute the immunoprecipitated proteins.

Western Blot Analysis

10. Load protein samples on 4% (homemade), 3%–8% Tris-acetate SDS–polyacrylamide precast gel, or 4%–12% Tris-glycine SDS–polyacrylamide precast gels (e.g., from Invitrogen). Run the gel and transfer to PVDF membrane (Bio-Rad).

Detection of the $PC1_{CTF}$ Subunit of $PC1^{cFL}$ and $PC1^{U}$

11. After blocking with a milk block (5% skimmed milk powder, TBS-Tween 20 buffer) or 4% BSA, incubate the membrane with rCC or E8 as primary antibody at 1:1000–2000 dilution overnight at 4°C.
12. Wash with TBS-Tween 20 buffer. Incubate with an HRP-conjugated anti-rabbit IgG antibody (GE Healthcare, NA934V) at 1:10,000 dilution for rCC or HRP-conjugated anti-rat IgG antibody (GE Healthcare, NA935) at 1:5000 dilution for E8, for 1 h. Wash with TBS-Tween 20 buffer.
13. Incubate with ECL Prime (GE Healthcare) reagent for detection of $PC1^{U}$ and $PC1_{CTF}$ bands on film, or with a chemiluminescence imaging system.

Detection of the $PC1_{NTF}$ Subunit of $PC1^{cFL}$ and $PC1^{U}$

14. Strip the membrane using Restore Western Blot Buffer (Pierce, VWR) for 5–10 min.
15. Reprobe with an anti-$PC1_{NTF}$ antibody (7e12, E3, or E4). The $PC1_{NTF}$ subunit of $PC1^{cFL}$ and $PC1^{U}$ can then be visualized as described above (steps 12 and 13).

Detection of the PC2 Protein within PC1/2 Complex

16. The cCC-IP products contain PC2 bound to PC1. This PC2 pool can be analyzed from the IP products by Western blot with PC2 antibody. This may be achieved by reprobing the same membrane as described above (steps 12 and 13).

1.6.3 Protocol II: Detection of $PC1^{deN}$ by Immunodepletion

Immunodepletion of $PC1^{cFL}$ and $PC1^{U}$

1. Perform immunoprecipitation from lysates with anti-cCC under nondenaturing conditions as described in Protocol I (steps 1–7). Collect the supernatant (i.e., flow through fraction) after capturing PC1-antibody complex with goat anti-chicken IgY-agarose.
2. Perform the immunoprecipitation for the supernatant and test for intact $PC1^{cFL}$ and $PC1^{U}$ in the resulting immunoprecipitate.
3. Repeat the immunoprecipitation for the supernatant until $PC1^{cFL}$ and $PC1^{U}$ are no longer detectable from its immunoprecipitate. They are "depleted" from the flow through. Depending on PC1 expression level in a given sample, several rounds (up to five) of immunoprecipitation reaction may be required to for successful immunodepletion.

Analysis of $PC1^{deN}$

4. The flow through fraction that is depleted in $PC1^{cFL}$ and $PC1^{U}$ ("depleted lysate," or L^{Δ}) can be used for evaluating $PC1^{deN}$ by *N*-glycosylation (see Protocol III) and Western blot analysis using a $PC1_{NTF}$-specific antibody.

1.6.4 Protocol III: *N*-Glycosylation Analysis of Polycystin-1 Products

N-glycosylation analysis of endogenous PC1 can be performed on immunoprecipitated polycystin complex (Protocol I), directly in depleted (L^{Δ}) (Protocol II), or on total protein lysates. Typical reaction conditions for immunoprecipitated polycystin complex are as follows:

1. Perform PC1 immunoprecipitation with anti-cCC from lysates (such as from kidney tissues) as described in Protocol I. Wash the final IgY-agarose which is bound with polycystin products (IP-IgY-agarose) thoroughly, and remove all wash buffer from the agarose beads.

2. Add 90 μL glycoprotein denaturing buffer (0.5% SDS, 40 mM DTT) to 30 μL of the IP-IgY-agarose beads.
3. Denature IP-IgY-agarose mixture by heating reaction at 100°C for 1 min.

Note: Do not heat the sample for prolonged time as the polycystin complex is prone to degradation.

4. Chill the denatured IP-IgY-agarose mixture on ice.
5. Divide the IP-IgY-agarose mixture evenly in three aliquots of 30 μL (plus 10 μL beads) for control, PNGase-F, and Endo-H reactions, respectively.
6. i. Control (Aliquot #1): Make a total reaction volume of 40 μL (plus 10 μL beads) by adding 4 μL 500 mM sodium phosphate (pH 7.5 at 25°C), 4 μL 10% NP-40, and 2 μL H_2O. Mix gently.
   ii. PNGase-F (Aliquot #2): Make a total reaction volume of 40 μL (plus 10 μL beads) by adding 4 μL 500 mM sodium phosphate (pH 7.5 at 25°C), 4 μL 10% NP-40, and 1 μL H_2O. Then add 1 μL of PNGase-F (500,000 units/mL, New England Biolabs, P0704), and mix gently.

Note: PNGase-F is inhibited by SDS, therefore it is essential to have NP-40 in the reaction mixture under denaturing conditions.

   iii. Endo-H (Aliquot #3): Make a total reaction volume of 40 μL (plus 10 μL beads) by adding 4 μL 500 mM sodium acetate (pH 6.0 at 25°C) and 5 μL H_2O. Then add 1 μL of Endo-H (500,000 units/mL, New England Biolabs, P0702), and mix gently.

Note: Endo-H is not inhibited by SDS.

7. Incubate all three aliquots at 37°C for 1 h. Add SDS-sample buffer to stop the reaction.
8. Assess the extent of deglycosylation by mobility shifts on SDS-PAGE and Western blot. The expected pattern of mobility shift is illustrated in Figure 1.2.

1.6.5 Expected Results and Interpretation of Polycystin-1 Cleavage Patterns

A typical result of biochemical analysis of endogenous PC1 immunoprecipitated with the cCC antibody is presented in Figure 1.3a for MEFs derived from *Pkd1*$^{+/+}$, *Pkd1*$^{-/-}$, and *Pkd1*$^{V/V}$ embryos, along with exogenously expressed recombinant PC1. Probing with rCC identifies the PC1^{U} band of ~520 kDa, and the more abundant PC1$_{CTF}$ band of ~150 kDa in WT MEFs. The rCC antibody recognizes only the ~520-kDa noncleavable PC1V band in *Pkd1*$^{V/V}$ MEFs without detecting PC1$_{CTF}$. Probing with anti-LRR detects the same PC1^{U} band and a strong PC1$_{NTF}$ doublet (~450 and ~370 kDa) that is associated with PC1$_{CTF}$ in WT samples, but only the noncleavable PC1V mutant band (~520 kDa) in *Pkd1*$^{V/V}$ MEFs. None of these bands are detectable in *Pkd1*$^{-/-}$ MEFs. Of note, the upper band of the PC1$_{NTF}$ doublet is more dominant than the lower band in the MEFs. However, the stoichiometry of the upper and lower bands varies among tissues.[9]

Figure 1.3b shows the *N*-glycosylation analysis of PC1 and PC2 products from WT MEFs captured by anti-cCC antibody. Treatment with PNGase-F shifts the $PC1_{CTF}$ from ~150 kDa to a slightly faster migrating band at ~140 kDa (approximating the predicted MW of the $PC1_{CTF}$). Endo-H digestion results in appearance of two bands at ~150 and ~140 kDa, revealing that the $PC1_{CTF}$ is composed of two distinct species differing in *N*-glycan types. Of note, the $PC1^{U}$ is exclusively sensitive to Endo-H in MEFs, as shown by its shift to ~460 kDa (the predicted MW of uncleaved PC1) upon treatment.

PNGase-F treatment reduces the $PC1_{NTF}$ doublet to a single one at ~320 kDa, the predicted MW of $PC1_{NTF}$. Analysis with Endo-H shows that the upper $PC1_{NTF}$ band is Endo-H resistant, while the lower one is Endo-H sensitive as revealed by its shift to ~320 kDa. Hence, the $PC1_{NTF}$ doublet bands result from differential *N*-glycosylation modification, as is found for the $PC1_{CTF}$ subunits. Collectively, these results provide evidence for one single endogenous $PC1^{cFL}$ form that could traffic from the Golgi compartment to the plasma membrane.

Probe with anti-PC2 antibody reveals the unique property of PC2 in the PC1/2 complex. In the whole-cell lysate, PC2 is detected as a single 120 kDa band in MEFs, as shown universally for various tissues and cells.[86,87] However, the fraction of PC2 that is complexed with PC1 appears as two distinct bands of 130 kDa ($PC2_{130}$) and 120 kDa ($PC2_{120}$) of similar intensity.[12] PNGase-F treatment reduces the PC2 doublet to a single one at ~110 kDa, the predicted MW of PC2.

Treatment with Endo-H reveals that $PC2_{130}$ is Endo-H resistant, whereas $PC2_{120}$ is Endo-H sensitive. These results indicate that a significant Endo-H-resistant pool of the polycystin complex exists *in vivo*. Together, these data provide evidence that PC1 and PC2 form a complex in the ER and that the cleaved PC1/2 complex reaches the Golgi apparatus and cilia.[12]

1.7 BIOCHEMICAL ANALYSIS OF CILIARY POLYCYSTINS

The primary cilium is an essential cellular organelle whose signaling function within a number of pathways has important connotations for polycystic kidney disease.[88] Defective cilia signaling, or mutations in proteins that localize to, or signal through, the cilium result in ciliopathies—a collection of diseases associated with cilia defects.[89] Cilia are thought to be involved in maintaining tubular diameter via several pathways, and therefore, as in ADPKD, many ciliopathies include cystic development of the renal architecture.[28,90] Despite the ciliary membrane being continuous with the plasma membrane, the cilium is in fact separated from the cell body, being gated at the base by structures associated with the basal body. Proteins that reside in the ciliary membrane or cilioplasm must be transported into the cilium by the intraflagellar transport system.[91,92]

Many efforts have been made to understand the route of polycystin trafficking through the cell and toward the ciliary membrane. Some studies suggest that PC2 is able to localize to cilia independently of PC1,[93,94] while others indicate that the polycystins localize only when in complex.[27,95,55] PC1 has been proposed to traffic to cilia from the trans-Golgi network (TGN) via post-Golgi vesicles in an Arf4-dependent manner,[65] whereas a more direct route has been suggested for cilia-destined PC2: this route avoids the TGN and PC2 travels directly from the cis-Golgi compartment.[94]

Analysis of the ciliary polycystin complex presents a unique challenge due to PC1 GPS cleavage and other posttranslational modifications such as *N*-glycosylation, both of which cannot be easily assessed directly by immunofluorescence. Biochemical analysis of the polycystin complex on isolated cilia provides an effective way to directly assess its molecular composition and modification in cilia. This is highlighted by the study performed by Kim et al.,[12] which showed that the cilia portion of polycystins is distinct from the cellular portions in terms of cleavage pattern and *N*-glycosylations. Cilia isolated from MDCK monolayers with stable and inducible expression of recombinant PC1 were shown to contain Endo-H resistant, PNGase-F sensitive PC1, which migrated to a size consistent with the $PC1_{NTF}$ or $PC1^{deN}$. The uncleaved full-length PC1 ($PC1^{U}$) was absent in cilia but present at a ~1:1 ratio with cleaved PC1 in the cell body. This indicates that PC1 cleavage is required for cilium entry. PC2 was also shown to be Endo-H resistant in the cilia fraction and is therefore consistent with the polycystins traveling to the cilium in complex.

Two hypotheses exist as to why only cleaved PC1 is found in the cilium: either PC1 cleavage rapidly occurs within the cilium at a highly efficient rate leaving the uncleaved species undetectable in the cilia preparations or the uncleaved form is specifically excluded from the cilium. The no-cleavable $PC1^{V}$ mutant was employed to elucidate the answer to this question.[12] Interestingly, not only was the uncleaved PC1 unable to enter the cilium, but PC2 was also undetectable within the cilium in MDCK cells with overexpression of $PC1^{V}$. This strongly suggests that cleavage of PC1 is essential for ciliary trafficking of both PC1 and PC2, and for proper ciliary function to maintain proper tubule architecture.

1.7.1 Protocol IV: Cilia Preparation of MDCK Cells for Biochemical Study

MDCK Cell Growth

The most effective way to obtain high numbers of long cilia is to grow MDCK cells on a 10 cm transwell plate (Corning) at 100% confluence for around 7–10 days (usually 10 days). Use DMEM + 10% FBS at the outside membrane and DMEM/NO FBS at the inside membrane of the transwell plate. Change the media every 2–3 days. This can be accomplished using a regular cell culture plate, although a transwell plate is more effective, by using low glucose DMEM + 10% FBS (this medium mimics a similar condition as serum starvation to cause longer cilia).

Cilia Preparation[12,14]

1. Wash ciliary-conditioned MDCK cells gently with DMEM media (without FBS) 3 times to remove any cell debris.
2. Incubate the monolayer in 30 mM ammonium sulfate for 3 h, which will induce shedding of intact cilia.
3. Collect the medium supernatant containing the cilia. Centrifuge at 2000 × g for 30 min to remove any floating cells.
4. Centrifuge again at 10,000 × g for 30 min to remove the cell debris.
5. Recentrifuge at 16,000 × g for 30 min; the pellet contains the cilia population.

6. Resuspend the cilia pellet in PBS for immunofluorescence or in lysis buffer (20 mM sodium phosphate, pH 7.2; 150 mM NaCl; 1 mM EDTA;10% (vol/vol) glycerol; 1% Triton X-100; and protease inhibitors) for Western blot.

1.7.2 Quality Control of Ciliary Preparation: Immunofluorescence

Prior to undertaking Western blot quality control for the cilia preparation, immunofluorescent screening can be used to assess the effectiveness of deciliation and cilia isolation.

Having isolated the cilia and cellular fraction by the method described in Protocol IV, take a small aliquot of both fractions for assessment by immunofluorescence following one of two protocols, as follows.

For Cilia Preparation

Method A

1. Resuspend the cilia fraction in PBS.
2. Pipette the cilia suspension onto a poly-L-lysine–coated slide and cover with a siliconized coverslip.
3. Chill the slide on dry ice.
4. Remove the coverslip.
5. Fix with 4% paraformaldehyde for 15 min.
6. Permeate with 0.1% Triton X 100 in PBS for 5 min.
7. Wash with PBS.
8. Block 3% BSA or FBS in PBS for 30 min.
9. Incubate with anti-acetylated tubulin antibody such as mouse anti-acetylated tubulin from Sigma-Aldrich (T7451, 1:4000) or with rabbit anti-Arl13b from Proteintech (17711-1-AP, 1:200) in 4% FBS for 30–60 min at room temperature.
10. Wash with PBS for 10 min. Repeat 3 times.
11. Incubate secondary Ab in a block solution for 60 min.
12. Wash cells with PBS for 10 min. Repeat 3 times.
13. Stain with DAPI.
14. Mount.

Method B

1. Resuspend the cilia fraction in 4% FBS for 30 min blocking period at room temperature.
2. Centrifuge at 16,000 × g for 30 min to pellet the cilia; remove the blocking solution.
3. Resuspend pellet with primary antibody such as mouse anti-acetylated tubulin from Sigma-Aldrich (T7451, 1:4000) or rabbit anti-Arl13b from Proteintech (17711-1-AP, 1:200) diluted in 4% FBS for 30–60 min at room temperature.

4. Centrifuge at 16,000 × g for 30 min to pellet the cilia; remove the primary antibody solution.
5. Carefully wash the pellet with PBS before resuspending in fluorescent secondary antibody solution directed toward the selected primary antibody. Dilute the secondary antibody in 4% FBS and incubate for 30 min at room temperature in the dark.
6. Centrifuge at 16,000 × g for 30 min to pellet the cilia; remove the secondary antibody solution.
7. Wash with PBS and resuspend in DAPI solution for 5–10 min. (1 μg/mL, a low concentration of DAPI will ensure that background staining is kept to a minimum. Do not use mounting medium containing DAPI.)
8. Centrifuge at 16,000 × g for 30 min to pellet the cilia; remove the DAPI solution.
9. Wash the pellet with PBS being careful not to disturb the pellet.
10. Resuspend the cilia in mounting medium before mounting on a glass slide and coverslipping. Visualize the slide on a fluorescent microscope.

Note that the cilia fractions should not be stained by DAPI but should stain positively with cilia markers.

For Cell Body Fraction

The cell body fraction can also be stained to determine efficiency of deciliation using Method B. Effective deciliation is indicated by little or no staining with cilia markers.

1.7.3 Quality Control of Ciliary Preparation: Western Blot

Assessing the purity of the cilia fraction is essential because cellular contamination will confound results. There should be no contamination from cell body cytoskeleton proteins such as β-Actin, secreted protein such as IGFBP-2, and exosomal protein such as Alix (an abundant component of exosome proteomes), nor from cellular organelles such as nuclear, Golgi, or ER proteins. Although immunofluorescence provides reassurance that whole cells are not contaminating the cilia portion, it is still possible that some plasma membrane fragments remain attached to the cilium preparation. The purity should thus be further validated by absence of plasma membrane protein such as c-Met (an apical plasma membrane protein).[12] Since protein levels are likely to be low, it may be favorable to probe one blot with a number of different antibodies, making use of different molecular weight, fluorescent colors, or stripping and sequential probing.

Once the purity of cilia fraction is ascertained, the sample can be used for biochemical analysis for the polycystin complex or other proteins by Western blot or other methods.

1.8 CONCLUDING REMARKS

Throughout this chapter, we have provided detailed protocols to enable detection and assessment of the three distinct species of PC1: $PC1^{U}$, $PC1^{cFL}$, and $PC1^{DeN}$. We highlighted the need for rigorous, genetically defined controls for the antibodies and the importance of understanding where the selected antibody binds and which PC1 species it is able to detect. We have described a number of useful antibodies that can be used to biochemically detect each of the three PC1 species and provided advice concerning the interpretation of results. However, there is still no suitable antibody that can reliably detect endogenous PC1 for immunofluorescence, immunoelectron microscopy, or immunohistochemistry. There is also demand to develop antibodies that can differentiate between uncleaved and cleaved forms of PC1 species in the native state by recognizing conformational changes caused by GPS cleavage. Finally, there is a need to develop PC1-specific antibodies that can modulate the activities of the polycystin complex and have therapeutic potential for ADPKD. Development of those antibodies and their standardized protocols will provide a plethora of long-overdue and critical information for the field.

ACKNOWLEDGMENTS

This work was supported by grants from the National Institute of Diabetes and Digestive and Kidney Diseases (NIDDK) (R01DK111611 to FQ) and the Baltimore Polycystic Kidney Disease Research and Clinical Core Center (P30DK090868). This work was also supported in part by a research Grant-in-Aid from the Polycystic Kidney Disease Foundation (Grant #195G14a to FQ) and a grant from the Living Legacy Foundation of Maryland (to FQ).

REFERENCES

1. Boletta, A, Germino, GG: Role of polycystins in renal tubulogenesis. *Trends Cell Biol*, 13: 484–492, 2003.
2. Boletta, A, Qian, F, Onuchic, LF, Bhunia, AK, Phakdeekitcharoen, B, Hanaoka, K, Guggino, W, Monaco, L, Germino, GG: Polycystin-1, the gene product of *PKD1*, induces resistance to apoptosis and spontaneous tubulogenesis in MDCK cells. *Mol Cell*, 6: 1267–1273, 2000.
3. Grimm, DH, Karihaloo, A, Cai, Y, Somlo, S, Cantley, LG, Caplan, MJ: Polycystin-2 regulates proliferation and branching morphogenesis in kidney epithelial cells. *J Biol Chem*, 281: 137–144, 2006.
4. Nickel, C, Benzing, T, Sellin, L, Gerke, P, Karihaloo, A, Liu, ZX, Cantley, LG, Walz, G: The polycystin-1 C-terminal fragment triggers branching morphogenesis and migration of tubular kidney epithelial cells. *J Clin Invest*, 109: 481–489, 2002.
5. Hateboer, N, v Dijk, MA, Bogdanova, N, Coto, E, Saggar-Malik, AK, San Millan, JL, Torra, R, Breuning, M, Ravine, D: Comparison of phenotypes of polycystic kidney disease types 1 and 2. European PKD1-PKD2 Study Group. *Lancet*, 353: 103–107, 1999.
6. Grantham, JJ, Cook, LT, Wetzel, LH, Cadnapaphornchai, MA, Bae, KT: Evidence of extraordinary growth in the progressive enlargement of renal cysts. *Clin J Am Soc Nephrol*, 5: 889–896, 2010.
7. Fedeles, SV, Gallagher, AR, Somlo, S: Polycystin-1: A master regulator of intersecting cystic pathways. *Trends Mol Med*, 20: 251–260, 2014.
8. Hopp, K, Ward, CJ, Hommerding, CJ, Nasr, SH, Tuan, HF, Gainullin, VG, Rossetti, S, Torres, VE, Harris, PC: Functional polycystin-1 dosage governs autosomal dominant polycystic kidney disease severity. *J Clin Invest*, 122: 4257–4273, 2012.

9. Kurbegovic, A, Kim, H, Xu, H, Yu, S, Cruanes, J, Maser, RL, Boletta, A, Trudel, M, Qian, F: Novel functional complexity of polycystin-1 by GPS cleavage in vivo: Role in polycystic kidney disease. *Mol Cell Biol*, 34: 3341–3353, 2014.
10. Wei, W, Hackmann, K, Xu, H, Germino, G, Qian, F: Characterization of cis-autoproteolysis of polycystin-1, the product of human polycystic kidney disease 1 gene. *J Biol Chem*, 282: 21729–21737, 2007.
11. Yu, S, Hackmann, K, Gao, J, He, X, Piontek, K, Garcia-Gonzalez, MA, Menezes, LF, Xu, H, Germino, GG, Zuo, J, Qian, F: Essential role of cleavage of polycystin-1 at G protein-coupled receptor proteolytic site for kidney tubular structure. *Proc Natl Acad Sci USA*, 104: 18688–18693, 2007.
12. Kim, H, Xu, H, Yao, Q, Li, W, Huang, Q, Outeda, P, Cebotaru, V, Chiaravalli, M, et al.: Ciliary membrane proteins traffic through the Golgi via a Rabepl/GGA1/Arl3-dependent mechanism. *Nat Commun*, 5: 5482, 2014.
13. Qian, F: *Polycystin-1*, Academic Press, Elsevier, 2012.
14. Freedman, BS, Lam, AQ, Sundsbak, JL, Iatrino, R, Su, X, Koon, SJ, Wu, M, Daheron, L, Harris, PC, Zhou, J, Bonventre, JV: Reduced ciliary polycystin-2 in induced pluripotent stem cells from polycystic kidney disease patients with *PKD1* mutations. *J Am Soc Nephrol*, 24: 1571–1586, 2013.
15. Chapin, HC, Caplan, MJ: The cell biology of polycystic kidney disease. *J Cell Biol*, 191: 701–710, 2010.
16. Trudel, M, Yao, Q, Qian, F: The role of G-protein-coupled receptor proteolysis site cleavage of polycystin-1 in renal physiology and polycystic kidney disease. *Cells*, 5, 2016.
17. Qian, F, Boletta, A, Bhunia, AK, Xu, H, Liu, L, Ahrabi, AK, Watnick, TJ, Zhou, F, Germino, GG: Cleavage of polycystin-1 requires the receptor for egg jelly domain and is disrupted by human autosomal-dominant polycystic kidney disease 1-associated mutations. *Proc Natl Acad Sci USA*, 99: 16981–16986, 2002.
18. Garcia-Gonzalez, MA, Jones, JG, Allen, SK, Palatucci, CM, Batish, SD, Seltzer, WK, Lan, Z, Allen, E, Qian, F, Lens, XM, Pei, Y, Germino, GG, Watnick, TJ: Evaluating the clinical utility of a molecular genetic test for polycystic kidney disease. *Mol Genet Metab*, 92: 160–167, 2007.
19. Arac, D, Boucard, AA, Bolliger, MF, Nguyen, J, Soltis, SM, Sudhof, TC, Brunger, AT: A novel evolutionarily conserved domain of cell-adhesion GPCRs mediates autoproteolysis. *EMBO J*, 31: 1364–1378, 2012.
20. Busch, T, Kottgen, M, Hofherr, A: TRPP2 ion channels: Critical regulators of organ morphogenesis in health and disease. *Cell Calcium*, 66: 25–32, 2017.
21. Tsiokas, L, Kim, S, Ong, EC: Cell biology of polycystin-2. *Cell Signal*, 19: 444–453, 2007.
22. Koulen, P, Cai, Y, Geng, L, Maeda, Y, Nishimura, S, Witzgall, R, Ehrlich, BE, Somlo, S: Polycystin-2 is an intracellular calcium release channel. *Nat Cell Biol*, 4: 191–197, 2002.
23. Qian, F, Germino, FJ, Cai, Y, Zhang, X, Somlo, S, Germino, GG: *PKD1* interacts with *PKD2* through a probable coiled-coil domain. *Nat Genet*, 16: 179–183, 1997.
24. Su, Q, Hu, F, Ge, X, Lei, J, Yu, S, Wang, T, Zhou, Q, Mei, C, Shi, Y: Structure of the human *PKD1-PKD2* complex. *Science*, 361, 2018.
25. Tsiokas, L, Kim, E, Arnould, T, Sukhatme, VP, Walz, G: Homo- and heterodimeric interactions between the gene products of *PKD1* and *PKD2*. *Proc Natl Acad Sci USA*, 94: 6965–6970, 1997.
26. Hanaoka, K, Qian, F, Boletta, A, Bhunia, AK, Piontek, K, Tsiokas, L, Sukhatme, VP, Guggino, WB, Germino, GG: Co-assembly of polycystin-1 and -2 produces unique cation-permeable currents. *Nature*, 408: 990–994, 2000.
27. Nauli, SM, Alenghat, FJ, Luo, Y, Williams, E, Vassilev, P, Li, X, Elia, AE, Lu, W, Brown, EM, Quinn, SJ, Ingber, DE, Zhou, J: Polycystins 1 and 2 mediate mechanosensation in the primary cilium of kidney cells. *Nat Genet*, 33: 129–137, 2003.

28. Ma, M, Gallagher, AR, Somlo, S: Ciliary mechanisms of cyst formation in polycystic kidney disease. *Cold Spring Harb Perspect Biol*, 9, 2017.
29. Delling, M, DeCaen, PG, Doerner, JF, Febvay, S, Clapham, DE: Primary cilia are specialized calcium signalling organelles. *Nature*, 504: 311–314, 2013.
30. Rowe, I, Chiaravalli, M, Mannella, V, Ulisse, V, Quilici, G, Pema, M, Song, XW, Xu, H, Mari, S, Qian, F, Pei, Y, Musco, G, Boletta, A: Defective glucose metabolism in polycystic kidney disease identifies a new therapeutic strategy. *Nat Med*, 19: 488–493, 2013.
31. Menezes, LF, Zhou, F, Patterson, AD, Piontek, KB, Krausz, KW, Gonzalez, FJ, Germino, GG: Network analysis of a *Pkd1*-mouse model of autosomal dominant polycystic kidney disease identifies HNF4alpha as a disease modifier. *PLOS Genet*, 8: e1003053, 2012.
32. Woodward, OM, Li, Y, Yu, S, Greenwell, P, Wodarczyk, C, Boletta, A, Guggino, WB, Qian, F: Identification of a polycystin-1 cleavage product, P100, that regulates store operated Ca entry through interactions with STIM1. *PLOS ONE*, 5: e12305, 2010.
33. Chauvet, V, Tian, X, Husson, H, Grimm, DH, Wang, T, Hiesberger, T, Igarashi, P, Bennett, AM, Ibraghimov-Beskrovnaya, O, Somlo, S, Caplan, MJ: Mechanical stimuli induce cleavage and nuclear translocation of the polycystin-1 C terminus. *J Clin Invest*, 114: 1433–1443, 2004.
34. Lin, CC, Kurashige, M, Liu, Y, Terabayashi, T, Ishimoto, Y, Wang, T, Choudhary, V, Hobbs, R, Liu, LK, Lee, PH, Outeda, P, Zhou, F, Restifo, NP, Watnick, T, Kawano, H, Horie, S, Prinz, W, Xu, H, Menezes, LF, Germino, GG: A cleavage product of Polycystin-1 is a mitochondrial matrix protein that affects mitochondria morphology and function when heterologously expressed. *Sci Rep*, 8: 2743, 2018.
35. Low, SH, Vasanth, S, Larson, CH, Mukherjee, S, Sharma, N, Kinter, MT, Kane, ME, Obara, T, Weimbs, T: Polycystin-1, STAT6, and P100 function in a pathway that transduces ciliary mechanosensation and is activated in polycystic kidney disease. *Dev Cell*, 10: 57–69, 2006.
36. Hughes, J, Ward, CJ, Peral, B, Aspinwall, R, Clark, K, San Millan, JL, Gamble, V, Harris, PC: The polycystic kidney disease 1 (*PKD1*) gene encodes a novel protein with multiple cell recognition domains. *Nat Genet*, 10: 151–160, 1995.
37. Moy, GW, Mendoza, LM, Schulz, JR, Swanson, WJ, Glabe, CG, Vacquier, VD: The sea urchin sperm receptor for egg jelly is a modular protein with extensive homology to the human polycystic kidney disease protein, PKD1. *J Cell Biol*, 133: 809–817, 1996.
38. Schroder, S, Fraternali, F, Quan, X, Scott, D, Qian, F, Pfuhl, M: When a module is not a domain: The case of the REJ module and the redefinition of the architecture of polycystin-1. *Biochem J*, 435: 651–660, 2011.
39. Qian, F, Wei, W, Germino, G, Oberhauser, A: The nanomechanics of polycystin-1 extracellular region. *J Biol Chem*, 280: 40723–40730, 2005.
40. Saigusa, T, Bell, PD: Molecular pathways and therapies in autosomal-dominant polycystic kidney disease. *Physiology (Bethesda)*, 30: 195–207, 2015.
41. Li, Y, Santoso, NG, Yu, S, Woodward, OM, Qian, F, Guggino, WB: Polycystin-1 interacts with inositol 1,4,5-trisphosphate receptor to modulate intracellular Ca2+ signaling with implications for polycystic kidney disease. *J Biol Chem*, 284: 36431–36441, 2009.
42. Manzati, E, Aguiari, G, Banzi, M, Manzati, M, Selvatici, R, Falzarano, S, Maestri, I, Pinton, P, Rizzuto, R, del Senno, L: The cytoplasmic C-terminus of polycystin-1 increases cell proliferation in kidney epithelial cells through serum-activated and Ca(2+)-dependent pathway(s). *Exp Cell Res*, 304: 391–406, 2005.
43. Kim, E, Arnould, T, Sellin, LK, Benzing, T, Fan, MJ, Gruning, W, Sokol, SY, Drummond, I, Walz, G: The polycystic kidney disease 1 gene product modulates Wnt signaling. *J Biol Chem*, 274: 4947–4953, 1999.
44. Shillingford, JM, Murcia, NS, Larson, CH, Low, SH, Hedgepeth, R, Brown, N, Flask, CA, Novick, AC, Goldfarb, DA, Kramer-Zucker, A, Walz, G, Piontek, KB, Germino, GG, Weimbs, T: The mTOR pathway is regulated by polycystin-1, and its inhibition

reverses renal cystogenesis in polycystic kidney disease. *Proc Natl Acad Sci USA*, 103: 5466–5471, 2006.
45. Parnell, SC, Magenheimer, BS, Maser, RL, Rankin, CA, Smine, A, Okamoto, T, Calvet, JP: The polycystic kidney disease-1 protein, polycystin-1, binds and activates heterotrimeric G-proteins in vitro. *Biochem Biophys Res Commun*, 251: 625–631, 1998.
46. Parnell, SC, Magenheimer, BS, Maser, RL, Zien, CA, Frischauf, AM, Calvet, JP: Polycystin-1 activation of c-Jun N-terminal kinase and AP-1 is mediated by heterotrimeric G proteins. *J Biol Chem*, 277: 19566–19572, 2002.
47. Parnell, SC, Magenheimer, BS, Maser, RL, Pavlov, TS, Havens, MA, Hastings, ML, Jackson, SF, Ward, CJ, Peterson, KR, Staruschenko, A, Calvet, JP: A mutation affecting polycystin-1 mediated heterotrimeric G-protein signaling causes PKD. *Hum Mol Genet*, 27: 3313–3324, 2018.
48. Delmas, P, Nomura, H, Li, X, Lakkis, M, Luo, Y, Segal, Y, Fernandez-Fernandez, JM, Harris, P, Frischauf, AM, Brown, DA, Zhou, J: Constitutive activation of G-proteins by polycystin-1 is antagonized by polycystin-2. *J Biol Chem*, 277: 11276–11283, 2002.
49. Zhang, B, Tran, U, Wessely, O: Polycystin 1 loss of function is directly linked to an imbalance in G-protein signaling in the kidney. *Development*, 145, 2018.
50. Ponting, CP, Hofmann, K, Bork, P: A latrophilin/CL-1-like GPS domain in polycystin-1. *Curr Biol*, 9: R585–588, 1999.
51. Krasnoperov, VG, Bittner, MA, Beavis, R, Kuang, Y, Salnikow, KV, Chepurny, OG, Little, AR, Plotnikov, AN, Wu, D, Holz, RW, Petrenko, AG: Alpha-Latrotoxin stimulates exocytosis by the interaction with a neuronal G-protein-coupled receptor. *Neuron*, 18: 925–937, 1997.
52. Promel, S, Langenhan, T, Arac, D: Matching structure with function: The GAIN domain of adhesion-GPCR and PKD1-like proteins. *Trends Pharmacol Sci*, 34: 470–478, 2013.
53. Bjarnadottir, TK, Fredriksson, R, Hoglund, PJ, Gloriam, DE, Lagerstrom, MC, Schioth, HB: The human and mouse repertoire of the adhesion family of G-protein-coupled receptors. *Genomics*, 84: 23–33, 2004.
54. Bjarnadottir, TK, Fredriksson, R, Schioth, HB: The Adhesion GPCRs: A unique family of G protein-coupled receptors with important roles in both central and peripheral tissues. *Cell Mol Life Sci*, 2007.
55. Overgaard, CE, Sanzone, KM, Spiczka, KS, Sheff, DR, Sandra, A, Yeaman, C: Deciliation is associated with dramatic remodeling of epithelial cell junctions and surface domains. *Mol Biol Cell*, 20: 102–113, 2009.
56. Lin, HH, Chang, GW, Davies, JQ, Stacey, M, Harris, J, Gordon, S: Autocatalytic cleavage of the EMR2 receptor occurs at a conserved G protein-coupled receptor proteolytic site motif. *J Biol Chem*, 279: 31823–31832, 2004.
57. Guo, HC, Xu, Q, Buckley, D, Guan, C: Crystal structures of Flavobacterium glycosylasparaginase. An N-terminal nucleophile hydrolase activated by intramolecular proteolysis. *J Biol Chem*, 273: 20205–20212, 1998.
58. Xu, Q, Buckley, D, Guan, C, Guo, HC: Structural insights into the mechanism of intramolecular proteolysis. *Cell*, 98: 651–661, 1999.
59. Lee, JJ, Ekker, SC, von Kessler, DP, Porter, JA, Sun, BI, Beachy, PA: Autoproteolysis in hedgehog protein biogenesis. *Science*, 266: 1528–1537, 1994.
60. Paulus, H: Protein splicing and related forms of protein autoprocessing. *Annu Rev Biochem*, 69: 447–496, 2000.
61. Cai, Y, Fedeles, SV, Dong, K, Anyatonwu, G, Onoe, T, Mitobe, M, Gao, JD, Okuhara, D, Tian, X, Gallagher, AR, Tang, Z, Xie, X, Lalioti, MD, Lee, AH, Ehrlich, BE, Somlo, S: Altered trafficking and stability of polycystins underlie polycystic kidney disease. *J Clin Invest*, 124: 5129–5144, 2014.
62. Gainullin, VG, Hopp, K, Ward, CJ, Hommerding, CJ, Harris, PC: Polycystin-1 maturation requires polycystin-2 in a dose-dependent manner. *J Clin Invest*, 125: 607–620, 2015.

63. Puertollano, R, Randazzo, PA, Presley, JF, Hartnell, LM, Bonifacino, JS: The GGAs promote ARF-dependent recruitment of clathrin to the TGN. *Cell*, 105: 93–102, 2001.
64. Schrick, JJ, Vogel, P, Abuin, A, Hampton, B, Rice, DS: ADP-ribosylation factor-like 3 is involved in kidney and photoreceptor development. *Am J Pathol*, 168: 1288–1298, 2006.
65. Ward, HH, Brown-Glaberman, U, Wang, J, Morita, Y, Alper, SL, Bedrick, EJ, Gattone, VH, 2nd, Deretic, D, Wandinger-Ness, A: A conserved signal and GTPase complex are required for the ciliary transport of polycystin-1. *Mol Biol Cell*, 22: 3289–3305, 2011.
66. Castelli, M, Boca, M, Chiaravalli, M, Ramalingam, H, Rowe, I, Distefano, G, Carroll, T, Boletta, A: Polycystin-1 binds Par3/aPKC and controls convergent extension during renal tubular morphogenesis. *Nat Commun*, 4: 2658, 2013.
67. Paavola, KJ, Stephenson, JR, Ritter, SL, Alter, SP, Hall, RA: The N terminus of the adhesion G protein-coupled receptor GPR56 controls receptor signaling activity. *J Biol Chem*, 286: 28914–28921, 2011.
68. Paavola, KJ, Hall, RA: Adhesion G protein-coupled receptors: Signaling, pharmacology, and mechanisms of activation. *Mol Pharmacol*, 82: 777–783, 2012.
69. Freeze, HH: Use of glycosidases to study protein trafficking. *Curr Protoc Cell Biol*, Chapter 15: Unit 15.2, 2001.
70. Kornfeld, R, Kornfeld, S: Assembly of asparagine-linked oligosaccharides. *Annu Rev Biochem*, 54: 631–664, 1985.
71. Nauta, J, Goedbloed, MA, van den Ouweland, AM, Nellist, M, Hoogeveen, AT: Immunological detection of polycystin-1 in human kidney. *Histochem Cell Biol*, 113: 303–311, 2000.
72. Palsson, R, Sharma, CP, Kim, K, McLaughlin, M, Brown, D, Arnaout, MA: Characterization and cell distribution of polycystin, the product of autosomal dominant polycystic kidney disease gene 1. *Mol Med*, 2: 702–711, 1996.
73. Geng, L, Segal, Y, Pavlova, A, Barros, EJ, Lohning, C, Lu, W, Nigam, SK, Frischauf, AM, Reeders, ST, Zhou, J: Distribution and developmentally regulated expression of murine polycystin. *Am J Physiol*, 272: F451–459, 1997.
74. Van Adelsberg, J, Chamberlain, S, D'Agati, V: Polycystin expression is temporally and spatially regulated during renal development. *Am J Physiol*, 272: F602–609, 1997.
75. Huan, Y, van Adelsberg, J: Polycystin-1, the *PKD1* gene product, is in a complex containing E-cadherin and the catenins. *J Clin Invest*, 104: 1459–1468, 1999.
76. Weston, BS, Jeffery, S, Jeffrey, I, Sharaf, SF, Carter, N, Saggar-Malik, A, Price, RG: Polycystin expression during embryonic development of human kidney in adult tissues and ADPKD tissue. *Histochem J*, 29: 847–856, 1997.
77. Ward, CJ, Turley, H, Ong, AC, Comley, M, Biddolph, S, Chetty, R, Ratcliffe, PJ, Gattner, K, Harris, PC: Polycystin, the polycystic kidney disease 1 protein, is expressed by epithelial cells in fetal, adult, and polycystic kidney. *Proc Natl Acad Sci USA*, 93: 1524–1528, 1996.
78. Roitbak, T, Ward, CJ, Harris, PC, Bacallao, R, Ness, SA, Wandinger-Ness, A: A polycystin-1 multiprotein complex is disrupted in polycystic kidney disease cells. *Mol Biol Cell*, 15: 1334–1346, 2004.
79. Roitbak, T, Surviladze, Z, Tikkanen, R, Wandinger-Ness, A: A polycystin multiprotein complex constitutes a cholesterol-containing signalling microdomain in human kidney epithelia. *Biochem J*, 392: 29–38, 2005.
80. Scheffers, MS, van der Bent, P, Prins, F, Spruit, L, Breuning, MH, Litvinov, SV, de Heer, E, Peters, DJ: Polycystin-1, the product of the polycystic kidney disease 1 gene, co-localizes with desmosomes in MDCK cells. *Hum Mol Genet*, 9: 2743–2750, 2000.
81. Geng, L, Segal, Y, Peissel, B, Deng, N, Pei, Y, Carone, F, Rennke, HG, Glucksmann-Kuis, AM, Schneider, MC, Ericsson, M, Reeders, ST, Zhou, J: Identification and localization of polycystin, the *PKD1* gene product. *J Clin Invest*, 98: 2674–2682, 1996.

82. Lu, W, Shen, X, Pavlova, A, Lakkis, M, Ward, CJ, Pritchard, L, Harris, PC, Genest, DR, Perez-Atayde, AR, Zhou, J: Comparison of *Pkd1*-targeted mutants reveals that loss of polycystin-1 causes cystogenesis and bone defects. *Hum Mol Genet*, 10: 2385–2396, 2001.
83. Ong, AC, Harris, PC, Davies, DR, Pritchard, L, Rossetti, S, Biddolph, S, Vaux, DJ, Migone, N, Ward, CJ: Polycystin-1 expression in PKD1, early-onset PKD1, and TSC2/PKD1 cystic tissue. *Kidney Int*, 56: 1324–1333, 1999.
84. Pema, M, Drusian, L, Chiaravalli, M, Castelli, M, Yao, Q, Ricciardi, S, Somlo, S, Qian, F, Biffo, S, Boletta, A: mTORC1-mediated inhibition of polycystin-1 expression drives renal cyst formation in tuberous sclerosis complex. *Nat Commun*, 7: 10786, 2016.
85. Newby, LJ, Streets, AJ, Zhao, Y, Harris, PC, Ward, CJ, Ong, AC: Identification, characterization, and localization of a novel kidney polycystin-1-polycystin-2 complex. *J Biol Chem*, 277: 20763–20773, 2002.
86. Cai, Y, Maeda, Y, Cedzich, A, Torres, VE, Wu, G, Hayashi, T, Mochizuki, T, Park, JH, Witzgall, R, Somlo, S: Identification and characterization of polycystin-2, the *PKD2* gene product. *J Biol Chem*, 274: 28557–28565, 1999.
87. Wu, G, D'Agati, V, Cai, Y, Markowitz, G, Park, JH, Reynolds, DM, Maeda, Y, Le, TC, Hou, H, Jr., Kucherlapati, R, Edelmann, W, Somlo, S: Somatic inactivation of *Pkd2* results in polycystic kidney disease. *Cell*, 93: 177–188, 1998.
88. Singla, V, Reiter, JF: The primary cilium as the cell's antenna: Signaling at a sensory organelle. *Science (New York, NY)*, 313: 629–633, 2006.
89. Hildebrandt, F, Benzing, T, Katsanis, N: Ciliopathies. *N Engl J Med*, 364: 1533–1543, 2011.
90. Ma, M, Tian, X, Igarashi, P, Pazour, GJ, Somlo, S: Loss of cilia suppresses cyst growth in genetic models of autosomal dominant polycystic kidney disease. *Nat Genet*, 45: 1004–1012, 2013.
91. Garcia-Gonzalo, FR, Reiter, JF: Open Sesame: How Transition Fibers and the Transition Zone Control Ciliary Composition. *Cold Spring Harb Perspect Biol*, 9, 2017.
92. Nachury, MV, Seeley, ES, Jin, H: Trafficking to the ciliary membrane: How to get across the periciliary diffusion barrier? *Annu Rev Cell Dev Biol*, 26: 59–87, 2010.
93. Geng, L, Okuhara, D, Yu, Z, Tian, X, Cai, Y, Shibazaki, S, Somlo, S: Polycystin-2 traffics to cilia independently of polycystin-1 by using an N-terminal RVxP motif. *J of Cell Sci*, 119: 1383–1395, 2006.
94. Hoffmeister, H, Babinger, K, Gurster, S, Cedzich, A, Meese, C, Schadendorf, K, Osten, L, de Vries, U, Rascle, A, Witzgall, R: Polycystin-2 takes different routes to the somatic and ciliary plasma membrane. *J Cell Biol*, 192: 631–645, 2011.
95. Chapin, HC, Rajendran, V, Caplan, MJ: Polycystin-1 surface localization is stimulated by polycystin-2 and cleavage at the G protein-coupled receptor proteolytic site. *Mol Biol Cell*, 21: 4338–4348, 2010.

2 Structural Determination of the Polycystin-2 Channel by Electron Cryo-Microscopy

Qinzhe Wang and Erhu Cao

CONTENTS

2.1 INTRODUCTION

2.1.1 Polycystins and Autosomal Dominant Polycystic Kidney Disease

Autosomal dominant polycystic kidney disease (ADPKD) is one of the most common genetic disorders, with an estimated occurrence of 1 in every 400–1000 individuals. ADPKD causes life-threatening complications that affect the kidneys, the cardiovascular system, and other organs such as the liver and pancreas.[1–6] ADPKD is typically characterized by progressive accumulation of fluid-filled bilateral renal cysts that leads to kidney enlargement, normal tissue loss, and, consequently, a decline in renal functions. Normal kidney architecture is ultimately destroyed in half of all ADPKD patients by the age of 60 years, leading to the end-stage renal disease that requires kidney dialysis or transplantation. A significant portion of ADPKD patients also develop extra-renal symptoms such as hypertension, intracranial aneurysms, and hepatic and pancreatic cysts.[2]

In the early 1990s, *PKD1* and *PKD2* were positionally cloned as two causative genes that are mutated in patients with ADPKD.[7–9] About 85% of clinical mutations map onto the *PKD1* gene, and the remaining 15% of clinical mutations locate on the *PKD2* gene.[10] Functional impairment resulting from, presumably, loss-of-function mutations of a single copy of either the *PKD1* or *PKD2* gene is sufficient to drive cyst formation.[8,11] Homozygous mutations result in embryonic lethality in various mouse models,[1] likely due to placenta defects,[12] heart development failure,[13] and pancreas and kidney abnormalities.[4,14] Of note, pathogenic missense mutations of *PKD1* or *PKD2* are not evenly distributed along their primary sequences and instead enrich in particular domains, so-called hotspots such as the polycystin-1 GAIN domain and the TOP domain in both polycystin proteins, likely reflecting the important roles of these domains in ligand recognition, channel regulation, or trafficking of polycystins. Besides abundant point mutations, many ADPKD patients have truncated polycystin proteins due to large deletions, frameshift, or premature termination mutations.[15,16]

Polycystin-1, encoded by the *PKD1* gene, is an 11-membrane-spanning receptor-like protein remarkable for harboring a large extracellular region (>3000 amino acids) that consists of multiple predicted ligand binding and/or adhesive modules.[17–19] It has been suggested that polycystin-1 co-assembles with polycystin-2,[20–23] forming a receptor/ion channel complex in the primary cilia that contributes to flow and mechanical sensing[24] and/or detects and responds to chemical ligands, such as Wnts.[25] A conserved amino acid sequence at the very carboxyl end of human polycystin-1 has been identified as a cilia localization signal.[26] Besides cilia, plasma membrane, and endoplasmic reticulum (ER) membrane,[27,28] the polycystin complex has also been identified in secreted exosomes in urine,[29] resembling LOV-1 and PKD-2, two worm polycystin homologues that were also found in exosomes involved in cellular communication.[30] Despite extensive efforts by many research groups, it remains incompletely understood how polycystin-1 is activated by ligands and modulated by cellular factors and, once activated, how it regulates downstream effectors and signaling cascades. Nevertheless, polycystin-1 was reported to regulate G-protein signaling,[31–38] the Hippo signaling pathway,[39] and cellular cAMP levels.[40] Polycystin-1 function can also be modulated by direct binding with calmodulin,[41] as well as posttranslational modifications such as phosphorylation and palmitoylation.[42,43]

Polycystin-2 is a member of the large tetrameric transient receptor potential (TRP) ion channel family.[44,45] Polycystin-2 shares a similar architecture with other TRPs and related voltage-gated ion channels (VGICs) within the transmembrane core where the last two transmembrane helices and the intervening pore loop (S5-P-S6) from all four subunits come together to form the central pore domain; this central pore is flanked by four voltage sensor-like domains encompassing the first four helices (S1-S4). Beyond this conserved transmembrane core, polycystin-2 exhibits multiple regulatory sites and functional domains, including glycogen synthase kinase phosphorylation sites,[46] an extracellular "tetragonal opening for polycystins" (TOP) domain, a calmodulin binding domain, EF-hands,[47,48] a cilia transport motif,[49] the endoplasmic reticulum retention sequence,[50] and the coiled-coil domain.[21] Attempts to record PKD2 currents have been challenging and have produced conflicting channel biophysical properties.[51–57] A gain-of-function PKD2 mutant revealed a nonselective cation current in *Xenopus* oocytes.[58] More recently, a breakthrough in patch clamp of primary cilia has enabled the recording of native PKD2 currents in primary cilia.[59,60] Collectively, polycystin-2 appears to function as a Na^+/K^+-conducting channel with lower permeability and smaller single-channel conductance to Ca^{2+}.

2.1.2 A Brief Primer on TRP Channels

The TRP channels conduct cations and are grouped into seven subfamilies according to the relatedness of their protein sequences: TRPC (canonical), TRPM (melastatin), TRPA (ankyrin), TRPV (vanilloid), TRPN (NOMPC-like), TRPML (mucolipin), and TRPP (polycystin).[44,61] All TRP channel subfamilies, except for TRPN, are present in mammals. In humans, 27 TRP proteins have been identified, making the TRP family the second largest ion channel family, only outnumbered by the potassium-channel family. The TRP channels are sensory proteins that detect and integrate numerous environmental and endogenous stimuli to elicit proper cellular responses. Some TRP channels operate downstream of, or retain the ability to be regulated by, phospholipase C (PLC)-coupled receptors, as first discovered in the ancestral fly TRPs.[62,63] However, for most mammalian TRP channels, the *in vivo* activation mechanisms are yet to be determined.[64]

TRP channels assemble as homo- or hetero-tetramers, sharing a similar membrane topology and subunit organization as the VGICs.[65] Six transmembrane helices (S1-S6) form the conserved membrane-embedded ion channel core, which organizes into two functional modules: an ion conduction pore module formed by two transmembrane helices (S5 and S6) and the intervening pore loop, and a voltage sensor-like domain (VSLD) formed by the remaining four transmembrane helices (S1-S4). The pore module and VSLD are connected by an S4-S5 linker, as observed in VGICs. Notably, the S4 helices of TRP channels lack regularly spaced arginine or lysine residues, a hallmark of VGICs that enables voltage-depending gating; this difference likely explains the only modest voltage sensitivity of TRPs as compared with VGICs. Each TRP subtype is defined by a diverse collection of soluble domains, including cytoplasmic N- and C-terminal domains and extracellular or luminal domains (e.g., in TRPP and TRPML). These domains associate with the transmembrane ion channel core and likely contribute to channel assembly or trafficking, or serve as sites for cellular regulatory factors and ligands.

2.1.3 Electron Cryo-Microscopy (Cryo-EM)

Electron cryo-microscopy, also called cryo-electron microscopy or cryo-EM, became an attractive structural method in the early 1980s when Dubochet and his colleagues invented a practical method for rapidly freezing biological samples within a thin layer of vitreous ice.[66] Preserving biomolecules in a frozen, hydrated state overcame two fundamental obstacles encountered when imaging biological samples by an electron microscope: biological samples rapidly decay upon electron radiation and also dehydrate in the high vacuum of a typical electron microscope. In its earlier applications, cryo-EM has been mostly employed to study structures of large biological assemblies, such as ribosomes and viruses.[67–70] Indeed, near-atomic resolution structures have been determined for favorable large and symmetric icosahedral virus specimens.[71] However, low-resolution cryo-EM reconstructions characterized by blobs of densities are more commonly obtained; these maps are useful for demarcating distinct domains of a protein or localizing individual subunits in a multicomponent complex, but they are inadequate to resolve secondary structures, let alone side-chain densities. Nevertheless, in the 1990s, Richard Henderson and also others predicted that single-particle cryo-EM should theoretically be able to resolve structures of proteins as small as 100 kD at 3 Å resolution.[67,72] These predictions were rather bold at a time when cryo-EM only occupied a special niche in structural biology and was often ridiculed as "blobology" by X-ray crystallographers, but now has proved to be incredibly prescient. Indeed, who could foresee that, two decades later, steady improvements in electron microscopes, detectors, and computational algorithms finally reached an inflection point, leading to the so-called "resolution revolution" that is rapidly transforming structural biology.[73] Application of a new generation of direct electron detectors has been arguably driving this "resolution revolution" since they exhibit drastically improved performance across all spatial frequencies, yielding much sharper images with signals potentially extending to the Nyquist frequency that can then be extracted by various computational software for constructing high-resolution maps.[74,75]

In 2013, TRPV1, the heat- and capsaicin-activated cation channel, was the first membrane protein for which the structures were determined by single-particle cryo-EM at a near-atomic resolution without crystallization.[76,77] Because TRPV1 exhibits rich pharmacology that can be used to trap TRPV1 in different functional states and because cryo-EM does not involve the time-consuming step of growing crystals, three distinct pore conformations of TRPV1 were elucidated by cryo-EM perhaps within 2 months, when microscope accessibility was less of a limiting factor as it is now in many institutions.[76,78] Breakthroughs in TRPV1 structural biology exemplified how technical advancement can drive discoveries in science.[79] Indeed, single-particle cryo-EM is now a predominant method for resolving structures of membrane proteins and large protein complexes. On a side note, the popularity of cryo-EM will continue to grow as its achievable resolutions start to rival that of X-ray crystallography[80] and its suitable targets expand into much smaller proteins.[81] So, there is an urgent need to establish new, and to support existing, national cryo-EM centers, in much the same way as investments in national synchrotron X-ray sources have democratized X-ray crystallography, making it accessible to a broad scientific community beyond mere X-ray crystallography practitioners. Such national cryo-EM centers will allow investigators who have no access to high-end microscopes in their home institutions to routinely collect data for determining structures of their favorite molecules.[82]

In the following sections, step-by-step protocols used for polycystin-2 structure determination are described.

2.2 EXPRESSION OF POLYCYSTIN-2 IN MAMMALIAN CELL LINE

Eukaryotic membrane proteins are usually harder to produce in functional form using prokaryotic expression systems, such as *E. coli*, because bacteria hosts lack sophisticated protein folding machinery, posttranslational modifications, and specific regulatory cellular factors, all of which are often required for maturation of these challenging membrane protein targets. For instance, membrane lipid compositions, which differ significantly between *E. coli* and mammalian cells, can be critical for the stability and activity of eukaryotic membrane receptors and ion channels. As a result, eukaryotic heterologous expression systems, such as insect cells and mammalian cells, are more commonly chosen to express eukaryotic membrane proteins. In particular, the BacMam system enables rapid generation of baculovirus particles in *Spodoptera frugiperda* insect cells, which can be subsequently used to transduce mammalian cells for protein production.[83] The BacMam system avoids generation of stable cell lines, which is not only time-consuming, but can also be extremely challenging if the expression of the protein target is toxic to the host cell. Large-scale transient transfection of mammalian cells is also commonly used to express mammalian membrane proteins, but it requires preparation of a large amount of plasmids and transfection reagents that can be prohibitively expensive. Compared with the insect cell expression system, the mammalian cell expression system can be advantageous because it supports more native-like posttranslational modifications, such as glycosylation.

Commonly used mammalian host cell lines include human embryonic kidney (HEK293) and Chinese hamster ovary (CHO) cell lines.[84] The HEK293 cell line is derived from original human embryo kidney cells transformed with sheared fragments of human adenovirus 5 (Ad5) DNA integrated into chromosome 19.[85] It is capable of producing transcriptionally incompetent (E1-deleted) human adenoviral vectors.[86] Several modified HEK293 cell lines, into which additional virus DNA components are integrated, are available. For example, HEK293E cells have Epstein–Barr virus nuclear antigen-1 (HEK293-EBNA1, or 293E) expressed constitutively, which offers a threefold improvement in recombinant protein yield with expression vectors bearing the Epstein–Barr virus origin of replication.[87]

Methods to transiently express target protein in mammalian cells include electroporation, calcium phosphate method, lipophilic polymer, liposome, and virus transduction. Electroporation is a physical method to introduce DNA into cells according to the electro-potential differences generated by an external electrical

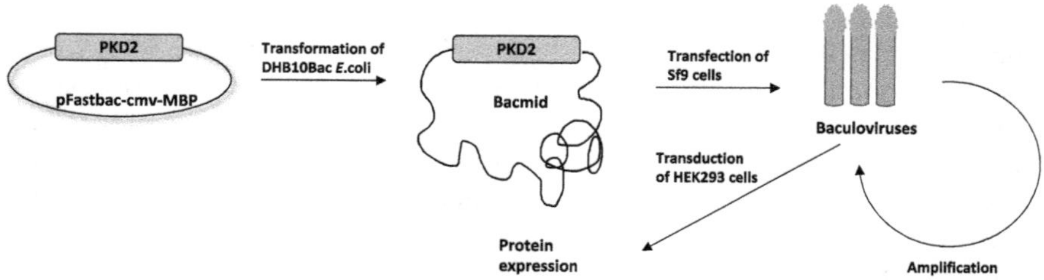

FIGURE 2.1 Expression of polycystin-2 using BacMam system. A full-length or truncated version of the *PKD2* gene is subcloned into a modified pFastbacl vector with an N-terminal MBP fusion tag. Bacmids are generated in DH10Bac *E. coli* strain. After successful packaging in Sf9 insect cells, the baculoviruses containing a PKD2 expression cassette are used to transduce human HEK293 cells for protein production.

power supply.[88] Calcium phosphate method exploits the fact that, in supersaturated solutions, DNA–calcium phosphate co-precipitates spontaneously form and can be efficiently taken up by mammalian cells via endocytosis.[89] Polyethylene glycol (PEG), DEAE-dextran, and polyethylenimine (PEI) all form complex with DNA and mediate transfection.[90–92] Liposomes are spherically shaped, lipid bilayer-enclosed structures that can be used as delivery vesicles.[93] Several commercial transfection reagents are liposome-based. Besides the methods mentioned above, hijacking virus's DNA injection machinery for transduction almost guarantees gene delivery to target cells, as demonstrated in the BacMam system.[83] Transduction efficiency could be further enhanced by baculovirus coated with polyethylenimine.[94]

The BacMam protocol includes four steps: subcloning into a BacMam compatible vector, recombinant bacmids production, baculovirus generation, and protein expression in mammalian cell culture (Figure 2.1). The target gene is first subcloned into a BacMam compatible vector with the transposon element for bacmid generation in DH10Bac *E. coli* cells. Positive clones are selected using blue–white screening, and extracted bacmids are introduced into an insect cell host to produce baculoviruses for protein expression in the mammalian host.

2.2.1 Subcloning into a BacMam Compatible Vector

The first component of the BacMam system (Invitrogen) is a bacmid donor vector that contains an expression cassette flanked by left and right arms of transposon Tn7, so the gene of interest can be integrated into the baculovirus genome bearing a Tn7 attachment site via site-specific transposition. The expression cassette contains a human cytomegalovirus immediate-early enhancer and promoter (CMV promoter), a multiple cloning site for cloning, a gentamicin resistance gene for later bacmid selection, and a polyadenylation signal from simian virus 40 (SV40) for efficient transcription termination. The donor vector also contains a pUC origin of replication and an ampicillin-resistant gene for vector propagation and selection in *E. coli*.

Similar to the general strategy mentioned above, polycystin-2 is cloned into a modified pFastbacl vector, in which a CMV-Kozak-8xHis-MBP sequence replaces the original polyhedrin promoter. A tobacco etch virus (TEV) proteolytic site is introduced between the maltose-binding protein (MBP) and the multiple cloning site. The modified vector produces a fusion protein with an amino terminal octa-histidine tag as well as an MBP that can be used for affinity purification.

2.2.2 Generation of Recombinant Bacmids

To generate the recombinant baculovirus containing a polycystin-2 expression cassette, DH10Bac *E. coli* strain is transformed with the donor pFastbac-PKD2 plasmid. The DH10Bac *E. coli* strain contains a baculovirus shuttle vector (bacmid with a Tn7 attachment site) and a helper plasmid expressing transposase. Upon transformation, the target gene from the donor vector transposes to the low copy baculovirus shuttle vector at a mini-attTn7 transposon recombination site to generate high molecular weight bacmids (>135 kb in size, 1–5 copies per cell). Positive clones are then selected on an LB plate with multiple antibiotics under blue-white selection. As a result, success in bacmid production can be accessed by a small-scale growth from white colonies on selection plates and PCR validation using M13 forward and reverse primers. Recombinant baculovirus is produced in *Spodoptera frugiperda* (Sf9) cells and harvested in the cell culture medium.

Reagents

Multiantibiotic selection plates for bacmids are freshly prepared and contain 50 μg/mL of kanamycin, 7 μg/mL of gentamycin, 10 μg/mL of tetracycline, 100 μg/mL of X-Gal, and 40 μg/mL of IPTG.

MAX Efficiency DH10Bac competent cells from ThermoFisher
S.O.C. medium from Invitrogen

Equipment

Water bath at 42°C
Incubator at 37°C

Step-by-Step Protocol

1. Add 40 ng of plasmid DNA into 100 μL of DH10Bac competent *E. coli.*
2. Keep sample on ice for 25 min.
3. Heat shock at 42°C for 45 s and add 900 μL of SOC medium. Shake at 37°C for 4 h.
4. Plate 200 μL per multiantibiotic plate.
5. Grow in 37°C for 48 h. Although white colonies are visible after 24 h, they become more easily identifiable from blue colonies later for accurate colony picking.
6. Pick two colonies for each plasmid and streak again on selection plate; grow at 37°C for 24 h.
7. After reselection using blue–white system, pick single colonies to grow overnight in liquid LB medium containing the same antibiotics.

2.2.3 Bacmids Extraction

Bacmids are extracted by isopropanol precipitation. Bacmids are large circular DNAs and should be handled with care. Do not vortex or pipette vigorously when working with bacmids.

Reagents

Resuspension buffer (stored at 4°C): 50 mM Tris; 10 mM EDTA, pH 8.0; 100 ug/mL RNase A.

Lysis buffer (stock at room temperature): 200 mM NaOH; 1% SDS
Neutralization buffer (stock at 4°C): 3 M potassium acetate, pH 5.5
Bacmid precipitation: Isopropanol
Bacmid wash: 70% ethanol
Bacmid storage: Sterile water

Equipment

Benchtop centrifuge providing >17,000 g centrifugal force
−20°C freezer

Step-by-Step Protocol

1. Pellet cells from 4 mL culture in 2 mL Eppendorf tube.
2. Add 300 μL resuspension buffer; pipette up and down to resuspend the cell pellet.
3. Add 300 μL lysis buffer; mix by inverting the tube 10 times.
4. Incubate sample at room temperature for 5 min.
5. Add 300 μL of neutralization buffer; mix immediately and thoroughly by inverting the tube ~10 times. A homogeneous suspension is visible.
6. After keeping the centrifuge tube on ice for 5 min, centrifuge sample at 4°C for 12 min.
7. Carefully transfer the supernatant to a new sterile Eppendorf tube and add 630 μL of isopropanol to the supernatant and mix thoroughly.
8. Bacmid DNA is precipitated at −20°C for 30 min and pelleted after centrifugation for 12 min. Alternatively, DNA can be precipitated overnight at −20°C.
9. Wash bacmid DNA with 1 mL of cold 70% ethanol 2 times and centrifuge for 1 min in between.
10. Dry bacmid DNA at room temperature for 4 min.
11. Dissolve bacmid DNA with 50 μL of sterilized water.
12. Confirm the presence of insertion by PCR using M13 forward and reverse primers. Recombinant bacmid migrates as a band at a molecular weight of 2300 bp plus the size of the insert on agarose gel, while untransposed bacmid shows one band at ~300 bp on agarose gel.

2.2.4 Virus Production

Sf9 (*Spodoptera frugiperda*) cells are common hosts for baculovirus production. To increase virus titer for high-level protein expression, two to three passages are usually required. P0 virus stock is defined as the virus progeny from a transfection or cotransfection. P1 virus is from the first expansion culture. P2 and P3 viruses are successively expanded viruses accordingly.

Reagents

ESF 921 Insect Cell Culture Medium, Protein Free from Expression Systems Inc.
Transfection reagent: *Cellfectin* II transfection reagent from Thermo Fisher Scientific

Equipment

Cell culture incubator or shaker with temperature control
6-well cell culture plates

Step-by-Step Protocol

1. Culture Sf9 cells at 27°C with routine passaging.
2. Allocate 1×10^6 cells into 6-well plates in a total of 2 mL ESF 921 medium. The cell density is 0.5 million per mL.
3. Let cells settle down for at least 30 min.
4. At the same time, mix 100 μL of ESF 921 medium with 5 uL of bacmids (solution A), and mix 100 μL of ESF 921 medium with 8 μL of transfection reagent (solution B).
5. After 5 min incubation at room temperature, add solution B to A, and then continue to incubate at room temperature for an additional 20 min.
6. Add 1 mL of fresh ESF 921 medium to the transfection mixture.
7. Remove 2 mL of culture medium from a 6-well plate and add transfection mixture prepared in the previous step. Be careful not to remove cells from the 6-well plate.
8. Incubate the 6-well plate for 6 h at 16°C.
9. Change the cell culture medium with 3 mL of fresh ESF 921 medium.
10. Incubate cells for 5 more days.
11. Collect supernatant after centrifugation at minimal speed (~1000 rpm) for 10 min. This is P0 stock.

2.2.5 Virus Expansion

The aim of virus expansion is to increase virus production and titer.

Reagent

ESF 921 Insect Cell Culture Medium, Protein Free from Expression Systems

Equipment

250 mL round bottle flasks
2.5 L round bottle flasks, without baffle

Step-by-Step Protocol

1. For P1 virus production, grow 25 mL of Sf9 cells in 250 mL round bottle flasks at 27°C to $<2 \times 10^6$ cells/mL.

2. Dilute cell culture with fresh ESF 921 medium and add P0 viruses and keep culturing for 2 days.
3. Check cell density and dilute cells to $<2 \times 10^6$. Split the cell culture if necessary.
4. Check the cell density again after 2–3 days. Properly infected cells are enlarged. Harvest virus if necessary.
5. For P2 virus expansion, 400–800 mL of Sf9 cells are cultured in a 2.5 L round bottle flask with shaking at 125 rpm at 27°C.
6. Add P1 viruses at 1/500–1000 volume ratio and keep culturing for 2 days. Add more ESF 921 medium if necessary. Note that the volume of virus stock added depends on the virus titer, which can be determined using the viral plaque assay[95] or the end-point dilution assay.[96]

2.2.6 Harvesting Virus

After expansion, baculoviruses are harvested from the supernatant of Sf9 cell culture using centrifugation.

Reagent

Bovine calf serum from Millipore Sigma

Equipment

Class II A B3 biological safety cabinet from Forma Scientific
Beckman Model TJ-6 Centrifuge
HERA Heraeus incubator
Metallized Hemacytometer Reichert Bright-Line
Sterilized 50 mL Falcon tubes

Step-by-Step Protocol

1. Check the morphology of insect cells. Swollen and larger cells are infected.
2. Transfer cell culture to Falcon tubes.
3. Low-speed centrifugation at 1200 rpm for 5 min to remove insect cells.
4. Transfer supernatant into new Falcon tubes.
5. Higher-speed centrifugation at 4000 rpm for 20 min.
6. Transfer the supernatant into new Falcon tubes and add bovine calf serum to 2% final concentration to increase long-term stability. Fetal bovine serum can substitute for bovine calf serum in this step.
7. Keep virus stock at 4°C for storage.

2.2.7 Mammalian Cell Growth and Protein Expression

HEK293S GnTI$^-$ is an *N*-acetylglucosaminyltransferase I-deficient, human embryonic kidney cell line.[97] Recombinant glycoproteins produced from this cell line have more a homogeneous *N*-glycosylation pattern, contrasted with wildtype HEK293 cells, thus the HEK293S GnTI$^-$ cell line becomes particularly suitable for structural biology study. Despite being an adherent cell line in nature, the HEK293S GnTI$^-$ is easy to adapt for suspension culture.

Reagents

HEK293S GnTI⁻(ATCC CRL-3022) human embryonic kidney cells
FreeStyle 293 expression medium is from Invitrogen, part of Thermo Fisher Scientific
Sodium butyrate from Alfa Aesar

Equipment

2.5 L baffled cell culture flasks
Cell culture incubator for large culture flasks

Step-by-Step Protocol

1. Culture 1 L of HEK293S GnTI⁻ cells in a 2.5 L baffled cell culture flask, shaking at 90 rpm at 37°C to 1 × 10^6 cells/mL.
2. Add 50 mL of P2 viruses. Note the exact volume of virus stock to be added depends on the individual protein being expressed and the virus titer.
3. After 8–12 hours, add sodium butyrate to 5–10 mM final concentration and decrease culture temperature to 30°C.
4. Continue culturing for 48 hours before cell harvesting. HEK293 cells that undergo multiple cell divisions over a period of 36–48 hours following transfection can form 2–4 cell clusters with individual cells attached to each other by gap junctions, which is common and have no effect on protein expression.

2.3 PURIFICATION OF POLYCYSTIN-2

Polycystin-2 is expressed as a fusion protein with an amino-terminal maltose binding protein affinity tag. Purification of polycystin-2 involves mammalian cell membrane preparation, protein solubilization, affinity purification, sample reconstitution, and size-exclusion purification. The solubilization and reconstitution steps, two key steps in membrane protein purification, are illustrated in Figure 2.2.

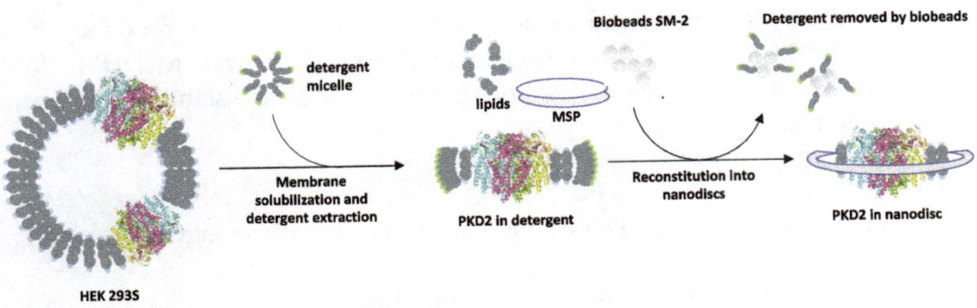

FIGURE 2.2 Solubilization and reconstitution of the polycystin-2 channel. The crude membrane prepared from the polycystin-2 expressing HEK293 cells is extracted with detergents. Purified PKD2 in detergent micelles is then incubated with purified membrane scaffold proteins and lipids, followed by removal of detergents with Bio-Beads. PKD2 protein is eventually reconstituted into an artificial lipid bilayer systems termed nanodisc.

2.3.1 Membrane Preparation

Reagents

Protease inhibitors:

Leupeptin 5 mg/mL in ethanol (1000× stock)
Pepstatin 2.8 mg/mL in ethanol (2000× stock)
PMSF 17.4 mg/mL in DMSO (1000× stock)
Aprotinin 2 mg/mL (1000× stock)
Hypotonic buffer: 10 mM HEPES, pH 8.0; 0.5 mM TCEP or DTT + 4 protease inhibitors
Resuspension buffer: 50 mM HEPES, 150 mM NaCl, 10% glycerol, 0.5 mM TCEP, pH 7.8 + 4 protease inhibitors

Equipment

Optima L-90 K ultracentrifuge from Beckman Coulter
Type 45 Ti rotor from Beckman Coulter
Dounce tissue grinder from VWR

Step-by-Step Protocol

1. Pellet cells at 4000 rpm for 10 min.
2. Resuspend cell pellet from 4 L cell culture in 180 mL of hypotonic buffer and stir in a beaker for 30 min at 4°C.
3. Separate the crude membrane from other intracellular components after ultracentrifugation at 40,000 rpm for 45 min.
4. Suspend membranes in 200 mL resuspension buffer and homogenized using a Dounce tissue grinder.
5. Allocate homogenized membrane suspension into ten 50 mL Falcon tubes and keep at −80°C after flash freezing in liquid nitrogen for storage. Each tube contains about 25 mL of homogenized membrane.

2.3.2 Protein Solubilization and Affinity Purification

Polycystin-2 can be solubilized in detergents mixtures such as n-dodecyl-β-D-maltoside (DDM)/cholesteryl hemisuccinate (CHS). Compared to DDM alone, adding CHS is beneficial since cholesterol has been reported as a stabilizer for several mammalian membrane proteins.[98]

Reagents

n-Dodecyl-β-D-maltoside (DDM) and cholesteryl hemisuccinate (CHS) from Anatrace
Soybean polar lipid extract from Avanti Polar Lipids, Inc.
Amylose resin from New England Biolabs
D-(+)-Maltose monohydrate from Millipore Sigma
Detergent stock solution (10×): 200 mM DDM/20 mM CHS in H_2O
Washing buffer: 50 mM HEPES (pH 7.4), 150 mM NaCl, 2 mM TCEP, 0.5 mM DDM, 0.1 mg/mL soybean polar lipid extract

Elution buffer: 50 mM HEPES (pH 7.4), 150 mM NaCl, 2 mM TCEP, 0.5 mM DDM, 20 mM maltose, 0.1 mg/mL soybean polar lipid extract

Equipment

Poly-Prep chromatography columns from Bio-Rad

Step-by-Step Protocol

1. Thaw 25 mL of homogenized membranes, which originated from 400 mL cell culture, in a warm water bath.
2. Add detergent stock solution into membrane suspension to make a final concentration of 20 mM DDM/2 mM CHS in a small beaker and stir the sample at room temperature for 1 h to extract polycystin-2 protein from the membrane preparation.
3. Discard insoluble membrane after centrifugation at 40,000 g for 30 min at 4°C.
4. Mix the resulting supernatant, which contains solubilized polycystin-2 protein, with 300 μL of prewashed amylose resin using head-over-head rotation at 4°C for 2 h to immobilize polycystin-2 onto amylose resin.
5. Pack the amylose resin with immobilized polycystin-2 protein into a Poly-Prep chromatography column and wash the resin with 3 mL, 10 column volume, of washing buffer.
6. Elute polycystin-2 proteins from amylose resin with 4-column volume of elution buffer.

2.3.3 Protein Reconstitution

Two reconstitution protocols are used for polycystin-2: nanodisc and amphipol. Bio-Beads SM-2 are nonpolar polystyrene adsorbent which removes detergents from the sample, thus facilitating the interactions between polycystin-2 and nanodisc or amphipol.

2.3.3.1 Nanodisc Reconstitution

Reagents

Membrane scaffold protein MSP2N2 is prepared in lab according to literature[99]
TEV protease from tobacco etch virus
Soybean polar lipid extract from Avanti Polar Lipids, Inc.
Bio-Beads SM-2 Resin (1523920, Bio-Rad)
Soybean lipid stock: 10 mM soybean polar lipid extract in 20 mM HEPES, 150 mM NaCl, 2 mM TCEP, pH 7.4

Step-by-Step Protocol

1. Mix purified polycystin-2 protein (2–3 mg/mL) solubilized in DDM/CHS/lipids with MSP2N2 (170–225 μM) and the soybean lipid stock at a 1:1:200 molar ratio and incubate on ice for 30 min.
2. Add three batches of Bio-Beads SM-2 (30 mg per 1 mL of reconstitution mixture) at 2–3 h intervals at 4°C with head-over-head rotation.

3. Add a fourth batch of Bio-Beads into the sample together with TEV protease (protease: polycystin-2 = 1:40 molar ratio) to remove the fusion tag and incubate the reconstitution mixture at 4°C overnight.
4. Next day, remove Bio-Beads SM-2 from the sample by passing the reconstitution mixture through an empty gravity column and reserve the sample for size-exclusion purification.

2.3.3.2 Amphipol Reconstitution

Reagents

100 mg/mL amphipols A8-35 from Anatrace

Step-by-Step Protocol

1. Incubate purified polycystin-2 protein (2–3 mg/mL) with TEV protease for 4 h at 4°C to remove the MBP fusion tag.
2. Mix the tag-removed sample with amphipols A8-35 at 1:3 (w/w) with head-over-head rotation for 4 h.
3. Remove detergent with Bio-Beads the same way as in the nanodisc reconstitution protocol.

2.3.4 Size-Exclusion Chromatography

To separate tag-removed tetrameric polycystin-2 channel from released MBP tag and soluble aggregates, size-exclusion chromatography is performed.

Reagents

Size-exclusion buffer: 20 mM HEPES, pH 7.4, 150 mM NaCl

Equipment

Superose 6, 10/300 GL column from GE Healthcare Life Sciences
A fast protein liquid chromatography system supporting system pressure equal to or higher than 2 MPa pressure and a flow rate of 1 mL/min
Vivaspin concentrator with 100 KNML

Step-by-Step Protocol

1. Centrifuge protein sample at 17,000 g at 4°C for 30 minutes to remove potential protein aggregation.
2. Load 0.5 mL of the supernatant to Superose 6 column using a syringe.
3. Elute proteins at 0.5 mL/min, and collect 0.5 mL fractions.
4. Concentrate fractions corresponding to tetrameric channel using a 100 kDa Vivaspin concentrator in a benchtop centrifuge operating at 4500 rpm.
5. Prepare EM grids using freshly purified polycystin-2 protein or flash freeze the sample in liquid nitrogen and keep at –80°C for storage.

2.4 STRUCTURAL DETERMINATION OF POLYCYSTIN-2

Structural determination of polycystin-2 involves preparation of cryo-EM grids, electron microscopy data collection, data analysis, and image processing, three-dimensional (3D) reconstruction, and model building.

2.4.1 Preparation of Cryo-EM Grids

Before mounting the sample onto a transmission electron microscope for imaging, the protein sample has to be spotted onto a solid support and preserved in a native-like, frozen hydrated state within a thin layer of vitreous ice. The cryo-EM sample grids, the solid support, is typically made from copper or gold. A thin layer of amorphous carbon is coated on one side of the copper grid, but often with regularly arrayed perforations (or holes) where a thin layer of vitreous ice with embedded biological molecules will form. As the carbon support film is hydrophobic in nature, a glow discharge treatment is needed to deposit negative charges on the carbon film, thus making it hydrophilic so that the applied protein sample can spread evenly across the grid.

Reagents

Quantifoil 1.2/1.3 holey carbon 400 mesh copper grids are from Quantifoil Micro Tools
Filter paper: 595 filter paper; 55/22 mm (Product #: 47000-100, Ted Pella)

Equipment

PELCO easiGlow Glow Discharge Cleaning System from Ted Pella, INC.
FEI Vitrobot Mark III Grid Plunging System from Thermo Fisher Scientific

Step-by-Step Protocol

1. Concentrate purified polycystin-2 protein to between 1 and 3.5 mg/mL. Glow discharge of copper grids at 25 mA for 25 s.
2. Apply 3.5 uL of sample on glow-discharged Quantifoil grid in a humidity controlled chamber on Vitrobot at 4°C with 75% relative humidity.
3. Plunge freezing sample in liquid ethane cooled by liquid nitrogen with the following parameters: 20 s wait time, ~−1 mm filter paper offset, 7 s blotting time. Note that the freezing parameters are protein specific and the parameters should be determined case by case.
4. The frozen grids are transferred to storage dewar filled with liquid nitrogen for further data collection.

2.4.2 Electron Microscopy Data Collection

Equipment

A transmission electron microscope (TEM) operating at 200 or 300 kV. To obtain better movies, the TEM typically is equipped with a field emission gun (FEG) as an electron source and a direct electron detector. Tecnai TF20 and Tecnai TF30 Polara microscopes from FEI Company and K2 Summit direct electron detector from Gatan Inc. were used to solve the 3D structure of the polycystin-2 protein.

A cryo-holder with temperature probe such as the Gatan 626 side entry cryo-holder from FEI.

An automated data collection software such as SerialEM.[100]

Step-by-Step Protocol

Transfer cryo-grids loaded with polycystin-2 protein on a cooled cryo-holder.

Insert cryo-holder into microscope and collect movie data using serialEM software. In the case of PKD2 on TF20 microscope, movies were recorded at a corrected magnification of 41,911× and at a dose rate of 8 electrons per pixel per second. Each movie, 60 subframes, was collected for 12 s at defocus ranges from 0.8 to 2.0 mm.

For the dataset collected on Tecnai TF30 Polara, movies were recorded using UCSFImage4[101] at a corrected magnification of 31,000× and at a dose rate of 8.2 electrons per pixel per second. Each movie, 40 subframes, was collected for 8 s at defocus ranges from 0.6 to 2.4 mm.

2.4.3 Image Processing and Three-Dimensional Reconstruction

Software

Movie alignment and motion correction software: MotionCor2.[102]
CTF estimation: CTFFIND4.[103]
Manual particle selection: SamViewer written by Maofu Liao, Harvard Medical School.
Automated particle selection: SPIDER.[104]
2D classification and 3D reconstruction: RELION.[105]
EM density map visualization: UCSF Chimera.[106]
Local Resolution estimation: ResMap.[107]
Pore radii: HOLE.[108]

Step-by-Step Protocol

1. Align, dose weigh, and sum movie frames into a single micrograph to correct potential electron beam induced specimen motion using MotionCor2, previously UcsfDfCorr, written by Shawn Zheng, UCSF.
2. Then determine the CTF parameters for aligned micrographs using the program CTFFIND4. CTF score of each micrograph is used as a criterion to exclude micrographs with low quality from further analyses.
3. Manually select 2000–3000 particles from micrographs in SamViewer to produce an initial set of 2D class averages, which will serve as templates for automated particle selection in SPIDER. Manual inspection of all micrographs after automated particle selection is recommended to remove ice and junk particles. In the case of human PKD2: 198–703 collected on TF30 Polara, 368,032 particles were selected from 1500 micrographs.[109]
4. Reject particles from poorly resolved 2D classes for 3D reconstruction. Several optional rounds of 3D classification, termed "*in silico* purification" further improves the quality of selected particles.
5. Perform 3D reconstruction and refinement in RELION. 93,805 particles of human PKD2: 198–703 were used for 3D reconstruction. Since PKD2 is a homotetramer, as are the other TRP family members, C4 symmetry was imposed. For the high-resolution reconstruction of PKD2 protein using the dataset collected at 300 kV, an initial medium resolution (4.2 angstrom) PKD2 structure generated from a dataset collected at 200 kV was low-pass filtered to lower resolution (40 angstroms) to minimize model bias and

served as the initial model. RELION auto-refinement procedure converged at an unmasked resolution of 3.38 Å and a 3.0 Å map. A B-factor sharpening step was applied to the EM density map to facilitate model building.[109]
6. Visualize and segment EM density maps using UCSF Chimera.
7. Compute local resolution of EM density map using ResMap.
8. For ion channel, the solvent accessible pathway of the ion conduction pore can be estimated using HOLE program.

2.4.4 Model Building

Equipment

Model building: Coot[110]
Global refinement and minimization: PHENIX[111]
Model assessment: MolProbity[112] and EMRinger[113]

Step-by-Step Protocol

1. Determine the necessity of a reference model. If the EM density map is of low resolution, one may need a reference model from a previously solved structure, or one may not be able to build an atomic model at all. In the case of PKD2, *de novo* model building was accomplished in Coot taking advantage of the high-resolution EM density map obtained for PKD2 protein.
2. Recognize domain boundaries and secondary structural elements, such as alpha helixes in the transmembrane region, and build polyglycine chains. Several disordered regions, where no clearly defined EM density was observed, were omitted in the model building: N-terminus (198–215), a loop (296–301) within the extracellular polycystin domain, S2-S3 loop (494–503), and C-terminus (695–703).[109]
3. Assign amino acid residues based on defined EM density for side chains of bulky aromatic amino acid residues (Trp, Phe, and Tyr) and Arg. Conversely, lack of side-chain density in a region with well-defined density is indicative of Gly residues.
4. Perform global real space refinement and energy minimization using the "phenix.real_space_refine" module.
5. Assess the refined geometries of the model using MolProbity and EMRinger.
6. Analyze the model for overfitting using the two half-map method.[114,115]

2.5 CONCLUDING REMARKS

The past several years saw significant progress in the biochemistry and structural biology of polycystin-1 and polycystin-2 proteins, the two key proteins mutated in ADPKD patients, largely due to the rapid improvement of techniques for membrane protein expression, purification, and structural determination. Taking advantage of the high transduction efficiency of baculovirus and the sophisticated protein folding and posttranslational modification machinery present in mammalian cells, the BacMam method is gaining popularity rapidly. With the exponential growth of our arsenal for membrane protein solubilization, purification, and reconstitution, the traditionally daunting tasks of membrane protein structural biology are undergoing a revolution, which has broadly expanded the scope of structurally "solvable" membrane proteins. Indeed, the previously deemed high-hanging fruits are becoming reachable. The evolution of methods in cryo-EM sample preparation, electron microscope detector technology, and data analysis software all contribute to near-atomic resolution structures of the PKD2 protein.[109,116–118] More recently, the transmembrane architecture of the unique PKD1/PKD2 complex has been determined.[119] Taken together, it is foreseeable that the mechanistic understanding of how PKD1 and PKD2 are regulated and how numerous disease-associated mutations affect structures and function of PKD proteins will be achieved.

ACKNOWLEDGMENTS

This work has been supported by the NIH R01 DK110575-01 Grant and the DoD W81XWH-17-1-0158 Discovery Award to E.C., and E.C. is a Pew Scholar supported by the Pew Charitable Foundation.

REFERENCES

1. Boulter, C., Mulroy, S., Webb, S., Fleming, S., Brindle, K., and Sandford, R. 2001. Cardiovascular, skeletal, and renal defects in mice with a targeted disruption of the *Pkd1* gene. *Proc Natl Acad Sci USA 98*, 12174–12179.
2. Harris, P.C. and Torres, V.E. 1993. Polycystic kidney disease, autosomal dominant. In *GeneReviews((R))*, M.P. Adam, H.H. Ardinger, R.A. Pagon, S.E. Wallace, L.J.H. Bean, K. Stephens, and A. Amemiya, eds. Seattle, WA: University of Washington, Seattle University of Washington, Seattle. GeneReviews is a registered trademark of the University of Washington, Seattle. All rights reserved.
3. Kim, K., Drummond, I., Ibraghimov-Beskrovnaya, O., Klinger, K., and Arnaout, M.A. 2000. Polycystin 1 is required for the structural integrity of blood vessels. *Proc Natl Acad Sci USA 97*, 1731–1736.
4. Lu, W., Peissel, B., Babakhanlou, H., Pavlova, A., Geng, L., Fan, X., Larson, C., Brent, G., and Zhou, J. 1997. Perinatal lethality with kidney and pancreas defects in mice with a targetted *Pkd1* mutation. *Nature Genet 17*, 179–181.
5. Ong, A.C.M. and Harris, P.C. 2015. A polycystin-centric view of cyst formation and disease: The polycystins revisited. *Kidney Int 88*, 699–710.
6. Wu, G.Q., Markowitz, G.S., Li, L., D'Agati, V.D., Factor, S.M., Geng, L., Tibara, S., Tuchman, J., Cai, Y.Q., Park, J.H. et al. 2000. Cardiac defects and renal failure in mice with targeted mutations in *Pkd2*. *Nature Genet 24*, 75–78.
7. Germino, G.G., Somlo, S., Weinstatsaslow, D., and Reeders, S.T. 1993. Positional cloning approach to the dominant polycystic kidney-disease gene, *Pkd1*. *Kidney Int 43*, S20–S25.
8. Mochizuki, T., Wu, G.Q., Hayashi, T., Xenophontos, S.L., Veldhuisen, B., Saris, J.J., Reynolds, D.M., Cai, Y.Q., Gabow, P.A., Pierides, A. et al. 1996. PKD2, a gene for polycystic kidney disease that encodes an integral membrane protein. *Science 272*, 1339–1342.
9. The European Polycystic Kidney Disease, C. 1994. The polycystic kidney disease 1 gene encodes a 14 kb transcript and lies within a duplicated region on chromosome 16. *Cell 77*, 881–894.

10. Rossetti, S., Consugar, M.B., Chapman, A.B., Torres, V.E., Guay-Woodford, L.M., Grantham, J.J., Bennett, W.M., Meyers, C.M., Walker, D.L., Bae, K. et al. 2007. Comprehensive molecular diagnostics in autosomal dominant polycystic kidney disease. *J Am Soc Nephrol 18*, 2143–2160.
11. The International Polycystic Kidney Disease, C. 1995. Polycystic kidney disease: The complete structure of the *PKD1* gene and its protein. *Cell 81*, 289–298.
12. Garcia-Gonzalez, M.A., Outeda, P., Zhou, Q., Zhou, F., Menezes, L.F., Qian, F., Huso, D.L., Germino, G.G., Piontek, K.B., and Watnick, T. 2010. *Pkd1* and *Pkd2* are required for normal placental development. *PLOS ONE 5*, e12821.
13. Balbo, B.E., Amaral, A.G., Fonseca, J.M., de Castro, I., Salemi, V.M., Souza, L.E., Dos Santos, F., Irigoyen, M.C., Qian, F., Chammas, R. et al. 2016. Cardiac dysfunction in *Pkd1*-deficient mice with phenotype rescue by galectin-3 knockout. *Kidney Int 90*, 580–597.
14. Wu, G. and Somlo, S. 2000. Molecular genetics and mechanism of autosomal dominant polycystic kidney disease. *Mol Genet Metab 69*, 1–15.
15. Hwang, Y.H., Conklin, J., Chan, W., Roslin, N.M., Liu, J., He, N., Wang, K., Sundsbak, J.L., Heyer, C.M., Haider, M. et al. 2016. Refining genotype-phenotype correlation in autosomal dominant polycystic kidney disease. *J Am Soc Nephrol 27*, 1861–1868.
16. Rossetti, S., Strmecki, L., Gamble, V., Burton, S., Sneddon, V., Peral, B., Roy, S., Bakkaloglu, A., Komel, R., Winearls, C.G. et al. 2001. Mutation analysis of the entire PKD1 gene: Genetic and diagnostic implications. *Am J Hum Genet 68*, 46–63.
17. Burn, T.C., Connors, T.D., Dackowski, W.R., Petry, L.R., Vanraay, T.J., Millholland, J.M., Venet, M., Miller, G., Hakim, R.M., Landes, G.M. et al. 1995. Analysis of the genomic sequence for the autosomal-dominant polycystic kidney-disease (*Pkd1*) gene predicts the presence of a leucine-rich repeat. *Hum Mol Genet 4*, 575–582.
18. Hughes, J., Ward, C.J., Peral, B., Aspinwall, R., Clark, K., Sanmillan, J.L., Gamble, V., and Harris, P.C. 1995. The polycystic kidney-disease-1 (*Pkd1*) gene encodes a novel protein with multiple cell recognition domains. *Nature Genet 10*, 151–160.
19. Qian, F., Boletta, A., Bhunia, A.K., Xu, H.X., Liu, L.J., Ahrabi, A.K., Watnick, T.J., Zhou, F., and Germino, G.G. 2002. Cleavage of polycystin-1 requires the receptor for egg jelly domain and is disrupted by human autosomal-dominant polycystic kidney disease 1-associated mutations. *Proc Natl Acad Sci USA 99*, 16981–16986.
20. Qian, F., Germino, F.J., Cai, Y.Q., Zhang, X.B., Somlo, S., and Germino, G.G. 1997. PKD1 interacts with PKD2 through a probable coiled-coil domain. *Nature Genet 16*, 179–183.
21. Tsiokas, L., Kim, E., Arnould, T., Sukhatme, V.P., and Walz, G. 1997. Homo- and heterodimeric interactions between the gene products of PKD1 and PKD2. *Proc Natl Acad Sci USA 94*, 6965–6970.
22. Yu, Y., Ulbrich, M.H., Li, M.H., Buraei, Z., Chen, X.Z., Ong, A.C.M., Tong, L., Isacoff, E.Y., and Yang, J. 2009. Structural and molecular basis of the assembly of the TRPP2/PKD1 complex. *Proc Natl Acad Sci USA 106*, 11558–11563.
23. Zhu, J., Yu, Y., Ulbrich, M.H., Li, M.H., Isacoff, E.Y., Honig, B., and Yang, J. 2011. Structural model of the TRPP2/PKD1 C-terminal coiled-coil complex produced by a combined computational and experimental approach. *Proc Natl Acad Sci USA 108*, 10133–10138.
24. Nauli, S.M., Alenghat, F.J., Luo, Y., Williams, E., Vassilev, P., Li, X., Elia, A.E., Lu, W., Brown, E.M., Quinn, S.J. et al. 2003a. Polycystins 1 and 2 mediate mechanosensation in the primary cilium of kidney cells. *Nat Genet 33*, 129–137.
25. Kim, S., Nie, H.G., Nesin, V., Tran, U., Outeda, P., Bai, C.X., Keeling, J., Maskey, D., Watnick, T., Wessely, O. et al. 2016. The polycystin complex mediates Wnt/Ca2+ signalling. *Nat Cell Biol 18*, 752–764.
26. Jerman, S., Ward, H., MacDougall, M., and Wandinger-Ness, A. 2013. A conserved ciliary signaling microdomain links craniofacial disorders and polycystic kidney disease. *Mol Biol Cell 24*.

27. Cai, Y.Q., Fedeles, S.V., Dong, K., Anyatonwu, G., Onoe, T., Mitobe, M., Gao, J.D., Okuhara, D., Tian, X., Gallagher, A.R. et al. 2014. Altered trafficking and stability of polycystins underlie polycystic kidney disease. *J Clin Invest 124*, 5129–5144.
28. Giamarchi, A., Padilla, F., Coste, B., Raoux, M., Crest, M., Honore, E., and Delmas, P. 2006. The versatile nature of the calcium-permeable cation channel TRPP2. *Embo Rep 7*, 787–793.
29. Pocsfalvi, G., Raj, D.A.A., Fiume, I., Vilasi, A., Trepiccione, F., and Capasso, G. 2015. Urinary extracellular vesicles as reservoirs of altered proteins during the pathogenesis of polycystic kidney disease. *Proteom Clin Appl 9*, 552–567.
30. Wang, J., Silva, M., Haas, L.A., Morsci, N.S., Nguyen, K.C., Hall, D.H., and Barr, M.M. 2014. C. elegans ciliated sensory neurons release extracellular vesicles that function in animal communication. *Curr Biol 24*, 519–525.
31. Hama, T. and Park, F. 2016. Heterotrimeric G protein signaling in polycystic kidney disease. *Physiol Genomics 48*, 429–445.
32. Kong, T.Q., Xu, D.S., Tran, M., and Denker, B.M. 2010. Regulation of integrin expression by G alpha 12 An additional potential mechanism modulating cell attachment. *Cell Adhes Migr 4*, 372–376.
33. Parnell, S.C., Magenheimer, B.S., Maser, R.L., Pavlov, T.S., Havens, M.A., Hastings, M.L., Jackson, S.F., Ward, C.J., Peterson, K.R., Staruschenko, A. et al. 2018. A mutation affecting polycystin-1 mediated heterotrimeric G-protein signaling causes PKD. *Hum Mol Genet 27*, 3313–3324.
34. Wu, Y., Xu, J.X., El-Jouni, W., Lu, T., Li, S.Y., Wang, Q.Y., Tran, M., Yu, W.F., Wu, M.Q., Barrera, I.E. et al. 2016. G alpha 12 is required for renal cystogenesis induced by *Pkd1* inactivation. *J Cell Sci 129*, 3675–3684.
35. Xu, J.X., Lu, T.S., Li, S.Y., Wu, Y., Ding, L., Denker, B.M., Bonventre, J.V., and Kong, T.Q. 2015. Polycystin-1 and G alpha 12 regulate the cleavage of E-cadherin in kidney epithelial cells. *Physiol Genomics 47*, 24–32.
36. Yu, W.F., Kong, T.Q., Beaudry, S., Tran, M., Negoro, H., Yanamadala, V., and Denker, B.M. 2010. Polycystin-1 protein level determines activity of the G alpha(12)/JNK apoptosis pathway. *J Biol Chem 285*, 10243–10251.
37. Yu, W.F., Ritchie, B.J., Su, X.F., Zhou, J., Meigs, T.E., and Denker, B.M. 2011. Identification of polycystin-1 and G alpha 12 binding regions necessary for regulation of apoptosis. *Cell Signal 23*, 213–221.
38. Zhang, B., Tran, U., and Wessely, O. 2018. Polycystin 1 loss of function is directly linked to an imbalance in G-protein signaling in the kidney. *Development 145*.
39. Cai, J., Song, X., Wang, W., Watnick, T., Pei, Y., Qian, F., and Pan, D. 2018. A RhoA-YAP-c-Myc signaling axis promotes the development of polycystic kidney disease. *Genes Dev 32*, 781–793.
40. Plouffe, S.W., Hong, A.W., and Guan, K.-L. 2015. Disease implications of the Hippo/YAP pathway. *Trends Mol Med 21*, 212–222.
41. Doerr, N., Wang, Y.D., Kipp, K.R., Liu, G.Y., Benza, J.J., Pletnev, V., Pavlov, T.S., Staruschenko, A., Mohieldin, A.M., Takahashi, M. et al. 2016. Regulation of polycystin-1 function by calmodulin binding. *PLOS ONE 11*.
42. Roy, K. and Marin, E.P. 2018. Polycystin-1, the product of the polycystic kidney disease gene *PKD1*, is post-translationally modified by palmitoylation. *Mol Biol Rep 45*, 1515–1521.
43. Wilson, P.D., Geng, L., Li, X., and Burrow, C.R. 1999. The PKD1 gene product, "polycystin-1," is a tyrosine-phosphorylated protein that colocalizes with alpha2beta1-integrin in focal clusters in adherent renal epithelia. *Lab Invest 79*, 1311–1323.
44. Ramsey, I.S., Delling, M., and Clapham, D.E. 2006. An introduction to TRP channels. *Annu Rev Physiol 68*, 619–647.
45. Semmo, M., Kottgen, M., and Hofherr, A. 2014. The TRPP subfamily and polycystin-1 proteins. *Handb Exp Pharmacol 222*, 675–711.

46. Streets, A.J., Moon, D.J., Kane, M.E., Obara, T., and Ong, A.C.M. 2006. Identification of an N-terminal glycogen synthase kinase 3 phosphorylation site which regulates the functional localization of polycystin-2 in vivo and in vitro. *Hum Mol Genet 15*, 1465–1473.
47. Allen, M.D., Qamar, S., Vadivelu, M.K., Sandford, R.N., and Bycroft, M. 2014. A high-resolution structure of the EF-hand domain of human polycystin-2. *Protein Sci 23*, 1301–1308.
48. Petri, E.T., Celic, A., Kennedy, S.D., Ehrlich, B.E., Boggon, T.J., and Hodsdon, M.E. 2010. Structure of the EF-hand domain of polycystin-2 suggests a mechanism for Ca2+-dependent regulation of polycystin-2 channel activity. *Proc Natl Acad Sci USA 107*, 9176–9181.
49. Geng, L., Okuhara, D., Yu, Z.H., Tian, X., Cai, Y.Q., Shibazaki, S., and Somlo, S. 2006. Polycystin-2 traffics to cilia independently of polycystin-1 by using an N-terminal RVxP motif. *J Cell Sci 119*, 1383–1395.
50. Cai, Z.Q., Maeda, Y., Cedzich, A., Torres, V.E., Wu, G.Q., Hayashi, T., Mochizuki, T., Park, J.H., Witzgall, R., and Somlo, S. 1999. Identification and characterization of polycystin-2, the *PKD2* gene product. *J Biol Chem 274*, 28557–28565.
51. Cai, Y.Q., Anyatonwu, G., Okuhara, D., Lee, K.B., Yu, Z.H., Onoe, T., Mei, C.L., Qian, Q., Geng, L., Wiztgall, R. et al. 2004. Calcium dependence of polycystin-2 channel activity is modulated by phosphorylation at Ser(812). *J Biol Chem 279*, 19987–19995.
52. Delmas, P., Nauli, S.M., Li, X.G., Coste, B., Osorio, N., Crest, M., Brown, D.A., and Zhou, J. 2004. Gating of the polycystin ion channel signaling complex in neurons and kidney cells. *FASEB J 18*, 740–742.
53. Gonzalez-Perrett, S., Kim, K., Ibarra, C., Damiano, A.E., Zotta, E., Batelli, M., Harris, P.C., Reisin, I.L., Arnaout, M.A., and Cantiello, H.F. 2001. Polycystin-2, the protein mutated in autosomal dominant polycystic kidney disease (ADPKD), is a Ca2+-permeable nonselective cation channel. *Proc Natl Acad Sci USA 98*, 1182–1187.
54. Hanaoka, K., Qian, F., Boletta, A., Bhumia, A.K., Piontek, K., Tsiokas, L., Sukhatme, V.P., Guggino, W.B., and Germino, G.G. 2000. Co-assembly of polycystin-1 and-2 produces unique cation-permeable currents. *Nature 408*, 990–994.
55. Koulen, P., Cai, Y.Q., Geng, L., Maeda, Y., Nishimura, S., Witzgall, R., Ehrlich, B.E., and Somlo, S. 2002. Polycystin-2 is an intracellular calcium release channel. *Nat Cell Biol 4*, 191–197.
56. Nauli, S.M., Alenghat, F.J., Luo, Y., Williams, E., Vassilev, P., Lil, X.G., Elia, A.E.H., Lu, W.N., Brown, E.M., Quinn, S.J. et al. 2003b. Polycystins 1 and 2 mediate mechanosensation in the primary cilium of kidney cells. *Nature Genet 33*, 129–137.
57. Vassilev, P.M., Guo, L., Chen, X.Z., Segal, Y., Peng, J.B., Basora, N., Babakhanlou, H., Cruger, G., Kanazirska, M., Ye, C.P. et al. 2001. Polycystin-2 is a novel cation channel implicated in defective intracellular Ca2+ homeostasis in polycystic kidney disease. *Biochem Bioph Res Co 282*, 341–350.
58. Arif Pavel, M., Lv, C.X., Ng, C., Yang, L., Kashyap, P., Lam, C., Valentino, V., Fung, H.Y., Campbell, T., Moller, S.G. et al. 2016. Function and regulation of TRPP2 ion channel revealed by a gain-of-function mutant. *Proc Natl Acad Sci USA 113*, E2363–E2372.
59. Kleene, S.J. and Kleene, N.K. 2017. The native TRPP2-dependent channel of murine renal primary cilia. *Am J Physiol Renal Physiol 312*, F96–F108.
60. Liu, X., Vien, T., Duan, J., Sheu, S.H., DeCaen, P.G., and Clapham, D.E. 2018. Polycystin-2 is an essential ion channel subunit in the primary cilium of the renal collecting duct epithelium. *Elife 7*.
61. Venkatachalam, K. and Montell, C. 2007. TRP channels. *Annu Rev Biochem 76*, 387–417.
62. Hardie, R.C. and Minke, B. 1995. Phosphoinositide-mediated phototransduction in Drosophila photoreceptors: The role of Ca2+ and trp. *Cell Calcium 18*, 256–274.

63. Montell, C. 1997. New Light on TRP and TRPL. *Mol Pharmacol 52*, 755–763.
64. Clapham, D.E., Montell, C., Schultz, G., and Julius, D. 2003. International union of pharmacology. XLIII. Compendium of voltage-gated ion channels: Transient receptor potential channels. *Pharmacol Rev 55*, 591–596.
65. Long, S.B., Campbell, E.B., and Mackinnon, R. 2005. Crystal structure of a mammalian voltage-dependent Shaker family K+ channel. *Science 309*, 897–903.
66. Adrian, M., Dubochet, J., Lepault, J., and McDowall, A.W. 1984. Cryo-electron microscopy of viruses. *Nature 308*, 32–36.
67. Frank, J. 2017. Single-Particle Reconstruction – Story in a Sample. In Nobel Lecture, https://www.nobelprize.org/uploads/2018/06/frank-lecture-slides.pdf. Retrieved 08/26/2019.
68. Grigorieff, N. and Harrison, S.C. 2011. Near-atomic resolution reconstructions of icosahedral viruses from electron cryo-microscopy. *Curr Opin Struct Biol 21*, 265–273.
69. Kaelber, J.T., Hryc, C.F., and Chiu, W. 2017. Electron cryomicroscopy of viruses at near-atomic resolutions. *Annu Rev Virol 4*, 287–308.
70. Zhou, Z.H. 2014. Structures of viral membrane proteins by high-resolution cryoEM. *Curr Opin Virol 5*, 111–119.
71. Zhang, X., Jin, L., Fang, Q., Hui, W.H., and Zhou, Z.H. 2010. 3.3 A cryo-EM structure of a nonenveloped virus reveals a priming mechanism for cell entry. *Cell 141*, 472–482.
72. Henderson, R. 1995. The potential and limitations of neutrons, electrons and X-rays for atomic resolution microscopy of unstained biological molecules. *Q Rev Biophy 28*, 171–193.
73. Callaway, E. 2015. The revolution will not be crystallized: A new method sweeps through structural biology. *Nature News 525*, 172.
74. Li, X., Mooney, P., Zheng, S., Booth, C.R., Braunfeld, M.B., Gubbens, S., Agard, D.A., and Cheng, Y. 2013. Electron counting and beam-induced motion correction enable near-atomic-resolution single-particle cryo-EM. *Nature Methods 10*, 584.
75. McMullan, G., Faruqi, A.R., Clare, D., and Henderson, R. 2014. Comparison of optimal performance at 300 keV of three direct electron detectors for use in low dose electron microscopy. *Ultramicroscopy 147*, 156–163.
76. Cao, E., Liao, M., Cheng, Y., and Julius, D. 2013. TRPV1 structures in distinct conformations reveal activation mechanisms. *Nature 504*, 113–118.
77. Liao, M., Cao, E., Julius, D., and Cheng, Y. 2013. Structure of the TRPV1 ion channel determined by electron cryo-microscopy. *Nature 504*, 107–112.
78. Gao, Y., Cao, E., Julius, D., and Cheng, Y. 2016. TRPV1 structures in nanodiscs reveal mechanisms of ligand and lipid action. *Nature 534*, 347–351.
79. Liao, M., Cao, E., Julius, D., and Cheng, Y. 2014. Single particle electron cryo-microscopy of a mammalian ion channel. *Curr Opin Struct Biol 27*, 1–7.
80. Merk, A., Bartesaghi, A., Banerjee, S., Falconieri, V., Rao, P., Davis, M.I., Pragani, R., Boxer, M.B., Earl, L.A., Milne, J.L.S. et al. 2016. Breaking cryo-EM resolution barriers to facilitate drug discovery. *Cell 165*, 1698–1707.
81. Khoshouei, M., Radjainia, M., Baumeister, W., and Danev, R. 2017. Cryo-EM structure of haemoglobin at 3.2 Å determined with the Volta phase plate. *Nat Commun 8*, 16099.
82. Stuart, D.I., Subramaniam, S., and Abrescia, N.G.A. 2016. The democratization of cryo-EM. *Nature Methods 13*, 607.
83. Goehring, A., Lee, C.-H., Wang, K.H., Michel, J.C., Claxton, D.P., Baconguis, I., Althoff, T., Fischer, S., Garcia, K.C., and Gouaux, E. 2014. Screening and large-scale expression of membrane proteins in mammalian cells for structural studies. *Nat Protoc 9*, 2574.
84. Khan, K.H. 2013. Gene expression in Mammalian cells and its applications. *Adv Pharm Bull 3*, 257–263.
85. Graham, F.L., Smiley, J., Russell, W., and Nairn, R. 1977. Characteristics of a human cell line transformed by DNA from human adenovirus type 5. *J Gen Virol 36*, 59–72.
86. Kovesdi, I. and Hedley, S.J. 2010. Adenoviral producer cells. *Viruses 2*, 1681–1703.

87. Tom, R., Bisson, L., and Durocher, Y. 2008. Culture of HEK293-EBNA1 cells for production of recombinant proteins. *CSH Protocols 2008*, pdb.prot4976.
88. Potter, H., Weir, L., and Leder, P. 1984. Enhancer-dependent expression of human kappa immunoglobulin genes introduced into mouse pre-B lymphocytes by electroporation. *Proc Natl Acad Sc USA 81*, 7161–7165.
89. Jordan, M., Schallhorn, A., and Wurm, F.M. 1996. Transfecting mammalian cells: Optimization of critical parameters affecting calcium-phosphate precipitate formation. *Nucleic Acids Res 24*, 596–601.
90. Kim, Y.K., Park, I.K., Jiang, H.L., Choi, J.Y., Je, Y.H., Jin, H., Kim, H.W., Cho, M.H., and Cho, C.S. 2006. Regulation of transduction efficiency by pegylation of baculovirus vector in vitro and in vivo. *J Biotechnol 125*, 104–109.
91. Kumar, P., Nagarajan, A., and Uchil, P.D. 2018. Transfection mediated by DEAE-Dextran. *Cold Spring Harb Protoc 2018*, pdb.prot095463.
92. Longo, P.A., Kavran, J.M., Kim, M.-S., and Leahy, D.J. 2013. Transient mammalian cell transfection with polyethylenimine (PEI). *Methods Enzymol 529*, 227–240.
93. Akbarzadeh, A., Rezaei-Sadabady, R., Davaran, S., Joo, S.W., Zarghami, N., Hanifehpour, Y., Samiei, M., Kouhi, M., and Nejati-Koshki, K. 2013. Liposome: Classification, preparation, and applications. *Nanoscale Res Lett 8*, 102–102.
94. Yang, Y., Lo, S.L., Yang, J., Yang, J., Goh, S.S., Wu, C., Feng, S.S., and Wang, S. 2009. Polyethylenimine coating to produce serum-resistant baculoviral vectors for in vivo gene delivery. *Biomaterials 30*, 5767–5774.
95. Dulbecco, R. and Vogt, M. 1953. Some problems of animal virology as studied by the plaque technique. *Cold Spring Harb Symp Quant Biol 18*, 273–279.
96. Hopkins, R. and Esposito, D. 2009. A rapid method for titrating baculovirus stocks using the Sf-9 Easy Titer cell line. *Biotechniques 47*, 785–788.
97. Reeves, P.J., Callewaert, N., Contreras, R., and Khorana, H.G. 2002. Structure and function in rhodopsin: High-level expression of rhodopsin with restricted and homogeneous N-glycosylation by a tetracycline-inducible N-acetylglucosaminyltransferase I-negative HEK293S stable mammalian cell line. *Proc Natl Acad Sci USA 99*, 13419–13424.
98. Burger, K., Gimpl, G., and Fahrenholz, F. 2000. Regulation of receptor function by cholesterol. *Cell Mol Life Sci 57*, 1577–1592.
99. Ritchie, T.K., Grinkova, Y.V., Bayburt, T.H., Denisov, I.G., Zolnerciks, J.K., Atkins, W.M., and Sligar, S.G. 2009. Chapter eleven - Reconstitution of membrane proteins in phospholipid bilayer nanodiscs. In *Methods in Enzymology*, N. Düzgünes, ed. Academic Press, pp. 211–231.
100. Mastronarde, D.N. 2003. SerialEM: A program for automated tilt series acquisition on tecnai microscopes using prediction of specimen position. *Microsc Microanal 9*, 1182–1183.
101. Li, X., Zheng, S., Agard, D.A., and Cheng, Y. 2015. Asynchronous data acquisition and on-the-fly analysis of dose fractionated cryoEM images by UCSFImage. *J Struct Biol 192*, 174–178.
102. Zheng, S.Q., Palovcak, E., Armache, J.-P., Verba, K.A., Cheng, Y., and Agard, D.A. 2017. MotionCor2: Anisotropic correction of beam-induced motion for improved cryo-electron microscopy. *Nature Methods 14*, 331.
103. Rohou, A. and Grigorieff, N. 2015. CTFFIND4: Fast and accurate defocus estimation from electron micrographs. *J Struct Biol 192*, 216–221.
104. Shaikh, T.R., Gao, H., Baxter, W.T., Asturias, F.J., Boisset, N., Leith, A., and Frank, J. 2008. SPIDER image processing for single-particle reconstruction of biological macromolecules from electron micrographs. *Nature Protoc 3*, 1941–1974.
105. Scheres, S.H.W. 2012. RELION: Implementation of a Bayesian approach to cryo-EM structure determination. *J Struct Biol 180*, 519–530.

106. Pettersen, E.F., Goddard, T.D., Huang, C.C., Couch, G.S., Greenblatt, D.M., Meng, E.C., and Ferrin, T.E. 2004. UCSF Chimera – a visualization system for exploratory research and analysis. *J Comput Chem 25*, 1605–1612.
107. Kucukelbir, A., Sigworth, F.J., and Tagare, H.D. 2014. Quantifying the local resolution of cryo-EM density maps. *Nature Methods 11*, 63–65.
108. Smart, O.S., Neduvelil, J.G., Wang, X., Wallace, B.A., and Sansom, M.S.P. 1996. HOLE: A program for the analysis of the pore dimensions of ion channel structural models. *J Mol Graph 14*, 354–360.
109. Shen, P.S., Yang, X.Y., DeCaen, P.G., Liu, X.W., Bulkley, D., Clapham, D.E., and Cao, E.H. 2016. The structure of the polycystic kidney disease channel *PKD2* in lipid nanodiscs. *Cell 167*, 763–773.
110. Emsley, P. and Cowtan, K. 2004. Coot: Model-building tools for molecular graphics. *Acta Cryst D, Biol Crystallogr 60*, 2126–2132.
111. Afonine, P.V., Poon, B.K., Read, R.J., Sobolev, O.V., Terwilliger, T.C., Urzhumtsev, A., and Adams, P.D. 2018. Real-space refinement in PHENIX for cryo-EM and crystallography. *Acta Cryst D, Struct Biol 74*, 531–544.
112. Chen, V.B., Arendall, W.B., 3rd, Headd, J.J., Keedy, D.A., Immormino, R.M., Kapral, G.J., Murray, L.W., Richardson, J.S., and Richardson, D.C. 2010. MolProbity: All-atom structure validation for macromolecular crystallography. *Acta Cryst D, Biol Crystallogr 66*, 12–21.
113. Barad, B.A., Echols, N., Wang, R.Y.-R., Cheng, Y., DiMaio, F., Adams, P.D., and Fraser, J.S. 2015. EMRinger: Side chain–directed model and map validation for 3D cryo-electron microscopy. *Nature Methods 12*, 943.
114. Amunts, A., Brown, A., Bai, X.C., Llacer, J.L., Hussain, T., Emsley, P., Long, F., Murshudov, G., Scheres, S.H.W., and Ramakrishnan, V. 2014. Structure of the yeast mitochondrial large ribosomal subunit. *Science 343*, 1485–1489.
115. Zhao, M., Wu, S., Zhou, Q., Vivona, S., Cipriano, D.J., Cheng, Y., and Brunger, A.T. 2015. Mechanistic insights into the recycling machine of the SNARE complex. *Nature 518*, 61–67.
116. Grieben, M., Pike, A.C., Shintre, C.A., Venturi, E., El-Ajouz, S., Tessitore, A., Shrestha, L., Mukhopadhyay, S., Mahajan, P., Chalk, R. et al. 2017. Structure of the polycystic kidney disease TRP channel polycystin-2 (PC2). *Nat Struct Mol Biol 24*, 114–122.
117. Wilkes, M., Madej, M.G., Kreuter, L., Rhinow, D., Heinz, V., De Sanctis, S., Ruppel, S., Richter, R.M., Joos, F., Grieben, M. et al. 2017. Molecular insights into lipid-assisted Ca(2+) regulation of the TRP channel polycystin-2. *Nat Struct Mol Biol 24*, 123–130.
118. Zheng, W., Yang, X., Hu, R., Cai, R., Hofmann, L., Wang, Z., Hu, Q., Liu, X., Bulkey, D., Yu, Y. et al. 2018. Hydrophobic pore gates regulate ion permeation in polycystic kidney disease 2 and 2L1 channels. *Nat Commun 9*, 2302.
119. Su, Q., Hu, F., Ge, X., Lei, J., Yu, S., Wang, T., Zhou, Q., Mei, C., and Shi, Y. 2018. Structure of the human PKD1-PKD2 complex. *Science 361.*

3 Recording Ion Channels in Cilia Membranes

Leo C.T. Ng, Amitabha Mukhopadhyay,**
Thuy N. Vien, and Paul G. DeCaen

CONTENTS

3.1 INTRODUCTION

3.1.1 Ciliary Diversity: Primary, Motile, and Sensory Types

Through the single lens of a homemade microscope, Antonie van Leeuwenhoek visualized the incredibly tiny feet of "animalcules." The function of these beating motile cilia for locomotion of protists in pond water was obvious to van Leeuwenhoek.[1]

* These individuals contributed to this work equally.

He also observed motile cilia and flagella in human cells, noting that their undulations may move fluid or propel the sperm. A few hundred years later, embryologist Alexander Kowalevsky identified single cilia projecting from a variety of vertebrate cells.[2,3] These "primary cilia" are immotile and are far more widespread than the motile cilia types within our organ systems. Besides motile and primary cilia, "sensory cilia" make up the third and most structurally diverse cilia type. Sensory cilia, as their name implies, are found in sensory tissues such as olfactory neurons and photoreceptor cells. In general, sensory and primary cilia are thought to function as cellular antennae—receiving and integrating stimuli (chemical and other) through the activation of molecular effectors such as G protein-coupled receptors (GPCRs) and ion channels which are concentrated in these projections.[4,5] However, beyond fluid motion and motility, some evidence suggests that motile cilia may also carry out a sensory function for cells.[6] Thus, there might be considerable functional overlap between these classifications of cilia. Below, we will discuss the structural, functional, and component differences among these three cilia types.

3.1.2 Structural Assembly of Cilia Types

Primary cilia formation occurs during the interphase of the cell cycle because ciliogenesis shares the same microtubule machinery as mitosis and, thus, is dynamically regulated by cell-cycle proteins (Figure 3.1b).[7–9] The mother centriole polarizes to the apical surface and becomes the basal body of the cilium, consisting of a PLK4–STIL–SAS-6 core and a ring of nine triplet microtubules, made of α and β tubulins.[10,11] During ciliogenesis, which occurs in the G_0 and G_1 phases of cell division, the centrosome microtubules polymerize to form an internal scaffolding structure called the axoneme. When complete, the cilium extends 2–12 μm from the

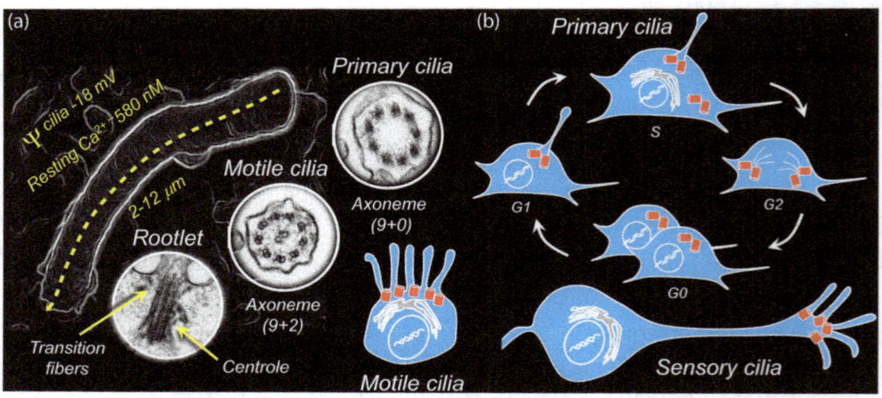

FIGURE 3.1 Types of cilia in mammalian cells. (a) A diagram depicting the cilia's dimensions and its unique ionic and electrical properties. *Right*: Negative EM staining of the ciliary rootlet and axoneme structures. Note the differences in the axonemes. (b) Three types of cilia and the temporal status of the primary cilia during cell division. Note that the primary cilium extends from the mother centriole and is absorbed or severed during active cell division. Centrioles are shown in red.

apical side of the plasma membrane with a diameter of less than 500 nm (Figure 3.1a). In general, motile cilia and primary cilia have unique axoneme structures. While both motile and primary cilia have a ring of nine doublets symmetrically arranged around the center (9+0 formation), most motile cilia have an extra pair of singlet microtubules in the center of the ring (9+2 formation, Figure 3.1a). The motile cilia also contain radial spokes and inner and outer dynein arm complexes to facilitate directional beating and bending of the cilia. The rhythmic beat is powered by the hydrolysis of ATP in dynein proteins anchored to the doublets. In sensory cilia (e.g., olfactory cilia), the axonemal shaft resembles the motile cilia but lacks the dynein arms required for movement—so, in this regard, its structure is similar to the primary cilia (Figure 3.1b). Mammalian cells exhibit only one primary cilium per cell. Motile cilia are most commonly found in multiples per cell—one exception being those in the embryonic node, which are solitary. When the cell divides, the centrosome redirects microtubule assembly to form the mitotic spindle and the primary cilia is absorbed by the cell. Ciliogenesis resumes in the two new cells after division. On the other hand, multiciliated cells contain dozens to hundreds of motile cilia, and ciliogenesis is nucleated by parental centrosomes and acentriolar structures called the deuterosomes.[12,13] Between the basal body and the axoneme is a junction created by Y-shaped fibers to form a transition zone.[14] This barrier, along with the intraflagellar transport protein (IFT) complexes, regulate the import and export of cargo in the cilia. This process involves two microtubular motor complexes, IFT-A and IFT-B, which are comprised of at least 7 and 16 different IFT subunits, respectively.[15,16] The IFT-B complex moves cargo from base to tip with kinesin-2 motor while the IFT-A complex recycles cargo back to the cell body with cytoplasmic dynein-2 motor.[17] The importance of the IFT system is highlighted by the failure of ciliogenesis and mitotic spindle orientation in *Chlamydomonas* and lethal phenotypes in mice when necessary components (e.g., KIF3a or IFT88) are genetically ablated.[18–20] In addition, ciliary protein localization and downstream signaling are crucial for tissue patterning during vertebrate embryogenesis.[20,21] Here, right-left symmetry is established by asymmetric Hedgehog activation in the embryonic node (also called Hensen's node or Kupffer vesical in fish) present in the gastrula-stage.[22,23] Beyond early development, Wnt and Hedgehog signaling is believed to regulate cytoskeletal cell polarity pathways and is dysregulated in forms of human cancer.[24,25] Briefly discussed in the next section are "ciliopathies"—genetic diseases that impact multiple organ systems which result from either abnormal formation of cilia or cilia-related signaling.

3.1.3 Ciliopathies

This section is a short primer on ciliopathy diseases. For those interested in a more in-depth review of ciliopathies and their disease mechanisms, we suggest reading reviews by Bisgrove and Yost[26] and Oh and Katsanis.[27] The importance of the primary cilia in human health is highlighted by more than 35 congenital diseases called "ciliopathies," which impact the development of organ systems such as the kidney, brain, heart, and eye. While some defects may only affect one single organ, the majority impact multiple organs with a combination of common ciliopathic phenotypes. For example, Joubert syndrome and Joubert-related disorders are caused by variants in the genes encoding for cilia assembly components—such as centrosomal CEP120 and BBS proteins, and the ciliary GTPase ARL13B.[28] These syndromic diseases primarily impact the development of the central nervous system but also have comorbidities which affect the renal and visual systems. In addition, dynein complexes are required for the coordinated and rhythmic motion of motile cilia and variants in these components result in impaired or immotile ciliary syndromes. For example, primary ciliary dyskinesis (PCD) is characterized by chronic respiratory tract infections and infertility because the cilia of airway epithelia and the sperm flagellum are immotile.[29] Hydrocephalus is a common comorbidity in PCD patients. Here, the beating of motile ependymal cilia is uncoordinated, resulting in an accumulation of cerebrospinal fluid in the ventricles of the brain. Developmental defects are common among individuals with ciliopathies—such as polydactyly (extra digits on hands and toes) and *situs inversus* or *totalis* (incorrect positioning of body organs).[27] However, the most common comorbidities shared by ciliopathies are polycystic kidney disease (PKD) and other renal defects but the reason for this is not known.[30] The most prevalent form of PKD is the autosomal dominant form (ADPKD), a ciliopathy which is caused by gene variants in *PKD1* and *PKD2*—which encode for two cilia-specific proteins polycystin-1 and polycystin-2, respectively.[29] ADPKD a common monogenetic disorder (~1:2000 people) characterized by kidney and liver cysts in adulthood. Embryonic knockout of both alleles of either *PKD1* or *PKD2* in mice causes renal disease *in utero* and results in embryonic death.[31,32] Homozygous ablation of either gene in mature mice results in progressive kidney cyst formation and recapitulates the ADPKD phenotype found in humans.[33,34] Importantly, gene variants which encode for downstream effectors of Ca^{2+} ciliary signaling are responsible for ciliopathies that share renal comorbidities. Thus, the significance of determining polycystin channel function and the impact of disease-causing variants will likely extend to other renal ciliopathies where localized ciliary Ca^{2+} dysregulation might be a shared disease-causing mechanism[35,36] The ciliary patch clamp configurations discussed in Section 3.3 provides a direct measurement of cilia-localized ion channel biophysics in native and heterologous expression systems.

3.1.4 Ion Channels in Cilia

The presence of ion channels in cilia membranes has been determined using localization studies, but their function within the cilia has been primarily supported by indirect means. Cilia are privileged organelles, which do not load with calcium fluorescent dyes. Thus, measuring changes in cytoplasmic calcium has led to confusion regarding the basic properties of ciliary ion channels.[5] In the next section, we describe electrophysiology methods for measuring ion channels directly from the cilia membrane.[37,38] Using these patch clamp configurations, we are able to determine the how, what, and why of cilia ion channel regulation.

The "how" of cilia ion channel regulation can be generalized in terms of biophysical regulation. Using the patch clamp configurations described in Section 3.3, we can determine the mechanistic regulation of cilia channels at the single-channel level, which is not possible with any other method. This includes, but is not limited to, determining the threshold of stimulus that regulate the states (e.g., voltage dependence of activation) and the type of ions that move through the channels (i.e., ion selectivity and conductance). The "what" refers to the genetic components which encode for ciliary ion channels, whereas the "why" describes their physiological function in teleological terms. By combining this approach with genetic ablation techniques, it is apparent that membranes of motile and primary cilia from disparate tissues contain distinct repertoires of ion channels. Outwardly rectifying currents from the primary cilia of different cells are conducted by distinct members of the polycystin subfamily of transient receptor potential ion channels (TRPPs). Here, polycystin-2 (TRPP2, encoded by the *PKD2* gene) forms an ion channel in the kidney collecting duct cilia whereas polycystin-2L1 (TRPP3, encoded by the *PKD2L1* gene) and polycystin-1L1 (encoded by the *PKD1L1* gene) are in retina pigmented epithelial cells (RPE) and mouse embryonic fibroblasts (MEFs).[37–40] Polycystin-2 is also reported to form heteromeric channels with polycystin-1; however, ion conduction by this complex is uncertain and is an area of active research. There are now high resolution cryo-EM structures of homomeric and heteromeric forms of the polycystin proteins, providing structural context for mutagenesis studies.[40–43] Both polycystin-2 and polycystin-2L1 expressing cells produce large conductance (96–180 pS), depending on the stoichiometry and charge carrier used. Polycystin-2L1 is more calcium-selective than polycystin-2, which is conferred by an additional aspartate (D525) residue found in the selectivity filter.[44] Both polycystin-2 and polycystin-2L1 are potentiated by high intraciliary calcium.[39,44] Under resting Ca^{2+} concentrations in the cilium (580 nM), polycystin channels will have a relatively low open probability as its voltage dependence of activation is positive relative to the resting membrane potential of the cilium (−18 mV).[45] However, when ciliary Ca^{2+} reaches μM levels, the voltage dependence shifts negatively, allowing polycystin-2 to spend more time in the open state at the ciliary resting membrane potential. Polycystin-2L1 channels exhibit irreversible desensitization by high levels of calcium, which may function as a negative-feedback mechanism in regulating ciliary calcium concentration.[44] In sperm flagella, calcium-selective channel CatSper and potassium channel Slo3 are required for hyperactivated motility—a process required for sperm fusion with the ovum.[46] In the sperm patch clamp method, the entire sperm membrane (including the flagellum and head) is voltage clamped by establishing a high resistance seal with the cytoplasmic droplet, which is located on the principle piece of the flagellum of the mouse and human sperm.[47] It appears that the membranes of the flagella/cilia and the cell body are electrically unified, and that the population of conductive channels are the same in ependymal cilia and its cell body. In contrast, the primary cilium is more electrically insulated from the cell body, as is evident by its unique membrane potential and resting Ca^{2+} concentration. In sensory olfactory cilia, the signal initiates with an odorant binding to a GTP-coupled receptor that increases production of cAMP by adenylyl cyclase catalysis. Luminal cAMP directly activates cyclic nucleotide–gated (CNG) channel, facilitating an influx of sodium and calcium ions into the cilium.[48] Calcium entry subsequently activates anoctamine-2 (ANO2), a member of the Ca-activated chloride channel family, which amplifies the initial ciliary membrane depolarization as $Cl-$ leaves the cell body of the olfactory neuron.[49] Thus, it appears that the cell body of sensory cells, like the cell body of motile ciliated cells, is electrically coupled to sensory ciliary membrane potential.

3.2 METHODS TO VISUALIZE CILIA FOR VOLTAGE CLAMP EXPERIMENTS

High electrical resistance seals from glass electrodes on cilia membranes for the purpose of measuring cilia-localized ion channels can be achieved with or without a cilia-specific fluorescent indicator.[38,50,51] For preparations in which fluorescence is not used as a guide, primary cilia can be visualized using phase contrast microscopy. This is achieved by culturing the cells on glass beads (or a rod-shaped surface) so that the cilium projects perpendicular to the light path on a standard microscope.[50] However, the dimensions of cilia are similar to that of filipodia. Thus, it is possible to confuse which cellular projection, cilia or filipodia, is being patched during an experiment without the aid of a ciliary target. Besides confirming that the cilia membrane is patched, fluorescence also provides a visual confirmation of the type of cilia patch configuration achieved. For example, establishing an "inside-out" patch configuration in conjunction with loading dyes would allow visual confirmation of liquid access to the internal compartment from the bath solution (see Section 3.3). Although overexpression of ARL13B-EGFP does not alter the morphology of the primary cilia, overexpression of some cilia-specific fluorophores may directly or indirectly alter the properties of ion channels present in the membrane. Thus, proper controls should be taken when deploying this technique. In unpublished work, we compared the biophysical properties of the polycystin-2 ion channel from inner medullary kidney collecting duct (IMCD) cells with and without ARL13B-GFP overexpression. We did not observe statistical differences in the number of polycystin-2 channels, conductance, or voltage dependence of channel opening between the two cell types (data not shown). Thus, these findings give us confidence in using fluorescently labeled cilia as our method of choice. Outlined below are two approaches to visualize primary cilia—immortalized cell lines and primary cells from transgenic mice—to be used in conjunction with the voltage clamp method to characterize ciliary ion channel biophysics.

3.2.1 Transgenic Mice for Recording Primary Cilia Ion Channels *In Situ* or from Primary Cultures

Primary cilia are visualized from live primary cells and tissues of transgenic mice that stably express cilia-specific fluorophores.[38] Previously, we stably expressed the ARL13B transgene conjugated to enhanced green fluorescent protein (*Arl13B-EGFP^{tg}*) in the genome of a founder C57BL/6 mouse. ARL13B (ADP-ribosylation factor-like protein 13B) is enriched but is not exclusively trafficked to motile and primary cilia. Importantly, *Arl13B-EGFP^{tg}* mice do not have a distinct behavioral or anatomical phenotype from C57BL/6 mice.[38,45] Furthermore, overexpression of the transgene does not alter the size or number of motile and primary cilia based on histological analyses of the lung, kidney, and eye. Heterozygous or homozygous expression of the transgene is sufficient for cilia visualization *in situ* or from cultured cells with green fluorescence protein (GFP) excitation under a standard widefield or confocal microscope. To facilitate cilia patch clamp experiments, we have successfully isolated ciliated hippocampal neurons, ependymal cells, RPE, MEF, and primary IMCD cells using tissue specific dissociation protocols to create primary cultures of each cell type.[52–54] When grown on a monolayer, the apical side with their projecting cilia invariably faces the top of the culture dish, which is easily accessible to a patch electrode (see Section 3.3.4).

3.2.2 Cell Lines to Record Native and Heterologous Ion Channels in Primary Cilia

Primary cilia from immortalized cultured cells can be used to study the function of ciliary ion channels and localization of various ciliary proteins.[38,39,45] Using the same strategy discussed previously, we have stably overexpressed several cilia-specific proteins (ARL13B, Smo, polycystin-2) conjugated to fluorescent proteins in several mammalian cell lines (IMCD, RPE) previously reported to have primary cilia. Thus, it is no surprise that cilia from these cells can be used for our electrophysiology experiments. However, when we stably express either GFP conjugated polycystin-2 (PKD2-GFP) or Smoothened (Smo-GFP), we observe that even human embryonic kidney (HEK) cells have primary cilia as well.[39] This is surprising since HEK cells are one of most common cell types used to study the function of heterologous ion channels. So, why has this observation not been reported before? The cilium is privileged and few types of overexpressed ion channels will localize to this compartment. Thus, the presence of cilia in HEK cells is easy to miss under light microscopy. Most fluorescently labeled channels—apart from polycystins—will not localize to the HEK cilia. HEK cells are easy to genetically manipulate, including overexpression and CRISPR/Cas9-mediated endogenous gene ablation.[55] Our lab is currently using heterologous mutagenesis strategies within these cell lines to assess structural regulation of ciliary ion channel biophysics.

3.3 PROTOCOL FOR RECORDING PRIMARY CILIA ION CHANNELS USING VOLTAGE CLAMP ELECTROPHYSIOLOGY

In this section, we describe strategies for measuring ciliary ion channels from cells expressing cilia-specific fluorescent proteins with several patch clamp configurations. This method is an extension of the voltage clamp method established many years ago by Neher and Sakmann,[56] albeit the membranes being recorded here are 10–100× smaller than most plasma membrane patches. Keep in mind that this section is only a primer, intended for investigators who already have experience with electrophysiology techniques. For those just starting out, we recommend reading *Single-Channel Recording*[57] and *The Axon™ Guide: A Guide to Electrophysiology and Biophysics Laboratory Techniques*.[58] These are important references to the theory and operation of the voltage clamp method. We also recommend gaining hands-on experience with measuring ion channels from plasma membranes prior to measuring them from cilia. This will provide an invaluable exercise and optimal starting point from which fundamental understanding of the method can be applied to the cilia voltage clamp technique.

3.3.1 Microscope and Cameras

We conduct our cilia electrophysiology experiments using either an inverted confocal or widefield microscope (Table 3.1). While it is possible to conduct these experiments

TABLE 3.1
List of Equipment Used with Manufacturer Part Numbers

Component	Part and/or Model Number	Manufacturer
Bright-Field Microscope Components		
Equipped with standard components	IX73	Olympus
Objective 20×, Air U Plan Fluorite, NA 0.5, WD 1.6 mm	1-U2B525	Olympus
Objective 40×, Air U Plan Fluorite, NA 0.75, WD 0.51 mm	1-U2B527	Olympus
Objective 60×, water emersion NA 1 2, WD 0.51 mm.	UPLSAPO	Olympus
Fluorescent light source	Lumen 200	Prior
Filter for EGFP, GCaMP3, mEmerald 470/20×, 485DM, 517/45 m	U-FGFP BX3 GFP/BLUE WIDE	Olympus
Filter for EYFP, Venus, Citrine 495/10×, 505DM, 537/45 m	U-FYFP; BX3 YFP/ YELLOW NARROW	Olympus
Filter for RFP, mCherry, DsRED 545/20×, 565DM, 597/55 m	U-FRFP; BX3 RFP/GREEN WIDE	Olympus
Electrophysiology and Imaging Components		
Patch clamp amplifier	Axopatch 200B	Molecular Devices
Digital to analog signal converter	Digidata 1440	Molecular Devices
CMOS camera	ORCA-FLASH 4.0LT	Hamamatsu
CCD camera	U-TV1X-2-7	Olympus
Rapid perfusion system	VC-77CS "Fast step system"	Warner Instruments
High-speed pressure clamp system	HSPC-1	ALA Scientific Instruments
Dual micromanipulator system (Stepper motor driven)	MP-300	Sutter Instruments
Dual micromanipulator system (Linear piezo driven)	Inchworm, uM 3-axis system 69-3432	Sensapex
Microelectrode Fabrication Components		
Flaming brown filament type micropipette electrode puller	P-97 3 mm box filament Settings: Heat = ramp test value; Pull = 44; Velocity = 65; Delay = 200; Pressure = 600	Sutter Instruments
Laser-based micropipette electrode puller	P-2000	Sutter Instruments
Capillary glass O.D. 1.5 mm, I.D. 0.84 mm	1B150F-4	World Precision Instruments

using an upright microscope, we prefer to keep the "air space" above the sample free from the objective to allow for other manipulations—such as excising the cilium and/or using a rapid exchange system. We typically use a 20× objective lens to locate cells which are projecting their cilia above the focal plane and have sufficient fluorescence. We then switch to a 60× water emersion objective with a 2× photomultiplier to visualize the cilia during seal formation. To capture images of the ARL13B-GFP fluorescence to confirm the cilia patch configuration, a standard CCD camera on a widefield microscope will suffice. However, we recommend using a CMOS camera (complementary metal-oxide semiconductor, Hammamatsu) if ratiometric calcium imaging experiments are being concurrently conducted (Table 3.1). This camera has sufficient frame speed and sensitivity to detect changes in calcium-dependent fluorescence in the cilium. However, to make a ratiometric measurement (GCaMP3:mCherry) necessary to quantify the calcium concentrations in the cilium with the cilia-localized calcium sensor Smo-mCherry-GCaMP3,[45] the camera will need to be coupled with an automated beam splitter to rapidly but independently excite GCaMP and mCherry fluorescence.

3.3.2 Glass Electrode Fabrication

In this section we will discuss the most important technical aspect of the protocol—the features of a glass electrode suitable for cilia electrophysiology. We typically use filamented glass with inner and outer diameters of 0.84 and 1.4 mm, respectively (Table 3.1). Previously, we fabricated our glass electrodes with a heated-filament type electrode puller but we now prefer a laser-heated puller because the resulting pipette geometry is more reproducible. In Table 3.1, we have listed the puller settings and the type of box filament used. The resistance of the "pulled" electrode is between 5–9 mΩ with an inner aperture of 0.5–1 μm in diameter (Figure 3.2a). Our experience, a priori, suggests that electrodes with a 5°–15° shank are optimal for seal formation. We then polished these electrodes using a commercially available electrode polisher (Table 3.1). This step smoothens the surface of the glass and narrows the aperture of the tip of the electrode, thereby increasing the resistance of the electrode. Recording saline is frontfilled and backfilled into these pipettes to ensure electrical continuity. The composition of the saline will depend on the patch configuration and type of ciliary ion channel to be tested (see Section 3.3.3). To measure polycystin-2 and polycystin-2L1 channel currents, we typically use a symmetric condition, where the pipette electrode and bath solutions both contain 140 mM NaCl and 10 mM HEPES. The solutions are pH and osmolarity balanced to 7.4 and 300 (±6) mOsm with NaOH and mannitol, respectively. Internal and external calcium can be added to either solution and chelated with BAPTA or EGTA to establish a resting free-calcium solution. To determine the optimal pipette resistance, we correlated the resistance with the success of achieving a high resistance seal with the cilia membrane (>10 GΩ, see Section 3.3.4.1) from many attempts (Figure 3.2a). Then, we used a scanning electron microscope to measure the inner diameter of each patch electrode used. Here, we determined electrodes with a 280–410 nm inner diameter would most frequently achieve high resistance seals suitable for measuring ciliary ion channels. These electrodes have a pipette resistance of 17–32 MΩ. While it is possible to record ciliary ion channels using different

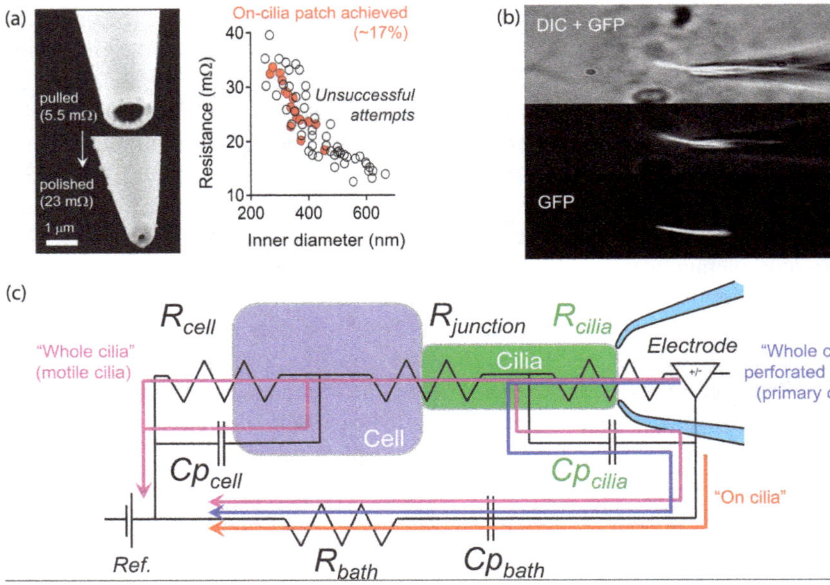

FIGURE 3.2 Establishing a ciliary giga seal is largely dependent on electrode fabrication. (a) Scanning electron microscope images of a cilia patch electrode before and after polishing the electrode tip. *Right*: Correlation between electrode resistance, lumen diameter and successful giga-seal formation. Red circles indicate successful trials. (b) Bright field microscope images of an electrode sealing onto the primary cilia membrane of an inner medullary kidney collecting duct cell expressing ARL13B-GFP. (c) Equivalency circuits cell membranes involved with the ciliary patch configurations.

electrode geometries than those described here (e.g., using electrodes with a wider shank), the inner diameter of the aperture must always be less than the diameter of cilium being recorded (<500 nm). Since the diameter of the electrode is not easily quantified using most microscopes, we use the 17–32 MΩ pipette resistance as our guiding metric for selecting suitable electrodes for these experiments.

3.3.3 Approaching the Cilia

Ciliated cells can either be cultured directly into recording chambers or onto 0.05 mm coverslips, which can then be placed into the recording chamber. Once in place, glass electrodes are placed into position using a micromanipulator. We use a traditional three-direction gear-driven manipulator commonly used for whole cell electrophysiology. However, most of these manipulators "drift" by 1–2 µm during an experiment. The amount of drift can be larger, especially with older manipulators or from those which were not well mechanically isolated. Drift can cause the loss of a high resistance seal, effectively ending the recording prematurely. Recently, we switched to an "Inchworm" manipulator driven by a piezoelectric actuator, which moves the shaft with nanometer precision and drift is undetectable (Table 3.1). Once we have an electrode in the bath solution with suitable resistance, we approach a ciliated cell using coarse manipulator speeds under a 20–40× objective magnification. At this point, we recommend turning off your fluorescent light source to minimize photobleaching of the sample. Once the electrode is 0.1–0.5 mm above the cell, switch to a higher magnification objective. We use a 60× water immersion objective and apply a drop of water in the interface of the recording chamber and the lens. Caution is needed because automated or manual switching of the objective or adding the liquid could invariably cause vibrations on the microscope stage. Keeping a safe distance between the electrode and sample will reduce the likelihood of causing damage during these vibrations.

3.3.4 Cilia Patch Configurations

In this section we describe several cilia patch configurations with distinct applications to assess ciliary channel biophysics (Figure 3.3a). To aid the discussion, we have added an equivalency circuit of the cilia patch configurations to keep track of which cellular membranes are involved in these configurations (Figure 3.2b). These configurations are based on traditional membrane patch configurations, with some subtle differences. Many of the suggestions here are made with the goal of measuring polycystin channels from the cilia of collecting duct cells. Because there are likely other ion channels present in cilia membranes from different cell types,

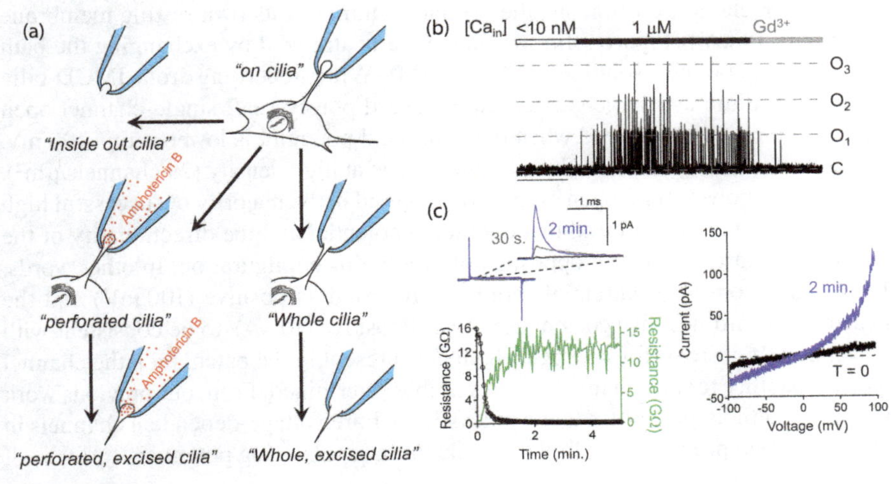

FIGURE 3.3 Cilia patch configurations and exemplar voltage clamp records. (a) Illustrations depicting the ciliary patch configurations. (b) Measuring polycystin-2 single channels in the inside-out configuration. The ciliary membrane was held at 100 mV. Note, the increase in the opening events (O_1–O_3) when internal calcium is increased from 10 nM to 1 μM. Polycystin channels were then blocked by 10 μM Gd^{3+}. (c) Establishing the perforated patch configuration with amphotericin B in the pipette electrode. *Top*: Increase capacitance two minutes after sealing onto the ciliary membrane. *Bottom*: The time course of amphotericin's effect on the ciliary membrane. Note that the capacitance and resistance are inversely proportional over time. *Right*: Activation of the total ciliary sodium current after perforating the cilia membrane and gaining electrical access to its capacitive membrane.

the experimentalist should alter this protocol accordingly. We typically collect our electrophysiology data using an Axopatch 200B patch clamp amplifier and the pClamp 10 software (Molecular Devices, Table 3.1). However, there are many other suitable amplifiers and acquisition/analysis software packages that can be used for cilia electrophysiology. As a rule of thumb, we recommend sampling the current 5× more frequently (25 KHz) than the low-pass Bessel filter rate. While this makes the size of the acquisition file larger, these settings allow for the sufficient sampling of both single- and whole-cilia currents, which can be re-sampled during the analysis phase. For data analysis, we use a combination of ClampFit (Molecular Devices) and IGOR Pro software (Wavemetrics).

3.3.4.1 On-Cilia Patch

All ciliary ion channel recordings begin in the on-cilia patch configuration (Figure 3.3a). Here, the electrode is lowered until it interfaces with the cilia tip, analogous to establishing the "on-cell configuration" with the plasma membrane. When approaching the cilia, it will deflect due to the outward pressure of the recording saline from the patch electrode, which indicates that the electrode and the cilia are at the same depth. Applying a gentle inward suction (2–10 mmHg, if you are using a pressure clamp system) causes the tip of the cilium to seal onto the pipette tip. At this point, the ciliary membrane resistance (R_{cilia}) will climb to levels above 10 GΩ, which is commonly known as a "giga-seal" (Figure 3.2c). Based on the results from current clamp experiments, the primary cilium has its own resting membrane potential (−18 mV) (Figure 3.1a), which can be neutralized by exchanging the bath with a high potassium solution (120–140 mM). When recording from IMCD cilia and using sodium as charge carrier, large inward polycystin-2 single-channel open events (−10 pA) are observed when the command potential is lowered to −100 mV. These channels are present in the cilia membrane at high density (29 channels/μm^2). In our hands, polycystin-2 open events are observed in the majority of successful high resistance seals. Keep in mind that command potential and the directionality of the currents measured are in the opposite polarity in this configuration. In other words, the negative command potential should be reported as positive (100 mV) and the measured inward/negative current as outward/positive (10 pA), to be consistent with convention. If current from a single channel is present in the patch, then the channel open probability for this channel can be reliably determined. From our previous work using this method, polycystin-2 and polcystin-2L1 are voltage-dependent channels in which the open probability is highest at the most depolarizing potentials.[38,39]

3.3.4.2 Inside-Out Cilia Patch

From the on-cilia configuration, the cilium membrane can be torn open to expose the intraciliary compartment to the external bath where it can be exchanged (Figure 3.3a,b). This is analogous to the inside-out membrane patch which is commonly achieved by quickly lifting the electrode. However, raising the cilia causes it to stretch and reseal on its severed end, which isolates the intraciliary membrane from the bath and limits the effect of bath exchange of the inside of the cilium. To avoid this, excised cilia patches are briefly pressed against the surface of a bead made of Sylgard 184 (Dow Corning). Visualization of the inside-out configuration is done by perfusing the bath solution with unconjugated Alexa 594 fluorescent dye (Invitrogen). If the inside-out cilia configuration is achieved, the dye indicator will localize to the inside of the cilia compartment. The reporting conventions regarding command and current directionality are the same as the on-cilia configurations, but there is no need to neutralize the resting membrane potential. Using this configuration, ion channel modifiers (e.g., lipids and GPCR effectors) can be applied to the inside of the cilia membrane to study how intraciliary signaling controls channel kinetics. An example of using the inside-out configuration is shown in Figure 3.3b. To study rapid effects of internal calcium on polycystin-2 channel biophysics, we use a Fast Step Perfusion System (Warner Instruments) to change internal calcium concentration ($[Ca_{in}]$) from ~10 nM to 1 μM within 250 μs. The solution change is triggered via the pClamp software by routing the Fast Step output and input signals through a digital-to-analog signal converter (Digidata 1440, Molecular Devices, Table 3.1). The impact of elevating internal calcium is obvious—the channels are almost always closed when it is low, then several open events are observed when it is raised (Figure 3.3b).

3.3.4.3 Whole, Perforated, and Excised Cilia Patch Configurations

From the on-cell configuration, the total population of ion channels within the ciliary membrane can be captured in the whole-cilia or perforated patch configuration (Figure 3.3a). To achieve this configuration, the membrane forming the high resistance seal must be broken or perforated by an antibiotic. These configurations allow for electrical continuity between the pipette solution and the total ciliary capacitor (Cp <1 femtofarad). The seal membrane is broken using 300 mV voltage pulse for 50–500 ms. We use a modified "Zap" function on the Axopatch 200B amplifier, where we have replaced resistor R143 and used a 1 KΩ trim pot to lower the standard input potential of this function to 1.3 V. This reduces the likelihood of the total cilium being over-charged, which can rupture the total membrane. In the whole-cilia configuration, currents from five to ten voltage ramps from −100 to 100 mV (at rate of 0.5 mV per millisecond) are averaged to assess the voltage dependence of the total ciliary current. Chemical continuity between the internal cilioplasm and the pipette can be visualized by pre-loaded unconjugated Alexa 594 fluorescent dye in the pipette under red fluorescent protein (RFP) excitation. The dye can be seen entering the cilioplasm of primary and motile ciliary patches. However, an important difference is that the Alexa fluorescence remains restricted to the primary cilia compartment but can diffuse into the cell body of motile cilia. These chemical diffusion differences agree with the cilia-to-cell electrical connectivity differences reported for cells with primary and motile cilia. Since the repertoire of ionic currents measured from cilia and plasma membranes are distinct, only the channels clamped in the cilia membrane are captured in whole-cilia recording of primary cilia (Figure 3.2c). Thus, the primary cilia and plasma membrane are separated by significant resistivity ($R_{junction}$), which is probably formed by the cilia rootlet structures (e.g., centrosomes, transition fibers) found at the cilia-cell junction (Figure 3.1a). In contrast, plasma-membrane currents (e.g., Ca_V) can be measured from the motile cilia membrane, suggesting that the motile cilia membrane is more electrically connected to the plasma membrane (Figure 3.2c).[59] Thus, when voltage clamping motile cilia, the electrical circuit is more complex and involves two capacitive (Cp_{cilia} and Cp_{cell})

and two resistive membranes (R_{cilia} and R_{cell}). From either the perforated (see next paragraph) or whole-cilia patch configurations, the cilia can be removed from the cell body by lifting the electrode with its connected cilia (Figure 3.3a). Here, the lifting motion stretches the cilia membrane until it breaks from the cilia-cell junction, effectively establishing a configuration where the cilia channels can be measured in physical isolation from the plasma membrane (Cp_{cilia}, R_{cilia}). We typically lose a percentage of the cilia membrane as it breaks off from the cell, and thus we also lose a portion (~35%) of the outwardly rectifying cilia current in RPE cells expressing the polycystin-1L1/2L1 heteromeric ion channel.[38] Because the cilia-cell connection is severed, this configuration demonstrates that the cilia membrane—and not the plasma membrane—is densely populated with the polycystin channels and provides a method to study their activity independent from the cell.

Alternatively, the total ciliary current can be measured by establishing a perforated patch configuration (Figure 3.3a). The perforated patch has several advantages over the whole-cilia configuration. First, the internal compartment is not dialyzed, meaning the levels of endogenous cilioplasmic factors, such as mediators for GPCR signaling, are not disturbed. A second advantage is that this configuration is typically easier to achieve than whole-cilia because it does not rely on breaking the ciliary membrane, which often can cause the loss of the electrical seal. Here, the tip of the patch electrode is "front-loaded" with standard intracellular saline, and "back-loaded" with saline containing an antibiotic. For recording polycystin channels in the cilia, we typically use amphotericin B, which permeates K^+, Na^+, and Cl^- ions. The antibiotic diffuses into the membrane and forms pores in high numbers—effectively reducing the ciliary resistance (R_{cilia}) over a time course of 0.1–5 minutes (Figure 3.3c). At this same time, the ciliary membrane capacitance increases to tens of femtofarads (Cp_{cilia}). Here, the total sodium current measured from cilium membrane is outwardly rectifying when activated with a voltage ramp—consistent with polycystin-2's single-channel properties measured in the on-cilia configuration. Keep in mind that the presence of amphotericin B in the pipette solution can prevent seal formation, and likewise any amphotericin B that leaks from the patch pipette near the cilia and cell can damage the membrane. "Front-loading" the pipette tip with antibiotic-free solution and establishing the on-cell configuration first before the amphotericin reaches the tip should prevent premature invasion of the membrane. How much antibiotic-free solution should be loaded into the patch pipette tip (i.e., how long the patch pipette should be held in the antibiotic-free solution) requires optimizing and depends on how quickly giga-ohm seals are formed. The more saline is front-loaded into the pipette, the longer it takes for the amphotericin to arrive at the aperture and perforate the membrane. The antibiotic-free solution should not go beyond 0.3 mm from the aperture. Typically, we formulate the antibiotic at 60 mg/mL of DMSO to make a stock solution. This is then diluted by 100–300× in recording saline to achieve a final concentration of 200–600 μg/mL. We recommend making both solutions fresh every experimental day, as the efficacy of antibiotic begins to decline after 24 hours.

3.3.4.4 Enveloped Cilia Patch Clamp Configurations

Like the on-cilia configuration, channel biophysics are measured at the single-channel level in the enveloped excised cilia patch clamp configuration.[50] This patch configuration was first adapted to record primary cilia currents by the Kleene laboratory from methods used to record ionic currents found in frog sensory olfactory cilia.[51] Importantly, the difference between the "enveloped cilia" and "on-cilia" configurations is where these seals are made with the glass electrode. Here, the entire cilium is enveloped within the patch electrode, and the physical seal is made at the junction between the cilia and cell membrane. The reporting conventions regarding command and current directionality are the same as the "on-cilia" configuration. The enveloped cilia patch can be torn off from the cell body to expose the inner membrane of the cilia. In this configuration, ion channels which might be electrically insulated from the primary cilia patch configuration—such as those localized to the membrane of the cilia-cell junction—can be measured. An advantage of this configuration is that these populations might be missed in the other cilia membrane patch clamp configurations. The Kleene group has determined that TRPM4 is at least in-part responsible for a 31 pS cation conductance using the enveloped cilia patch configuration.[60] In our hands, we do not observe this conductance in the on-cilia patch configuration. However, we regularly see the TRPM4 and polycystin-2 conductance in the enveloped cilia patch configuration from IMCD cells. These findings suggest that these two channel populations may reside at different locations within the membrane of the cilium or at the ciliary-cell junction.

3.4 CONCLUDING REMARKS

In this chapter, we offered our approach to measuring ion channel function in primary cilia using the voltage clamp method. As discussed, the ion channels found in cilia membranes are distinct between cilia types and across disparate tissues, which has been determined using direct cilia electrophysiology methods. In conjunction with cilia-specific calcium fluorescence, these methods have defined the primary cilium as a Ca^{2+} privileged organelle with its own membrane potential. Naturally progressing from this work, we propose to integrate molecular structure with ion channel biophysics with the goal of defining polycystin regulation of ciliary Ca^{2+}, which is proposed to initiate cystogenic signaling in ADPKD.[61–64] Importantly, ciliopathies with kidney cyst comorbidities are linked to several gene variants that encode for downstream effectors of Ca^{2+} signaling. Thus, the significance of understanding calcium conducting ion channels in the cilium will likely extend to other renal ciliopathies, where localized ciliary Ca^{2+} dysregulation is a shared disease-causing mechanism. These methods will be a mainstay for answering questions regarding cilia channel biophysics and cilia-cell electrical coupling. However, we envision advancements in designing cilia-specific optogenetics and designer receptors exclusively activated by designer drug (DREADD) methods to be used independently or in conjunction with these patch clamp methods to provide further insights into cilia-cell electrical communication and the role of cilia in cell biology.[65,66]

REFERENCES

1. Satir, P. Landmarks in cilia research from Leeuwenhoek to us. *Cell Motility and the Cytoskeleton* 32, 90–94, 1995.
2. Wheatley, D. N. Landmarks in the first hundred years of primary (9+0) cilium research. *Cell Biology International* 29, 333–339, 2005.

3. Kowalevsky, A. Entwickelungsgeschichte des Amphioxus lanceolatus. *Memoires de l'Academie Imperiale des Sciences de St-Petersbourg VII.* VII, 1–17. 1867.
4. Mykytyn, K. and Askwith, C. G-protein-coupled receptor signaling in cilia. *Cold Spring Harbor Perspectives in Biology* 9, 2017.
5. Pablo, J. L., DeCaen, P. G., and Clapham, D. E. Progress in ciliary ion channel physiology. *Journal of General Physiology* 149, 37–47, 2017.
6. Bloodgood, R. A. Sensory reception is an attribute of both primary cilia and motile cilia. *Journal of Cell Science* 123, 505–509, 2010.
7. Pugacheva, E. N., Jablonski, S. A., Hartman, T. R., Henske, E. P., and Golemis, E. A. HEF1-dependent Aurora A activation induces disassembly of the primary cilium. *Cell* 129, 1351–1363, 2007.
8. Qin, H., Wang, Z., Diener, D., and Rosenbaum, J. Intraflagellar transport protein 27 is a small G protein involved in cell-cycle control. *Current Biology : CB* 17, 193–202, 2007.
9. Parker, J. D. et al. Centrioles are freed from cilia by severing prior to mitosis. *Cytoskeleton* 67, 425–430, 2010.
10. Kollman, J. M. et al. Ring closure activates yeast gammaTuRC for species-specific microtubule nucleation. *Nature Structural & Molecular Biology* 22, 132–137, 2015.
11. Banterle, N. and Gonczy, P. Centriole biogenesis: From identifying the characters to understanding the plot. *Annual Review of Cell and Developmental Biology* 33, 23–49, 2017.
12. Meunier, A. and Azimzadeh, J. Multiciliated cells in animals. *Cold Spring Harbor Perspectives in Biology* 8, 2016.
13. Spassky, N. and Meunier, A. The development and functions of multiciliated epithelia. *Nature Reviews. Molecular Cell Biology* 18, 423–436, 2017.
14. Mizuno, N., Taschner, M., Engel, B. D., and Lorentzen, E. Structural studies of ciliary components. *Journal of Molecular Biology* 422, 163–180, 2012.
15. Bhogaraju, S., Engel, B. D., and Lorentzen, E. Intraflagellar transport complex structure and cargo interactions. *Cilia* 2, 10, 2013.
16. Taschner, M. and Lorentzen, E. The intraflagellar transport machinery. *Cold Spring Harbor Perspectives in Biology* 8, 2016.
17. Sung, C. H. and Leroux, M. R. The roles of evolutionarily conserved functional modules in cilia-related trafficking. *Nature Cell Biology* 15, 1387–1397, 2013.
18. Mueller, J., Perrone, C. A., Bower, R., Cole, D. G., and Porter, M. E. The FLA3 KAP subunit is required for localization of kinesin-2 to the site of flagellar assembly and processive anterograde intraflagellar transport. *Molecular Biology of the Cell* 16, 1341–1354, 2005.
19. Delaval, B., Bright, A., Lawson, N. D., and Doxsey, S. The cilia protein IFT88 is required for spindle orientation in mitosis. *Nature Cell Biology* 13, 461–468, 2011.
20. Taulet, N. et al. IFT proteins spatially control the geometry of cleavage furrow ingression and lumen positioning. *Nature Communications* 8, 1928, 2017.
21. Bangs, F. and Anderson, K. V. Primary Cilia and Mammalian Hedgehog Signaling. *Cold Spring Harbor Perspectives in Biology* 9, 2017.
22. Essner, J. J., Amack, J. D., Nyholm, M. K., Harris, E. B., and Yost, H. J. Kupffer's vesicle is a ciliated organ of asymmetry in the zebrafish embryo that initiates left-right development of the brain, heart and gut. *Development* 132, 1247–1260, 2005.
23. Drummond, I. A. Cilia functions in development. *Current Opinion in Cell Biology* 24, 24–30, 2012.
24. Goetz, S. C. and Anderson, K. V. The primary cilium: A signalling centre during vertebrate development. *Nature Reviews. Genetics* 11, 331–344, 2010.
25. Takebe, N. et al. Targeting Notch, Hedgehog, and Wnt pathways in cancer stem cells: Clinical update. *Nature Reviews. Clinical Oncology* 12, 445–464, 2015.
26. Bisgrove, B. W. and Yost, H. J. The roles of cilia in developmental disorders and disease. *Development* 133, 4131–4143, 2006.

27. Oh, E. C. and Katsanis, N. Cilia in vertebrate development and disease. *Development* 139, 443–448, 2012.
28. Roosing, S. et al. Mutations in CEP120 cause Joubert syndrome as well as complex ciliopathy phenotypes. *Journal of Medical Genetics* 53, 608–615, 2016.
29. Mirra, V., Werner, C., and Santamaria, F. Primary ciliary dyskinesia: An update on clinical aspects, genetics, diagnosis, and future treatment strategies. *Frontiers in Pediatrics* 5, 135, 2017.
30. Waters, A. M. and Beales, P. L. Ciliopathies: An expanding disease spectrum. *Pediatric Nephrology* 26, 1039–1056, 2011.
31. Lu, W. et al. Perinatal lethality with kidney and pancreas defects in mice with a targetted *Pkd1* mutation. *Nature Genetics* 17, 179–181, 1997.
32. Wu, G. et al. Trans-heterozygous *Pkd1* and *Pkd2* mutations modify expression of polycystic kidney disease. *Human Molecular Genetics* 11, 1845–1854, 2002.
33. Ma, M., Tian, X., Igarashi, P., Pazour, G. J., and Somlo, S. Loss of cilia suppresses cyst growth in genetic models of autosomal dominant polycystic kidney disease. *Nature Genetics* 45, 1004–1012, 2013.
34. Shibazaki, S. et al. Cyst formation and activation of the extracellular regulated kinase pathway after kidney specific inactivation of Pkd1. *Human Molecular Genetics* 17, 1505–1516, 2008.
35. Harris, P. C. and Torres, V. E. Genetic mechanisms and signaling pathways in autosomal dominant polycystic kidney disease. *Journal of Clinical Investigation* 124, 2315–2324, 2014.
36. Chebib, F. T., Sussman, C. R., Wang, X., Harris, P. C., and Torres, V. E. Vasopressin and disruption of calcium signalling in polycystic kidney disease. *Nature Reviews. Nephrology* 11, 451–464, 2015.
37. Kleene, S. J. and Kleene, N. K. The native TRPP2-dependent channel of murine renal primary cilia. *American Journal of Physiology. Renal Physiology* 312, F96–F108, 2017.
38. DeCaen, P. G., Delling, M., Vien, T. N., and Clapham, D. E. Direct recording and molecular identification of the calcium channel of primary cilia. *Nature* 504, 315–318, 2013.
39. Liu, X. et al. Polycystin-2 is an essential ion channel subunit in the primary cilium of the renal collecting duct epithelium. *eLife* 7, 2018.
40. Wilkes, M. et al. Molecular insights into lipid-assisted Ca(2+) regulation of the TRP channel polycystin-2. *Nature Structural & Molecular Biology* 24, 123–130, 2017.
41. Shen, P. S. et al. The structure of the polycystic kidney disease channel PKD2 in lipid nanodiscs. *Cell* 167, 763–773 e711, 2016.
42. Su, Q. et al. Structure of the human PKD1-PKD2 complex. *Science* 361, 2018.
43. Grieben, M. et al. Structure of the polycystic kidney disease TRP channel polycystin-2 (PC2). *Nature Structural & Molecular Biology* 24, 114–122, 2017.
44. DeCaen, P. G., Liu, X., Abiria, S., and Clapham, D. E. Atypical calcium regulation of the PKD2-L1 polycystin ion channel. *eLife* 5, 2016.
45. Delling, M., DeCaen, P. G., Doerner, J. F., Febvay, S., and Clapham, D. E. Primary cilia are specialized calcium signalling organelles. *Nature* 504, 311–314, 2013.
46. Zeng, X. H., Navarro, B., Xia, X. M., Clapham, D. E., and Lingle, C. J. Simultaneous knockout of Slo3 and CatSper1 abolishes all alkalization- and voltage-activated current in mouse spermatozoa. *Journal of General Physiology* 142, 305–313, 2013.
47. Lishko, P., Clapham, D. E., Navarro, B., and Kirichok, Y. Sperm patch-clamp. *Methods in Enzymology* 525, 59–83, 2013.
48. Li, R. C. et al. Ca(2+)-activated Cl current predominates in threshold response of mouse olfactory receptor neurons. *Proceedings of the National Academy of Sciences of the United States of America* 115, 5570–5575, 2018.
49. Li, R. C., Ben-Chaim, Y., Yau, K. W., and Lin, C. C. Cyclic-nucleotide-gated cation current and Ca2+-activated Cl current elicited by odorant in vertebrate olfactory receptor neurons.

Proceedings of the National Academy of Sciences of the United States of America 113, 11078–11087, 2016.
50. Kleene, N. K. and Kleene, S. J. A method for measuring electrical signals in a primary cilium. *Cilia* 1, 17, 2012.
51. Kleene, S. J. and Gesteland, R. C. Transmembrane currents in frog olfactory cilia. *Journal of Membrane Biology* 120, 75–81, 1991.
52. Faust, D. et al. Culturing primary rat inner medullary collecting duct cells. *Journal of Visualized Experiments: JoVE* 76, e50366, 2013.
53. Jozefczuk, J., Drews, K., and Adjaye, J. Preparation of mouse embryonic fibroblast cells suitable for culturing human embryonic and induced pluripotent stem cells. *Journal of Visualized Experiments: JoVE* 64, e3854, 2012.
54. Heller, J. P., Kwok, J. C., Vecino, E., Martin, K. R., and Fawcett, J. W. A method for the isolation and culture of adult rat retinal pigment epithelial (RPE) cells to study retinal diseases. *Frontiers in Cellular Neuroscience* 9, 449, 2015.
55. Wang, H. et al. One-step generation of mice carrying mutations in multiple genes by CRISPR/Cas-mediated genome engineering. *Cell* 153, 910–918, 2013.
56. Hamill, O. P., Marty, A., Neher, E., Sakmann, B., and Sigworth, F. J. Improved patch-clamp techniques for high-resolution current recording from cells and cell-free membrane patches. *Pflugers Archiv: European Journal of Physiology* 391, 85–100, 1981.
57. Sakmann, B. S. a. E. *Single-Channel Recording.* second ed., Springer, 2009. LLC, M. D. (ed. Rivka Sherman-Gold), 2012.
58. Sherman-Gold, R (Ed.). *The Axon™ Guide: A guide to Electrophysiology and Biophysics Laboratory Techniques*, 2012. Molecular Devices, LLC.
59. Doerner, J. F., Delling, M., and Clapham, D. E. Ion channels and calcium signaling in motile cilia. *eLife* 4, 2015.
60. Flannery, R. J., Kleene, N. K., and Kleene, S. J. A TRPM4-dependent current in murine renal primary cilia. *American Journal of Physiology. Renal Physiology* 309, F697–F707, 2015.
61. Winyard, P. and Jenkins, D. Putative roles of cilia in polycystic kidney disease. *Biochimica et Biophysica Acta* 1812, 1256–1262, 2011.
62. Novarino, G., Akizu, N., and Gleeson, J. G. Modeling human disease in humans: The ciliopathies. *Cell* 147, 70–79, 2011.
63. Somlo, S. and Ehrlich, B. Human disease: Calcium signaling in polycystic kidney disease. *Current Biology: CB* 11, R356–R360, 2001.
64. Mangolini, A., de Stephanis, L., and Aguiari, G. Role of calcium in polycystic kidney disease: From signaling to pathology. *World Journal of Nephrology* 5, 76–83, 2016.
65. Fenno, L., Yizhar, O., and Deisseroth, K. The development and application of optogenetics. *Annual Review of Neuroscience* 34, 389–412, 2011.
66. Roth, B. L. DREADDs for neuroscientists. *Neuron* 89, 683–694, 2016.

4 Electrophysiological Recording of a Gain-of-Function Polycystin-2 Channel with a Two-Electrode Voltage Clamp

Courtney Ng, Zhifei Wang, Bin Li, and Yong Yu

CONTENTS

4.1 INTRODUCTION

4.1.1 Polycystin-2 (PC2) and Autosomal Dominant Polycystic Kidney Disease (ADPKD)

The activity of ion channels and ion pumps, which enables the transport of ions across the cell membranes, governs the membrane potential of the cell and initiates or regulates cell signaling. Transient receptor potential (TRP) channels, a group of nonselective cation channel proteins, play prominent roles in sensory physiology, such as detecting and responding to temperature, mechanical stress, pH, osmotic pressure, and pain, as well as contributing to olfaction and taste.[1,2] In accordance with their diverse sensory responses, expression of TRP channels are widely distributed to different cell and tissue types where they perform the specialized roles. This makes TRP channels crucial for sensory signal transduction, cell homeostasis, neuronal excitability, tissue development, and disease development. All TRP channels are intrinsic membrane proteins with six transmembrane domains (S1–S6), intracellular N- and C-termini, and a channel pore region between S5 and S6.[1–3] The intracellular termini vary significantly in length and the functional domains included among all TRP members and play crucial roles in the assembly of the channels and regulation of channel function.[3]

Mutations in one of the members of the TRP channel polycystin subfamily, polycystin-2 (PC2, also called TRPP2 or PKD2) and its related protein, polycystin-1 (PC1, also called PKD1), result in autosomal dominant polycystic kidney disease (ADPKD), one of the most common, monogenetic human diseases.[4–9] PC2 is known to have six transmembrane domains and intracellular N- and C-termini and functions as a nonselective and potentially Ca^{2+}-permeable cation channel.[4,10–13] Similar to other TRP proteins, four PC2 proteins can form a homotetrameric channel.[14–18] At the same time, it can associate with PC1 to form a heterotetrameric ion channel complex, which contains three PC2 subunits and one PC1 subunit.[19–22] With or without association with PC1, PC2 has been found to localize to the subcellular compartments including the plasma membrane, endoplasmic reticulum (ER), and primary cilium.[4,11] Disruption in Ca^{2+} signaling through the PC1/2 heteromeric complex, and maybe also the PC2 homomeric channel, is speculated to be the cause of renal cyst formation and progressive failure of renal function.

4.1.2 PC2_F604P, a Gain-of-Function (GOF) PC2 Mutant

Genetic studies have revealed that the linker region between S4 and S5 (S4-S5 linker) and the first half of S5 helix is a hot spot for GOF mutations of many TRP channels.[23] Mutations at this region have been linked to over activity-induced channelopathies in human beings. For example, N855S in TRPA1 causes familial episodic pain syndrome,[24] G573S in TRPV3 causes Olmsted syndrome,[25] multiple mutations in this region of TRPV4 cause various skeletal dysplasias and motor/sensory neuropathies,[26,27] and multiple mutations of TRPML1 have been found to be the underlying cause of mucolipidosis type IV.[28] Some other GOF mutations of TRP channels have been generated by doing mutagenesis screens, which also fall into this region.[29] These GOF mutations paved the way for scientists to study the function and regulation of these channels, especially for those that are difficult to be activated extrinsically or intrinsically.

Study of the channel function of the PC2 homomeric channel and the PC1/PC2 complex has been greatly delayed since a direct activation mechanism for these channels remains undetermined, leading to the absence of a functional readout for channel activity analysis. In this situation, GOF mutants of these channels, especially constitutively-activated mutants, which would bypass the unknown activation process, can facilitate the functional study of PC2, as well as the PC1/PC2 complex, and aid in our understanding of these functionally crucial proteins and their roles in ADPKD. Using a proline-scanning mutagenesis method to screen proline mutations around the S4-S5 linker and the S5 helix, we have recently generated a constitutively active GOF mutant of PC2, PC2_F604P.[14] When PC2_F604P was expressed in *Xenopus laevis* oocytes, it trafficked to the plasma membrane and exhibited robust current. Our result has been successfully repeated by other labs,[30,31] although it was reported this mutation did not lead to significant current when expressing PC2_F604P in human embryonic kidney (HEK) cells,[32] which is most likely caused by low surface expression of PC2 in these cells in the absence of PC1. The current from *Xenopus* oocytes can be easily recorded using the two-electrode voltage clamp (TEVC), a widely used electrophysiology method for studying ion channel function and regulation, which is described below.

The recently published cryo-EM structure of PC2_F604P shows the F604P mutation causing twisting and bending of the distal S6 helix and opening of the lower pore gate.[30] Interestingly, the structure of PC2_F604P is very similar to an open structure of PC2L1, a PC2 homologue,[33] suggesting that the F604P mutation may result in a conformational change, reflecting the natural gating of PC2. Recently, several other mutations that remove the hydrophobic restriction at the lower gate of the channel have been reported to also lead to leak activity of the PC2 channel.[30] These mutants open PC2 in a completely different manner compared to the GOF effect of F604P.

The GOF mutants of PC2 bypass the gating of this channel and provide an open channel for functional readout. It opens a new avenue for studying the function and regulation of PC2 and potentially the PC1/PC2 complex.

4.1.3 Two-Electrode Voltage Clamp (TEVC)

The voltage difference across a cell membrane is called membrane potential (V_m). Resting membrane potential, the electrical potential across a membrane prior to signaling, is determined by the concentration of ions (their gradient across the cell membrane) and by the channels that are essentially always open/active.[34] When primarily closed channels are activated (or gated) by stimuli, such as changes in the V_m or binding of ligands, ion conductance through the channels and the shift of V_m will happen.

The voltage clamp method allows electrophysiologists to measure the electric current caused by the movement of ions across a cell membrane while holding the V_m at a certain level. When a voltage clamp experiment is done, membrane potential (voltage) will first be measured and then changed (clamped) to the desired value by injecting current. This current is equal, although opposite in direction, to the current passing across the cell membrane through ion channels at the set voltage. Thus, by recording the injected current, we are able to measure the current passing through ion channels. This method was first designed by Cole and Marmont in 1949,[35] and was utilized and improved into the two-electrode voltage clamp (TEVC) method by Hodgkin, Huxley, and Katz in 1952 for measuring current from giant squid axons.[36] Today, TEVC is a widely used electrophysiological technique for investigating the properties of ion channels and electrogenic transporters expressed in large cells such as *Xenopus* oocytes. The applications of this technique include characterization of the ion selectivity, activation and desensitization, structural-functional analysis, and screening for agonists or activators and antagonists or inhibitors.

The control of membrane voltage is accurately achieved by a high-gain clamping amplifier. Gain is the measure of how much the device can amplify or increase the amplitude of the incoming signal. Two microelectrodes are inserted into the membrane of the oocyte; one is for voltage sensing, and the other for current injection, to artificially manipulate the membrane potential. The voltage-sensing electrode monitors the V_m, and that input is transmitted to the amplifier, as illustrated in Figure 4.1. Typically, the amplifier has a potentiometer to offset, or control, the holding potential, prior to clamping. The amplifier will bring the V_m to the command potential by injecting current through the current electrode that provides the access resistance, which is the resistance through the pipette tip to the ground. Increasing the gain will allow V_m to approach command potential accurately and reduce the effect of the access resistance. When the cell is clamped at a constant membrane potential, the current in response to the command voltage is observed and recorded.

Voltage clamping allows the separation of ionic and capacitative currents, whereby the basis can be understood through the equation of a cell circuit,

$$I_m = I_i + C\frac{dV}{dt}$$

where the total current, I_m, flowing across the membrane, is the sum of the ionic current, I_i, and membrane capacitance changing with respect to voltage and time, C(dV/dt), or capacitative current. When the step command is initiated, there is a brief moment of capacitative current. The current passing through the cell membrane is recorded at the electrode that grounds the bath solution, via the virtual-ground circuit, as shown in Figure 4.1. Conventionally for whole-cell voltage clamping, positive current corresponds to outward membrane current and negative current corresponds to inward current. Inward current is canceled by an equal and opposite current flowing into the headstage from the current electrode. Conversely, the current flowing from the headstage through the micropipette tip is positive current. For more reading on the voltage clamp method and ion channels, please refer to some excellent books.[34,35,37,38]

An analog-digital converter, or digitizer, is needed to take the analog signals from the amplifier and convert them to digital signals. Acquisition of the digital data is performed using a computer and software such as Axon pCLAMP (Molecular Devices). TEVC amplifier or the whole recording system are commercially available

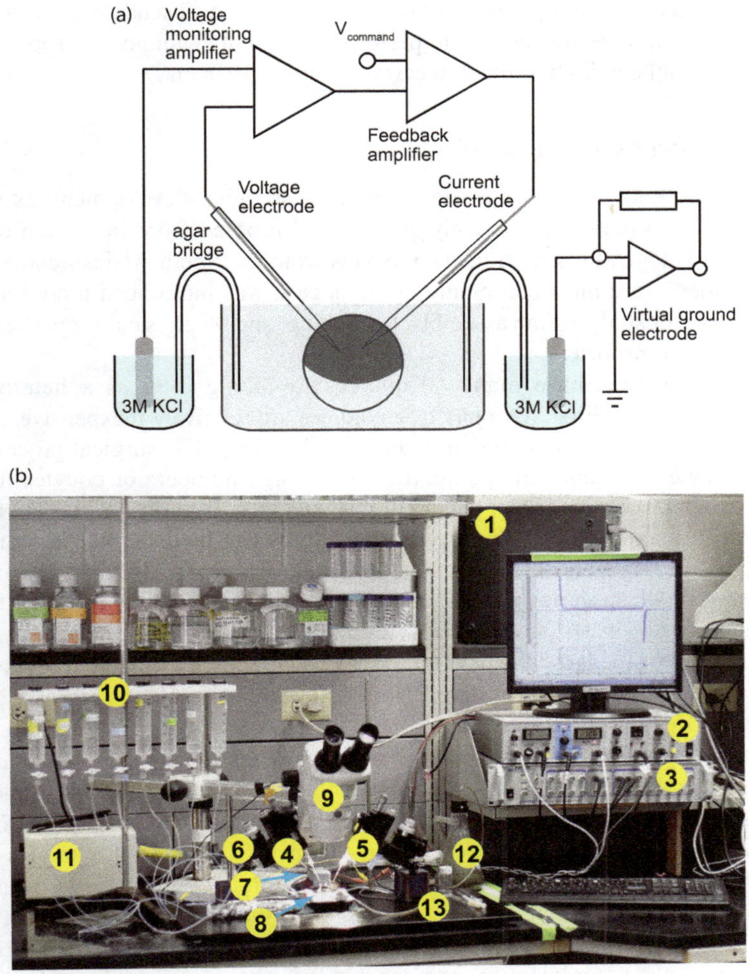

FIGURE 4.1 Two-electrode voltage clamp (TEVC) system. (a) Schematics of the TEVC method. (b) A typical setup of the TEVC recording system. 1. Computer; 2: amplifier; 3: analog-digital converter (digitizer); 4: voltage electrode; 5: current electrode; 6: magnetic stand and micromanipulator; 7: bath electrodes; 8: recording chamber; 9: stereomicroscope; 10: perfusion system; 11: cold light source; 12: vacuum system; 13: steel base plate or vibration isolation table. Vibration isolation equipment is optional based on the experimental environment.

from companies such as Molecular Devices (AxoClamp 900 A) and Warner Instruments (TEV-700).

The application of the TEVC method is extremely useful for measuring and controlling large currents, which single patch clamping is not ideal for. It is also known for a fast clamp settling time with low noise. This electrophysiological technique has been shown to be especially useful for studying the function of channel proteins, such as transient receptor potential (TRP) ion channels, as discussed in this chapter. TEVC recording is relatively easy to perform so students with no electrophysiology experience can be quickly trained to carry out the experiment.

4.1.4 *Xenopus* Oocyte System

Oocytes have been used as a model system for studying development, as well as an expression system that is widely applied and profited from in the ion channel and pharmacological research fields. Oocytes from the South African clawed frog, *Xenopus laevis*, are most commonly used, since it was introduced more than four decades ago, for studying the assembly, trafficking, and biophysical properties of ion channels and transporters.

Xenopus oocytes have many advantages for being used as a heterologous expression system.[39] First, *Xenopus* frog colonies are relatively inexpensive, and the housing system is easy to maintain in an animal facility. The surgical procedure to extract oocytes is simple and straightforward. Large numbers of oocytes (usually >20,000) can be obtained from one mature female frog. If fewer oocytes are needed each time, a maximum of four surgical procedures can be done on the same frog. Collection from right and left ovaries should be alternated with 2–3 months of resting time between two surgeries. Frogs can be kept a long time in the lab since, after the completion of oogenesis, stage VI oocytes can stay in the resting state for years in female frogs. For the labs which do not want to deal with housing the frogs, oocytes and ovaries can be directly ordered from commercial vendors such as EcoCyte Bioscience, Xenopus One, or Nasco.

Second, the large cell size (~1.0 mm in diameter) of oocytes makes it easy for manual microinjection and electrophysiological recordings, and efficient for expression of cell membrane channels and transporters from cRNA, cDNA, or mRNA, though cRNA is most used for ion channel study. Multiple samples can be injected simultaneously when expression of more than one protein is needed for protein complex assembly. This also comes with the convenience that the ratios of the RNA or DNA samples can be easily adjusted to optimize their relative expression levels. After injection, oocytes survive and maintain well *ex vivo/in vitro* for about a week and a half, as they are mainly self-sufficient. The self-sufficient capacity of *Xenopus* oocytes means that they have relatively low expression of endogenous membrane channels and receptors since there is no reliance on uptake of nutrients from its environment.[40] Additionally, it is easy to control the surrounding environment of the oocyte in order to study specific responses. Therefore, we can easily express our ion channels and membrane receptors of interest in *Xenopus* oocytes and study their functional characteristics and even molecular mechanisms of activation with TEVC and other voltage clamp methods, such as patch clamp.[41] The large size of the oocytes and the high degree of expression of exogenous mRNA or cRNA also lead to another advantage of studying ion channel proteins, which is that a single oocyte can be recorded for current and analyzed biochemically for protein expression and interaction in order to match the biochemical data with functional results. In many cases, we were able to detect enough protein collected from one oocyte via Western blot.

Third, compared with whole-cell recording with cultured mammalian cells, TEVC recording from *Xenopus* oocytes can be much faster, as well as more efficient. Data from many oocytes can be collected within a relatively short period of time. Depending on the research target, the same oocyte can be recorded for minutes or longer with TEVC. Commercially available automated systems have already been able to accomplish RNA microinjection and electrophysiological measurements on oocytes fully automatically. These new technical advances allow for a high-throughput screen of potential drugs that target membrane transport proteins.

4.1.5 Recording the Gain-of-Function PC2_F604P Channel

When wildtype (WT) human PC2 was expressed in *Xenopus* oocytes, although it can traffic to the plasma membrane, no current was recorded with TEVC due to the low open probability of this channel. However, when PC2_F604P was expressed, robust whole-oocyte current can be recorded with TEVC.[14] A significant feature of this current is that the presence of 2 mM Ca^{2+} or Mg^{2+} at the extracellular side greatly inhibits the inward current carried by Na^+ influx.

In this chapter, we will describe how to collect and inject *Xenopus* oocytes and record the PC2_F604P channel with the TEVC method. Figure 4.2 shows the workflow of this experiment.

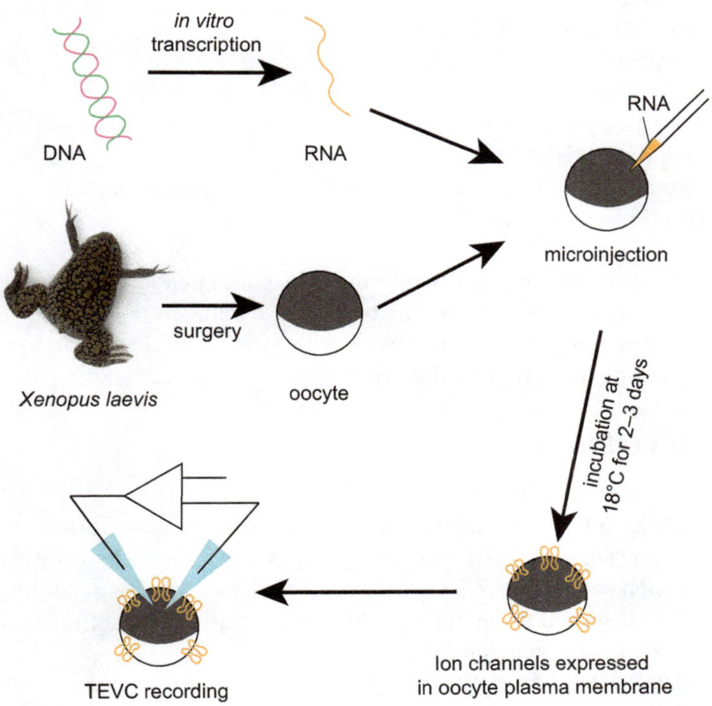

FIGURE 4.2 The flowchart of the experiments for recoding the channel current of PC2_F604P with the TEVC method.

4.2 MATERIALS

4.2.1 Frog Surgery, Oocyte Removal, and Preparation

1. Tricaine (ethyl 3-aminobenzoate methanesulfonate).
2. Female frogs of *Xenopus laevis.*
3. Collagenase A (Roche).
4. OR2 solution: 82.4 mM NaCl, 2.5 mM KCl, 1 mM $MgCl_2$, 5 mM HEPES, pH 7.6.
5. ND96 solution: 96 mM NaCl, 2.5 mM KCl, 1 mM $MgCl_2$, 1.8 mM $CaCl_2$, 5 mM HEPES, pH 7.6; add 1/100 penicillin/streptomycin (P/S) before use (Gibco 10,000 U/mL stock).
6. Sharp forceps.
7. Dissecting scissors.
8. Nylon sutures (3-0, NFS-1).
9. Glass bead sterilizer.
10. Benchtop shaker.

4.2.2 cRNA Injection

1. Microinjector (Drummond Nanoinject II).
2. Stereomicroscope (Nikon SMZ-1B).
3. Glass capillaries (3.5″, Drummond).
4. Pipette puller (Narishige PC-10).
5. cRNA (diluted stocks).
6. RNase-free pipette tips.
7. Mineral oil (Fisher).
8. 35 mm tissue culture or Petri dishes (Falcon) or parafilm.
9. 24-well plates (Falcon).
10. ND96 solution.
11. Oocyte transfer pipette, which is made by cutting and fire polishing a glass Pasteur pipette with an opening of about 1.5 mm.
12. Oocyte moving tool, which is made with a glass Pasteur pipette by bending and sealing the tip over a Bunsen burner.

4.2.3 TEVC Recording

1. Stereomicroscope (Nikon SMZ-745).
2. Amplifier (Oocyte Clamp OC-725C, Warner Instrument Corp.).
3. Digitizer (Digidata 1440 A Digitizer Acquisition System, Axon Instruments).
4. Perfusion cell chamber for holding oocyte in place during recording.
5. Perfusion system using gravity (slow flow) or air pressure (fast flow). Here we use a gravity-fed perfusion system.
6. Aspiration system or vacuum pump for suction the solution out of the recording chamber.
7. Pipette puller such as Narishige PC-10.
8. Micromanipulators (MM 33, Marzhauser Wetzlar).

9. Borosilicate glass capillaries (1.0 mm OD × 0.75 mm ID). Capillaries with fibers such as those from FHC, Inc. are much easier to fill.
10. Filtered 3 M KCl.
11. Glass capillaries, agarose, KCl, and Bunsen burner for making agar bridge.
12. Ag/AgCl electrode wires and pellet.
13. Recording software (Clampex 10.3 version from Axon Instruments or similar software).
14. Solutions used in this study:
    Divalent ion-free solution: 100 mM NaCl, 3 mM HEPES, pH 7.5.
    2 mM Ca^{2+} solution: 100 mM NaCl, 2 mM $CaCl_2$, 3 mM HEPES, pH 7.5.

4.3 PROTOCOLS FOR OOCYTE PREPARATION AND INJECTION

4.3.1 Frog Surgery Preparation

1. Tricaine is kept at 4°C. Warm up the bottle to room temperature before opening the lid. Prepare a fresh 0.1% tricaine solution (1 g tricaine in 1 L deionized water), with the pH adjusted to 7.0 using $NaHCO_3$.
2. When handling frog, wear powder-free sterile gloves moistened with deionized water at all times.
3. Submerge female *Xenopus laevis* in the tricaine solution, keeping the head elevated and out of the solution to prevent drowning.
4. At the same time, sterilize the tools necessary for surgery with the glass bead sterilizer.
5. It usually takes approximately 15–20 minutes for the frog to be anesthetized. A surgical level of anesthesia is confirmed by gently pinching the fleshy part of the rear foot with forceps and ensuring that the frog is nonresponsive to painful stimuli.

4.3.2 Frog Surgery and Oocyte Collection

1. Place the frog on a clean, moist surface and keep the skin moistened during surgery.
2. Remove contaminants from the skin by gently cleaning with sterile deionized water.
3. Make a small incision (1–1.5 cm) on the abdomen above the groin, lateral to the midline.
4. Use scissors to dissect through the fascia and muscle to visualize the oocytes.
5. Pick up several ovary sacks with forceps and cut.
6. The incision is closed by suturing both the fascia and skin layer in two layers. Skin suture must be nonabsorbable. It will be removed 10–14 days after surgery if it does not fall off by itself.
7. Recover the frog in shallow tank water with the head elevated and the rest of the body submerged.
8. The animal will be monitored until the voluntary movement is observed. Once the frog is moving freely, she can be replaced in the recovery tank.

4.3.3 Oocyte Treatment

1. The ovary sacks can be divided into smaller clumps using sharp scissors and transferred into a 50 mL Falcon tube. Rinse the oocytes several times with OR2 solution.
2. Replace the OR2 wash with 1 mg/mL collagenase in OR2 solution for defolliculation. A volume of 20–30 mL of OR2 solution is used for 5–10 mL of oocytes. The oocytes are incubated with frequent agitation (shaker at approximately 200 rpm) for 60–90 min and checked under a stereomicroscope for adequate enzymatic digestion to remove the ovarian follicles. Avoid overdigestion by checking after 30–45 min and repeat every additional 10 min until about half of the oocytes are defolliculated.
3. The reaction is stopped by washing out the collagenase/OR2 solution with OR2 several times until the solution is clear.
4. Rinse dissociated oocytes with ND96 solution 3 times with gentle shaking at low speed (<100 RPM) for 10 min each.
5. Oocytes are stored in fresh ND96 solution and incubated in an 18°C incubator for at least 30 min prior to selection. When selecting, the follicular layer left on any oocytes can be manually removed using forceps.
6. Select stage VI or V oocytes under a stereomicroscope. Selected oocytes should have a clear division of the animal and vegetal pole and be free of blotchy spots.
7. Incubate selected oocytes in ND96 buffer at 18°C for a couple of hours or overnight before injecting cRNA.

4.3.4 cRNA Injection

1. Glass capillaries are pulled (Narishige PC-10), and the tips are manually broken using sharp forceps under a microscope to create a tip diameter of approximately 15 μm. Gloves should always be worn when working with cRNA and injection materials.
2. The injection needles are backfilled with mineral oil using a syringe needle and mounted into the microinjector (Drummond Nanoinject II).
3. The cRNA sample can be centrifuged briefly before filling into the injection needle to prevent any precipitate from blocking the needle. A 1 μL droplet of the cRNA sample can be placed on a piece of parafilm or on the underside of the lid of a 35 mm Petri dish (use a marker to divide the dish lid into 4–8 pie sections for multiple samples) for filling into the needle under the stereomicroscope.
4. Transfer a group of oocytes with oocyte transfer tool into an ND96 solution–filled 35 mm Petri dish lined with nylon mesh (G: 0.8 mm), which is used to prevent oocyte movement during the injection.
5. The position of the oocytes can be adjusted with the oocyte moving tool to line up the oocytes and make sure their animal pole is facing up (this can also be done with the injection needle later).

6. Guide the needle to inject each oocyte using the micromanipulator. A good injection point on oocyte will be on the animal pole side close to the separation of the animal and vegetal poles.
7. Adjust the volume of the injector so that it delivers 30–50 nL of solution into each oocyte, which contains ~30–50 ng of cRNA (1 μg/μL concentration, see note below). To judge whether the injection is successful, pay attention to the change in the shape of the oocyte. The needle insertion will form a dimple on the membrane and that would swell back when injected.

Note: For injection of PC2 cRNA, a concentration of 1 μg/μL is appropriate for expression of the homomeric channel and recording 2 days after injection. For expressing different subunits of a heteromeric channel, the stoichiometry of the subunits should be considered, with the length of the cRNAs taken into account for calculating the molar ratio.

8. After injection, the oocytes are placed into a Petri dish or divided into wells of a multi-well plate (e.g., 24-well plate with ~5 oocytes/well) filled with ND96 solution containing antibiotics, and stored in 18°C. Damaged oocytes should be removed on a daily basis postinjection before electrophysiological applications. The oocytes are typically incubated between 2 and 5 days before using TEVC to observe electrophysiological function, depending on the rate of translation and trafficking (see note below).

Note: Larger proteins tend to take longer time to express and/or traffic to the plasma membrane in oocytes. Proteins that are coexpressed and form complex channels may also require additional time due to their assembly and trafficking. Typically, the electrophysiological activity of PC2 can be observed and recorded after a 2-day postinjection incubation, as described above.

9. This injection procedure can be used for changing the intracellular environment by treating the oocyte with substances such as BAPTA, a Ca^{2+} chelator, or dyes.

Note: Injection of 25 mM BAPTA can be used to chelate intracellular Ca^{2+}. Oocytes will typically stay stable for 30 min after injection of this concentration.

4.4 TEVC RECORDING

4.4.1 VOLTAGE PROTOCOLS FOR RECORDING

1. Generating the voltage protocol for recording, and the details in amplifier and software operation, can be variable based on the recording system and the recording software used. Below we will briefly describe the two voltage protocols used in our recording. A similar voltage protocol can be generated in any recording software.
2. The voltage ramp protocol. With this protocol, oocytes are held at −60 mV, and the voltage ramp is run from −80 mV to +80 mV in 160 ms. We use this protocol to monitor the currents of PC2_F604P in different solutions. In some cases, we use it to monitor the reversal potential shift.

3. The voltage step protocol. With this protocol, oocytes are held at −60 mV, and 50-ms voltage steps from −80 to +60 mV in 10-mV increments are applied. Current in each step is recorded and used for extracting the I-V curves. If the channel has voltage-dependent activation or inactivation, this protocol will help to observe it.

4.4.2 TEVC Setup Preparation

1. Make the KCl-agar bridge.
   Bend the capillary tubes with appropriate length (judge based on your recording chamber) with forceps on a Bunsen burner. Dip the bent capillary into 10 mL hot 1% agarose-KCl solution, which is prepared by dissolving agarose in water first then adding KCl to a 3 M final concentration. Tip: Dip one end of the capillary into the solution first, then let the whole capillary slowly fall in. This will help to push out all air in the capillary to avoid air bubbles. After taking the filled capillary out, remove the extra agarose outside of it and store at room temperature in a 3 M KCl solution.
2. Turn on the digitizer and amplifier, and open the recording software (Clampex 10.3 is used here).
3. Rinse out perfusion tubes and tubing with deionized water several times. Fill perfusion tubes with recording solutions and drain through the tubing. Make sure the aspirator or vacuum pump is working properly.
4. Pull glass capillaries for the electrodes using the pipette puller (Narishige PC-10) with a two-step pulling protocol. Adjust the heating parameters to make sure the electrodes have proper openings at the tips. The resistance should range from 2–4 MΩ for the voltage electrode and should be ≤1 MΩ for the current electrode. The resistance can be determined using the electrode test function on the oocyte clamp amplifier OC-725C (refer to the manual).
5. The electrode needles are backfilled with 3 M KCl with a syringe and fine tubing/needle, avoiding any bubbles in the tip of the pipettes. The glass pipettes are inserted into the electrode holders (with Ag/AgCl wire electrodes) and mounted to micromanipulators, making sure that the Ag/AgCl wires are in contact with the KCl solution. The Ag/AgCl wires should be bleached prior to recording.
6. The OC-725C system has two bath clamp electrodes (I_{sense} and I_{out}) designed to maintain a virtual ground in the oocyte recording. Proper placement of these electrodes is important. One can refer to the manual of OC-725C for more details. In our recording, I_{sense} is placed on the same side as the voltage recording electrode and is connected to the recording chamber through an agar bridge (made above). The connection is achieved by submerging both the I_{sense} electrode pellet and one end of the agar bridge in a well filled with 3 M KCl. The other end of the agar bridge is placed in recording chamber on the same side as the voltage recording electrode. I_{out} electrode is positioned in the recording chamber on the same side of the current electrode and

submerged in the flow downstream of the oocyte. It should be placed at a greater distance from the oocyte.

7. At this point, the setup is ready for recording current of the PC2 GOF channel.

4.4.3 TEVC Recording

1. After the equipment and materials are set up, carefully transfer one oocyte into the recording chamber. Adjust the position of the oocyte using the oocyte moving tool.
2. Run the recording solution for about 10 s to rinse out the ND96 solution.
3. The following steps can be variable based on the amplifier used in the experiment. We briefly describe it based on the OC-725C system.
4. Manually guide the tips of the microelectrodes into the bath solution in the chamber and adjust the voltage offsets of both the voltage and current electrodes to "0 mV" reference value. This adjusts the junction potential to zero value from the voltage electrode and a null reference for resting potential readings from the current electrode.
5. The electrode resistance can be checked by pressing the electrode test buttons. Ideally, the resistance should range from 2–4 MΩ for the voltage electrode and should be $\leq$1 MΩ for the current electrode.
6. Impale the animal pole of the oocyte with the electrodes one at a time starting with the voltage electrode, and then the current electrode. The readings on the meters will show the membrane potential, which is typically around −20 mV or more negative. The voltage reads the difference between membrane potential and the bath solution. One can judge whether the oocyte is healthy by looking at this number. If the reading is very close to zero or even becomes positive, then the oocyte membrane is very leaky. If it is not expected, one should remove the current oocyte and change to another one.
7. Switch the clamp from "OFF" to "FAST," and the GAIN to maximum by turning the dial clockwise. Oocytes are typically clamped in the fast mode, and in order to achieve the fastest clamp speed, the resistance of the current electrode should be kept as low as possible (suggested $\leq$1 MΩ).
8. Record currents via step protocol with the holding potential starting at 0 mV and commanded to generate the clamped membrane potential to step from −80 to +80 mV in 10 mV increments.
9. Perfuse the next solution (2 mM Ca^{2+}) into the chamber and observe the current via ramp protocol until the current becomes steady. Click STOP and start the step protocol again.
10. Clampex software has a function of setting sequence keys for quickly starting a single or series of protocols, which is very handy if more than one protocol is needed during one recording.
11. After data are acquired, return the GAIN to "OFF" by turning counter clockwise, and switch the clamp to "OFF."

4.4.4 Data Analysis

Data collected in Clampex can be directly analyzed with Clampfit software in the same software package from Axon Instruments or transferred out and analyzed with third-party software such as Excel or Prism. Figure 4.3 shows a typical recording protocol and current of PC2_F604P in divalent ion-free bath solution and 2 mM Ca^{2+} bath solution. Extracellular Ca^{2+} inhibits the inward current of PC2_F604P, which is carried mainly by Na^+ influx.[14]

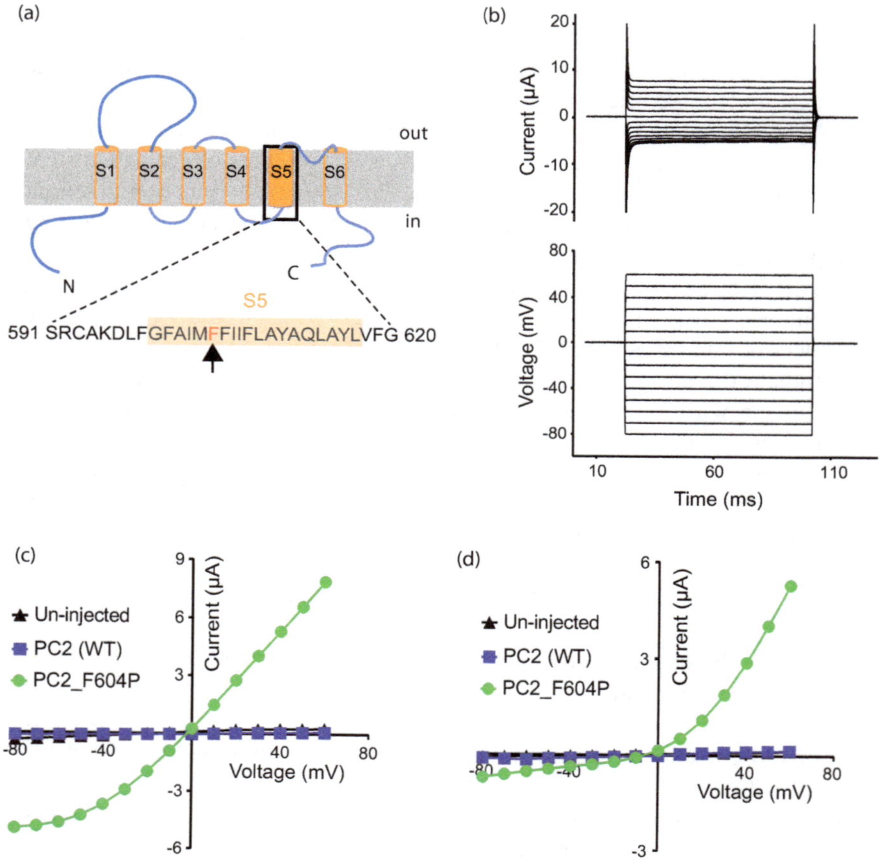

FIGURE 4.3 Recording data of the PC2_F604P channel. (a) Transmembrane topology of PC2 protein, showing the sequence of S5 and the position of F604. Mutation F604P leads to a GOF channel with constitutive activity. (b) A typical recording protocol of voltage steps (*lower chart*) and the corresponding current traces (*upper chart*). (c) Representative currents of uninjected oocyte and oocytes expressing wildtype PC2 or PC2_F604P in a bath solution containing 100 mM NaCl, 2 mM HEPES, pH 7.5. (d) Representative currents of uninjected oocyte and oocytes expressing wildtype PC2 or PC2_F604P in a bath solution containing 100 mM NaCl, 2 mM $CaCl_2$, 2 mM HEPES, pH 7.5. Extracellular Ca^{2+} blocks the inward current carried by Na^+ influx.

4.5 CONCLUDING REMARKS

The genetic connection between the mutations in PC1 and PC2 to ADPKD has been known for more than two decades. However, there is still no adequate treatment for ADPKD. The available treatments are not efficient enough and have systemic side effects.[6] Efforts to develop more effective therapies have been delayed due to the lack of a precise understanding of polycystins' key physiological functions. It has been generally believed that PC2 forms a cation channel and PC1 plays a role in regulating the PC2 channel function, or is directly involved in forming a heteromeric channel pore with PC2. Although the recording of PC2 or PC1/PC2 channel function has been reported in the previous publications,[15–18,22,31,42–47] a reliable and easy-to-perform method for recording PC2 or the PC1/PC2 channel is still missing. The combination of the GOF PC2 channel and the robust expression system of the *Xenopus* oocyte provide a nice platform for studying the channel properties of PC2 and how pathogenic mutations affect PC2 channel activity.[14] Our data also show that the GOF PC2 exhibited better rescue of defects in kidney caused by downregulation of the endogenous PC2 expression in zebrafish,[14] suggesting a potential application of the GOF mutant in disease treatment. The structural data of PC2_F604P further validate its value by showing that F604P leads to conformational changes that mimic what occurs in the natural gating process of this channel.[30] In a recent study, we have generated a GOF PC1/PC2 complex by mutating the lower gate of PC2 protein and illuminated the ion channel function of PC1 in this complex.[48] These studies has provided critical insights into the function and regulatory mechanism of ion channel function of the PC1/PC2 complex and its critical role in PKD pathogenesis. They will also aid the development of new therapeutic strategies for ADPKD treatment.

Since the GOF mutation causes the constitutive opening of the channel, it can bypass the difficulty caused by the lack of a known activation mechanism and provide a powerful tool to study ion channel function of polycysitns. However, we also need to be aware of the limits of its application. A significant drawback of using a GOF channel is that the channel gating process cannot be easily studied. Thus, if a pathogenic mutation affects channel gating, we may not be able to detect it with the GOF channel. Similarly, we may not be able to use it in testing a potential ligand of the channel if its binding does not lead to further conformational change. Finally, the open pore of the GOF channel may not be exactly the same as a naturally gated channel. Thus, caution needs to be taken when using the GOF mutant to study particular channel properties such as ion permeability.

ACKNOWLEDGMENTS

This work is supported by NIH Grants DK102092 (to Y.Y.) and a PKD Foundation Research Grant (to Y.Y.).

REFERENCES

1. Ramsey, I. S., Delling, M., and Clapham, D. E. An introduction to TRP channels. *Annual Review of Physiology* 2006; **68**, 619–647.
2. Montell, C. The TRP superfamily of cation channels. *Science's STKE: Signal Transduction Knowledge Environment* 2005; 2005, re3.

3. Venkatachalam, K. and Montell, C. TRP channels. *Annu Rev Biochem* 2007; **76**, 387–417.
4. Semmo, M., Kottgen, M., and Hofherr, A. The TRPP subfamily and polycystin-1 proteins. *Handbook of Experimental Pharmacology* 2014; **222**, 675–711.
5. Harris, P. C. and Torres, V. E. Polycystic kidney disease. *Annual Review of Medicine* 2009; **60**, 321–337.
6. Bergmann, C. et al. Polycystic kidney disease. *Nature Reviews Disease Primers* 2018; **4**, 50.
7. Chapin, H. C. and Caplan, M. J. The cell biology of polycystic kidney disease. *Journal of Cell Biology* 2010; **191**, 701–710.
8. Zhou, J. Polycystins and primary cilia: Primers for cell cycle progression. *Annual Review of Physiology* 2009; **71**, 83–113.
9. Wu, G. and Somlo, S. Molecular genetics and mechanism of autosomal dominant polycystic kidney disease. *Molecular Genetics and Metabolism* 2000; **69**, 1–15.
10. Tsiokas, L., Kim, S., and Ong, E. C. Cell biology of polycystin-2. *Cellular Signalling* 2007; **19**, 444–453.
11. Kottgen, M. TRPP2 and autosomal dominant polycystic kidney disease. *Biochimica et Biophysica Acta* 2007; **1772**, 836–850.
12. Giamarchi, A. et al. The versatile nature of the calcium-permeable cation channel TRPP2. *EMBO Reports* 2006; **7**, 787–793.
13. Mochizuki, T. et al. PKD2, a gene for polycystic kidney disease that encodes an integral membrane protein. *Science* 1996; **272**, 1339–1342.
14. Arif Pavel, M. et al. Function and regulation of TRPP2 ion channel revealed by a gain-of-function mutant. *Proceedings of the National Academy of Sciences of the United States of America* 2016; **113**, E2363–E2372.
15. Luo, Y., Vassilev, P. M., Li, X., Kawanabe, Y., and Zhou, J. Native polycystin-2 functions as a plasma membrane Ca2+-permeable cation channel in renal epithelia. *Molecular and Cellular Biology* 2003; **23**, 2600–2607.
16. Koulen, P. et al. Polycystin-2 is an intracellular calcium release channel. *Nature Cell Biology* 2002; **4**, 191–197.
17. Vassilev, P. M. et al. Polycystin-2 is a novel cation channel implicated in defective intracellular Ca(2+) homeostasis in polycystic kidney disease. *Biochemical and Biophysical Research Communications* 2001; **282**, 341–350.
18. Gonzalez-Perrett, S. et al. Polycystin-2, the protein mutated in autosomal dominant polycystic kidney disease (ADPKD), is a Ca2+-permeable nonselective cation channel. *Proceedings of the National Academy of Sciences of the United States of America* 2001; **98**, 1182–1187.
19. Su, Q. et al. Structure of the human PKD1-PKD2 complex. *Science* 2018; **361**.
20. Zhu, J. et al. Structural model of the TRPP2/PKD1 C-terminal coiled-coil complex produced by a combined computational and experimental approach. *Proceedings of the National Academy of Sciences of the United States of America* 2011; **108**, 10133–10138.
21. Yu, Y. et al. Structural and molecular basis of the assembly of the TRPP2/PKD1 complex. *Proceedings of the National Academy of Sciences of the United States of America* 2009; **106**, 11558–11563.
22. Hanaoka, K. et al. Co-assembly of polycystin-1 and -2 produces unique cation-permeable currents. *Nature* 2000; **408**, 990–994.
23. Myers, B. R., Saimi, Y., Julius, D., and Kung, C. Multiple unbiased prospective screens identify TRP channels and their conserved gating elements. *Journal of General Physiology* 2008; **132**, 481–486.
24. Kremeyer, B. et al. A gain-of-function mutation in TRPA1 causes familial episodic pain syndrome. *Neuron* 2010; **66**, 671–680.
25. Lin, Z. et al. Exome sequencing reveals mutations in TRPV3 as a cause of Olmsted syndrome. *American Journal of Human Genetics* 2012; **90**, 558–564.

26. Nilius, B. and Owsianik, G. Channelopathies converge on TRPV4. *Nature Genetics* 2010; **42**, 98–100.
27. Nilius, B. and Voets, T. The puzzle of TRPV4 channelopathies. *EMBO Reports* 2013; **14**, 152–163.
28. Wang, W., Zhang, X., Gao, Q., and Xu, H. TRPML1: An ion channel in the lysosome. *Handbook of Experimental Pharmacology* 2014; **222**, 631–645.
29. Hofmann, L. et al. The S4–S5 linker - gearbox of TRP channel gating. *Cell Calcium* 2017; **67**, 156–165.
30. Zheng, W. et al. Hydrophobic pore gates regulate ion permeation in polycystic kidney disease 2 and 2L1 channels. *Nature Communications* 2018; **9**, 2302.
31. Liu, X. et al. Polycystin-2 is an essential ion channel subunit in the primary cilium of the renal collecting duct epithelium. *Elife* 2018; **7**.
32. Shen, P. S. et al. The structure of the polycystic kidney disease channel PKD2 in lipid nanodiscs. *Cell* 2016; **167**, 763–773.
33. Su, Q. et al. Cryo-EM structure of the polycystic kidney disease-like channel PKD2L1. *Nature Communications* 2018; **9**, 1192.
34. Kandel, E. R., Schwartz, J. H., Jessell, T. M., Siegelbaum, S. A., and Hudspeth, A. J. *Principles of Neural Science*. 5th ed., 2013; McGraw-Hill.
35. Hille, B. *Ion Channels of Excitable Membranes*. 3rd ed., 2001; Sinauer Associates.
36. Moore, J. W. A personal view of the early development of computational neuroscience in the USA. *Front Comput Neurosci* 2010; **4**.
37. Sakmann, B. and Neher, E. *Single-Channel Recording*. 2nd ed., 2009; Springer.
38. Zheng, J. and Trudeau, M. C. *Handbook of Ion Channels*. 2015; CRC Press.
39. Lin-Moshier, Y. and Marchant, J. S. The Xenopus oocyte: A single-cell model for studying Ca2+ signaling. *Cold Spring Harb Protoc* 2013; **2013**.
40. Wagner, C. A., Friedrich, B., Setiawan, I., Lang, F., and Broer, S. The use of *Xenopus laevis* oocytes for the functional characterization of heterologously expressed membrane proteins. *Cell Physiol Biochem* 2000; **10**, 1–12.
41. Brown, A. L., Johnson, B. E., and Goodman, M. B. Patch clamp recording of ion channels expressed in Xenopus oocytes. *Journal of Visualized Experiments* 2008; **20**, e936.
42. Ma, R. et al. PKD2 functions as an epidermal growth factor-activated plasma membrane channel. *Molecular and Cellular Biology* 2005; **25**, 8285–8298.
43. Cai, Y. et al. Calcium dependence of polycystin-2 channel activity is modulated by phosphorylation at Ser812. *Journal of Biological Chemistry* 2004; **279**, 19987–19995.
44. Chen, X. Z. et al. Transport function of the naturally occurring pathogenic polycystin-2 mutant, R742X. *Biochemical and Biophysical Research Communications* 2001; **282**, 1251–1256.
45. Kleene, S. J. and Kleene, N. K. The native TRPP2-dependent channel of murine renal primary cilia. *American Journal of Physiology. Renal physiology* 2017; **312**, F96–F108.
46. Kim, S. et al. The polycystin complex mediates Wnt/Ca(2+) signalling. *Nature Cell Biology* 2016; **18**, 752–764.
47. Delmas, P. et al. Gating of the polycystin ion channel signaling complex in neurons and kidney cells. *FASEB Journal: Official Publication of the Federation of American Societies for Experimental Biology* 2004; **18**, 740–742.
48. Wang, Z. et al. The ion channel function of polycystin-1 in the polycystin-1/polycystin-2 complex. *EMBO Reports* 2019; **20**, e48336.

5 Functional Studies of PKD2 and PKD2L1 through Opening the Hydrophobic Activation Gate

Wang Zheng, Lingyun Wang, Jingfeng Tang, Ji-Bin Peng, and Xing-Zhen Chen

CONTENTS

5.1 INTRODUCTION

Transient receptor potential (TRP) channels are a superfamily of cation channels that have emerged as cellular sensors responding to a broad range of intracellular or extracellular stimuli.[1,2] The 28 TRP members in mammals have been divided into six subfamilies: TRPC (canonical), TRPM (melastatin), TRPV (vanilloid), TRPML (mucolipin), TRPA (ankyrin), TRPP (polycystin).[3] During the past decades, TRP channels have been extensively studied, which greatly deepened our understanding of their channel functions and physiological functions. With the recent exciting achievements on the determination of high-resolution structures of TRPs,[4] we have now gained insights into their gating and regulation mechanisms on the molecular level. However, compared with other TRP channels, our understanding of TRPP channel functions lags far behind to date, largely because of lack of reliable current readout and unknown agonist. In this chapter, we discuss the strategy and logic to constitutively open the TRPP channels through mutation to the hydrophobic activation gate.

The TRPP subfamily contains three members: PKD2 (also called polycystin-2 or TRPP2), PKD2L1 (also called polycystin-L or TRPP3) and PKD2L2 (also called TRPP5). PKD2, the founding member of the TRPP subfamily, is mutated in about 15% of the autosomal dominant polycystic kidney disease (ADPKD) cases; ADPKD is the most common monogenetic disorder of the kidney, affecting over 12.5 million people worldwide.[5] The disease is featured by the progressive formation of fluid-filled, enlarged renal cysts, which lead to a decline in the renal function.[6] The remaining 85% of the ADPKD cases are caused by mutations in PKD1, which forms a heterotetrameric complex with PKD2 in a stoichiometry of 1:3.[7–9] To date, PKD1 and PKD2 have been well characterized to be involved in many signaling pathways related to cell proliferation and apoptosis, two hallmarks of renal cystic cells.[10] Dysfunctions of PKD1 and/or PKD2 presumably result in abnormal signaling pathways, which together initiate and promote cyst growth. However, it remains largely unclear as to whether and how the channel function of the PKD2 homotetramer or the PKD1/PKD2 heterotetramer is involved in the disease pathogenesis. Also, there have been controversies about the channel function of PKD2 and the PKD1/PKD2 complex. For instance, PKD2 was initially proposed to be a Ca^{2+} release channel in the endoplasmic reticulum,[11] but recently, increasing evidence suggested that PKD2 has a low permeability, if not impermeable, to Ca^{2+}.[12,13] The PKD1/PKD2 complex was previously hypothesized to act as a fluid flow sensor in primary cilia of renal epithelial cells[14] and coexpression of PKD1 and PKD2 in CHO cells was reported to give rise to distinct cation currents.[15] However, with genetically engineered Ca^{2+} sensors in the primary cilia, no Ca^{2+} influx was detected at physiological or even supraphysiological levels of fluid flow.[16] Also, by directly measuring currents across primary cilia, PKD2, but not PKD1, was shown to be an essential ion channel subunit in the primary cilium of renal collecting duct epithelial cells.[13] In addition, a gain-of-function (GOF) PKD2 mutant F604P was recently found to induce robust cation currents when expressed in *Xenopus* oocytes,[12] but no current was detected when this mutant was expressed in mammalian cells.[17] Although high-resolution structures of PKD2 homotetramers and PKD1/PKD2 heterotetramers are now available[9,18–20] and they were both revealed to be pore-forming complexes, their channel functions have still remained elusive. Therefore, a reliable and well-recognized function readout is essential to understanding PKD2 channel function.

PKD2L1 shares 54% amino acid identity with PKD2 but is not involved in ADPKD. In contrast to PKD2, PKD2L1 has been more rigorously studied by electrophysiological approaches, and all the evidence support that PKD2L1 functions as a Ca^{2+} permeable nonselective cation channel.[21–23] Recently determined PKD2L1 structures[24,25] provided plausible explanations on ion permeation differences between PKD2L1 and PKD2, i.e., Ca^{2+} permeates through PKD2L1 but not PKD2. PKD2L1 forms a Ca^{2+} channel complex with PKD1L1, a homologue of PKD1, in primary cilia to regulate ciliary Ca^{2+} concentration and hedgehog signaling.[26,27] PKD2L1 also co-localizes with PKD1L3 in a subset of mouse taste receptor cells,[28–30] and they form a channel complex that responds to acid stimulation.[29,31] On the other hand, we showed that PKD2L1 alone in oocytes is sufficient to respond to acid.[32] PKD2L1/PKD1L3 was proposed to be the sour taste receptor,[28,29] but this was challenged by subsequent studies in animal models. While PKD2L1 knockout mice exhibited mildly reduced response to sour, no corresponding phenotype was observed in PKD1L3 knockout mice.[33,34] To date, the physiological function of PKD2L1 has remained elusive. Generating a PKD2L1 GOF mouse model through knockin of a GOF PKD2L1 point mutant would be an interesting approach to investigate PKD2L1 *in vivo* function. However, no GOF PKD2L1 mutation has been reported so far.

PKD2L2 shares 48% amino acid identity with PKD2. Little is known about the PKD2L2 channel and physiological function. Whether PKD2L2 actually forms a functional channel remains an open question.

5.1.1 Hydrophobic Gate Theory

Biological ion channels generally form transmembrane pores with diameters falling into the nanometer scale. Within such a narrow confine, water and ion molecules could behave in an unusual way not seen in bulk, macroscopic solutions. Molecular dynamics (MD) simulations on model nanopores[35–37] found that if the pore is lined by a hydrophobic surface, the unfavorable environment for water molecules may lead the pore to stochastically alternate between the liquid and vapor states (Figure 5.1a). Under the vapor state, the pore is effectively devoid of water molecules, resulting in a high energetic barrier to ion permeation, while under the liquid state, the pore is largely occupied by water molecules, allowing ions to pass through. This observation is reminiscent of closing-opening events of an ion channel in single-channel recordings and leads to the hydrophobic gating hypothesis, which was strongly supported by subsequent structural, functional, and computational studies.

Based on the hydrophobic gate theory, a hydrophobic gate allows the pore to close through keeping the low probability of the liquid state rather than through physical, steric occlusion of the pore. MD simulation studies showed that there are two main ways of opening a hydrophobic pore: increase the pore diameter or hydrophilicity in the constricting region of the pore[38] (Figure 5.1b). The first way is represented by the activation of biological ion channels, during which conformational changes of pore-lining helices (e.g., following binding of an agonist) increase the pore diameter,

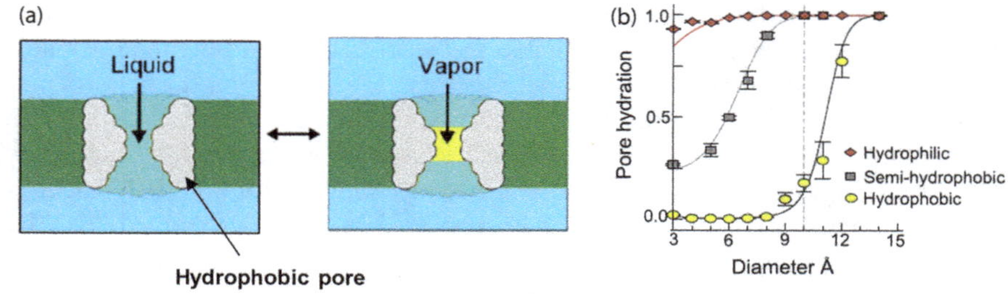

FIGURE 5.1 Hydrophobic gating principles. (a) Schematic illustration of liquid-vapor transitions within a hydrophobic nanopore. The membrane is indicated in green and the water solution is indicated in cyan. The area highlighted in yellow is devoid of water molecules. (b) The dependence of the pore hydration rate on the pore diameter or the hydrophilicity of the pore. (Adapted from Aryal, P et al. *J. Mol. Biol.* 427, 121–130, 2015.)

thereby resulting in abolishment of hydrophobic barrier and pore opening. MD simulations on model nanopores showed that the pore hydration rate is close to zero when a hydrophobic pore has a physical diameter of less than 9 Å (i.e., no ion is eligible to pass through) and dramatically increases to the maximum in a strong sigmoidal dependence when the pore diameter reaches 12–13 Å[39] (Figure 5.1b), which well explains the observed pore sizes of crystallized bacterial small conductance mechanosensitive channel MscS in closed (8 Å) and open (13 Å) states.[40,41] The other way to open a hydrophobic pore, through increasing the hydrophilicity of the pore, is presented by mutagenesis in biological ion channels. Replacement of the hydrophobic gate residue by a hydrophilic residue would drastically increases the pore hydration rate, resulting in a constitutive open pore.[39] This has led to the identification of hydrophobic gates in the TWIK-1 K2P potassium channel and the MscL mechanosensitive channel in which channel activity strongly correlated with the gate residue hydrophilicity.[42–44] In this chapter, we discuss the application of hydrophobic gate theory in the identification and characterization of the PKD2L1 and PKD2 pore gates, which will provide a valuable tool to study their channel functions.

5.2 IDENTIFICATION OF HYDROPHOBIC GATE RESIDUES OF PKD2L1 AND PKD2 BY HYDROPHILIC SUBSTITUTIONS TO CONSTITUTIVELY OPEN THE PORE

According to published structures of TRP channels, there is one or more pore constriction seals/rings separated by about one helical turn (usually 3–4 aa), mainly formed by a hydrophobic residue(s) in the cytoplasmic part of the 4 pore-lining S6 helices.[4,45] In some cases, a pore constriction may also be formed by a hydrophilic residue, for example, in TRPC4, -A1 and -V5 (Figure 5.2). Currently, these residues are all recognized as physical gate residues. However, it remains to be determined whether or not a physical gate actually controls the opening and closing of the pore (i.e., whether or not it actually acts as a functional gate). By functional studies on TRPV6, -V5, -V4, -C4 and -M8, we showed that a functional gate is always to be a hydrophobic gate, that TRPV6 and -V5 possess a double-consecutive residue gate, and

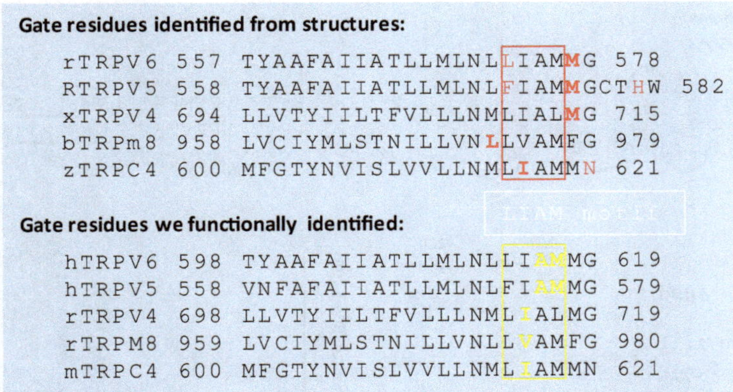

FIGURE 5.2 Physical and functional gate residues in TRP channels. *Upper panel*: Gate residues (red) that were identified from reported structures. Bold red residues indicated hydrophobic residues that define the narrowest constriction. The LIAM motif is shown within the box. *Lower panel*: Gate residues (yellow) that we functionally identified.

that all identified functional gate residues are within the "LIAM" motif (Figure 5.2).[46] By comparing with the corresponding physical gate residue, a significant discrepancy is noticed (Figure 5.2). We think that it could be due, at least in part, to different experimental conditions for cryo-EM and electrophysiology. We proposed that PKD2L1 and PKD2 also possess a similar hydrophobic gate formed by a residue(s) in the cytoplasmic part of S6. Based on the hydrophobic gate theory, increasing the hydrophilicity of the gate would result in the hydrophobic barrier collapse and an elevated hydration rate of the pore, thereby giving rise to a constitutively open pore.[39] Therefore, we expect to detect constitutive channel activities of PKD2L1 and PKD2 when the hydrophobic gate residue is replaced by a hydrophilic residue. To identify the hydrophobic gate in PKD2L1, we replaced each of the eight candidate hydrophobic residues in the lower part of S6 with hydrophilic asparagine (N) (Figure 5.3a) and examined channel activities of the resulting mutant channels in *Xenopus* oocytes with two-electrode voltage clamp, which was well-established in our laboratory.[22,47–49] We indeed found that mutants L557N and A558N exhibit large constitutive channel activity (Figure 5.3b), suggesting that L557 and A558 may together form a hydrophobic gate, which we called a two-residue gate. We examined the plasma membrane expression of the two mutants with biotinylation and immunofluorescence assays (see detailed protocols below in Sections 5.5 and 5.6) and found no significant difference from that of wildtype (WT) PKD2L1 (see Figure 1 in Zheng et al.[45]), suggesting that the increased channel activity of L557N or A558N is not due to the increased surface expression. Since asparagine has a similar size as leucine and is larger than alanine, the constitutive activity of the two mutants is unlikely due to an increase in the pore size. Therefore, these data suggest that L557 and A558 together form a hydrophobic gate in PKD2L1 and that the L557N or A558N mutation abolishes the hydrophobic barrier, resulting in the constitutive channel activity. Considering the existing controversies with respect to PKD2 channel function, it will be valuable to obtain a functional readout through providing an open

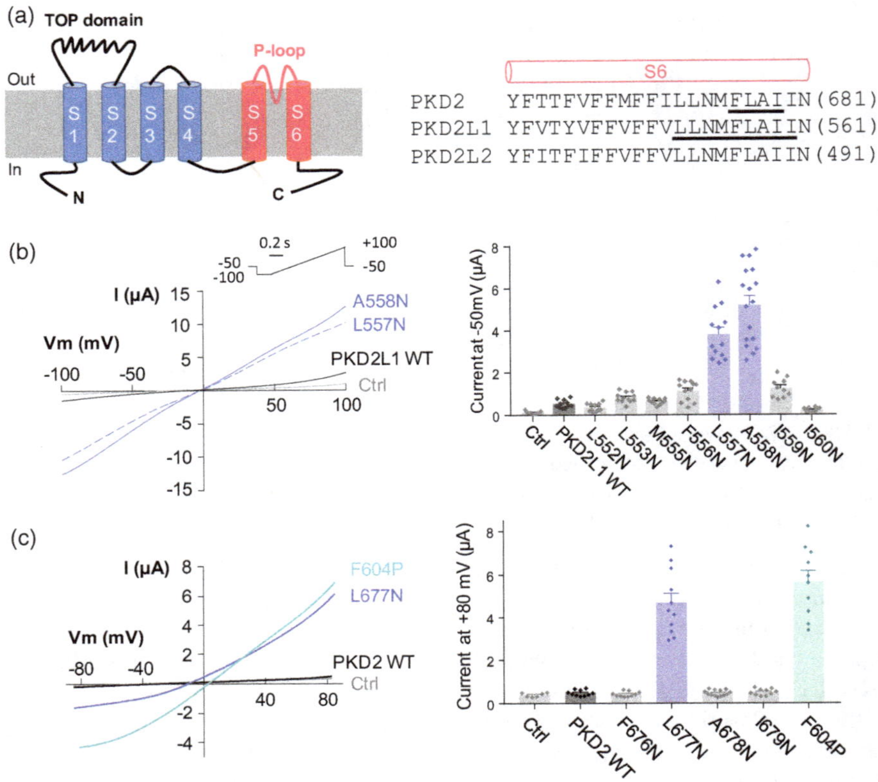

FIGURE 5.3 Identification of hydrophobic gate inhuman PKD2L1 and PKD2. (a) *Left panel*: Membrane topology of PKD channels. *Right panel*: Amino acid sequence alignment of S6 helix of human PKD channels. The underlined amino acids were mutated to N. (b) *Left panel*: Representative current-voltage curves obtained from Ctrl (H_2O-injected oocytes), PKD2L1 WT, L557N or A558N expressing oocytes obtained using the indicated voltage ramp protocol. *Right panel*: Averaged currents at –50 mV. (c) *Left panel*: Representative current-voltage curves obtained from oocytes expressing WT or a mutant PKD2, as indicated. The F604P mutant[12] serves as a positive control. *Right panel*: Averaged current at +80 mV. (Adapted from Zheng, W et al. *Nat. Commun.* 9, 2302, 2018.)

mutant channel by the same strategy. Since PKD2 possesses a cytoplasmic part of S6 that is identical to PKD2L1 (Figure 5.3a), we wondered whether PKD2 would also share the same pore gate residues with PKD2L1. We carried out similar asparagine substitution at each site in the F676-I679 fragment (corresponding to the PKD2L1 gate-containing fragment F556-I559). Using the *Xenopus* oocyte expression system, two-electrode voltage clamp, immunofluorescence, and surface biotinylation, we found that the L677N mutant (corresponding to L557N of PKD2L1), but not the other mutants including A678N, exhibits large constitutive channel activity, suggesting that L677 acts as the single-residue hydrophobic pore gate, which is consistent with the resolved PKD2 structures[18–20] showing that L677 faces the pore and forms a constriction while A678 points away from the pore.

5.2.1 Documentation on the Hydrophobic Gate by Further Amino Acid Substitutions

Based on the hydrophobic gate theory, the channel activity (or pore opening) would strongly depend on the hydrophilicity and size of the gate residue. To further support that L557/A558 in PKD2L1 and L677 in PKD2 functionally form the pore gate, we obtained a series of mutant channels by substituting each of these residues with amino acids of different hydrophilicity and sizes. We indeed found that the channel activity generally remains high with a hydrophilic substitution and negatively correlates with the size of a hydrophobic gate residue (i.e., positively correlates with the size of a hydrophobic pore) (Figure 5.4) (by biotinylation we found that the surface membrane expression of these mutants are not significantly altered), in strong agreement with the hydrophobic gate theory. Thus, a hydrophobic gate residue not only allows to open the channel via substitution with a hydrophilic or smaller hydrophobic residue but also allows to close the channel via substitution with a larger hydrophobic residue, which could also be used as a screening method in the identification of the gate residue(s). Of note, because either L557N or A558N is sufficient to open the channel, it would "logically" require that both sites be occupied by large hydrophobic residues (e.g., L557W or -F and A558W or -F) to close the channel. This is presumably because L557 and A558 together form only one gate, rather than two separate gates, and that changes at either site will affect the gate. In summary, the substitutions at gate

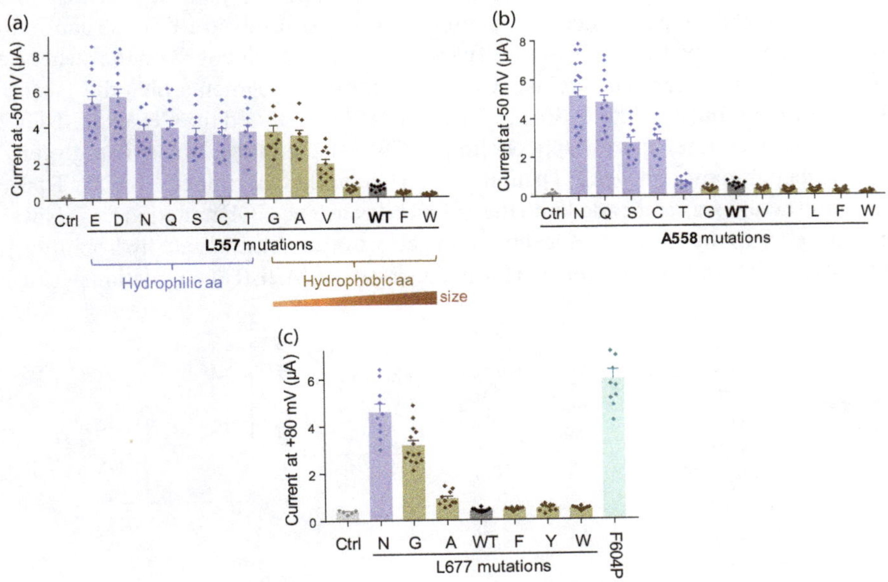

FIGURE 5.4 Characterization of hydrophobic gate residues in human PKD2L1 and PKD2. (a) and (b) Averaged currents at −50 mV obtained from oocytes expressing PKD2L1 L557 mutant (a) or A558 mutant (b), as indicated. (c) Averaged currents at +80 mV obtained from oocytes expressing PKD2 F604P mutant or indicated L677 mutants. (Adapted from Zheng, W et al. *Nat. Commun.* 9, 2302, 2018.)

residues allow obtaining a series of gain-of-function (GOF) and loss-of-function (LOF) mutant channels.

5.2.2 Documentation on the Hydrophobic Gate by Channel Activity Rescue

While we think that LOF of PKD2L1 gate mutants L557W and A558W is due to pore gate closure, it is also possible that the W substitution results in substantial conformational changes due to altered inter- or intra-subunit interactions, which abolishes the channel function. Because the channel activity depends on both the hydrophobicity and size of the gate residue(s), we wondered whether increasing hydrophilicity of site 557 (or 558) would compensate for LOF mutant A558W (or L557W). Indeed, we detected robust channel activity for double mutants L557N/A558W and L557W/A558N (Figure 5.5), indicating that the N substitution rescues the LOF mutants A558W and L557W, and that the LOF is unlikely due to conformational changes associated with the W substitution. Therefore, the rescue experiments were in support of our conclusion that L557 and A558 residues together form the hydrophobic gate in PKD2L1.

5.2.3 Molecular Dynamics Simulations

In order to further support the concept of the hydrophobic gate in PKD2L1 and PKD2 channels, we performed MD simulations on the resolved PKD2 structure to examine the pore hydration. To simplify the simulation, we focused on the structure of the PKD2 pore region (the last two transmembrane segments plus the P-loop, S5-loop-S6, aminoacid M590-S689; PDB: 5T4D). We first used the CHARMM-GUI member builder to set up a phosphocholine—POPC (1-palmitoyl,2-oleoyl-sn-glycero-3-phosphocholine)—bilayer.[50] Then, the experimentally determined PKD2WT pore region structure was embedded into the POPC bilayer. The GOF mutation L677N was introduced into the structure to examine the effect of increasing gate hydrophilicity. Two 40 ns MD simulations were performed using the AMBER14 simulation package

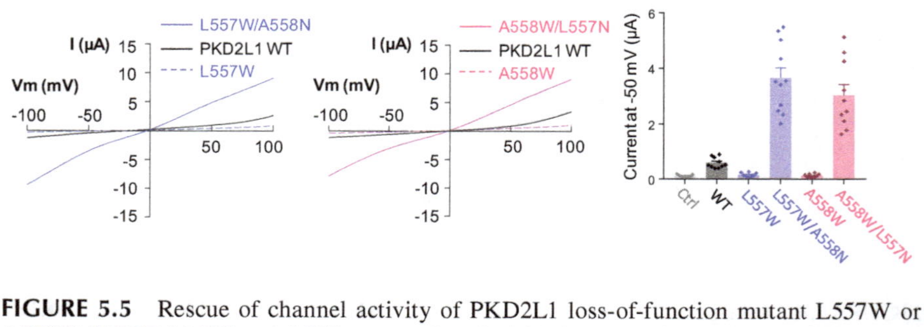

FIGURE 5.5 Rescue of channel activity of PKD2L1 loss-of-function mutant L557W or A558W. PKD2L1 L557 and A558 were replaced with W or N. *Left panel* and *middle panel*: Representative current-voltage curves of PKD2L1 WT or indicated mutant expressed in oocytes. *Right panel*: Averaged currents at −50 mV. (Adapted from Zheng, W et al. *Nat. Commun.* 9, 2302, 2018.)

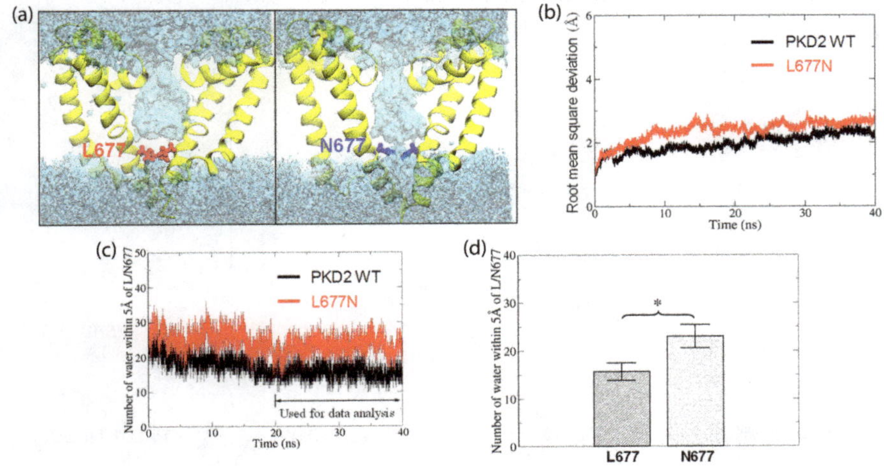

FIGURE 5.6 Hydrophobic cuff formed by L677 in a human PKD2 channel pore. (a) MD simulations of PKD2 pore region structure (yellow). Shown are the average densities of water molecules (cyan) within the pore of the PKD2 WT and L677N mutant during simulations. (b) Root mean square deviation for the C_α atoms of PKD2 pore region through simulations. (c) The number of water molecules within 5 Å of the gate residue during simulations. (d) Averaged number of water molecules within 5 Å of the gate residue during simulations. *$P < 0.05$ by student *t*-test.

based on a previously established protocol.[51] The root means square deviations for the C_α atoms of the PKD2 pore region were calculated to assess the equilibration of the simulations (Figure 5.6b). Interestingly, we observed that the pore constriction region formed by the L677 is largely devoid of water molecules during the simulation while this region remains hydrated in the GOF mutant L677N (Figure 5.6a). When the hydration rate was calculated within the last 20 ns simulations, the L677N mutant was found to exhibit a significantly higher water density within the 5 Å of the gate than the WT channel (Figure 5.6c and d). These simulation data strongly suggested that L677 residues from four PKD2 subunits form a hydrophobic cuff to prevent ion permeation, and that the L677N mutation abolishes this hydrophobic cuff, thus resulting in constitutive PKD2 channel activity.

5.3 COMPARISON BETWEEN FUNCTIONAL AND STRUCTURAL STUDIES

By functional studies using oocytes electrophysiology, we identified a double-residue hydrophobic gate in PKD2L1 (L557/A558) and a sing-residue hydrophobic gate in PKD2 (L677). Recently, structures of PKD2 and PKD2L1 in distinct states were reported by different groups,[18–20,24,25,45] which allowed us to compare the functional and structural data and better understand the single- and double-residue gate. Three PKD2 structures in the closed state were independently determined (see Figure 5.7 with PDB accession numbers 5T4D, 5K47, and 5MKE), and they all revealed that

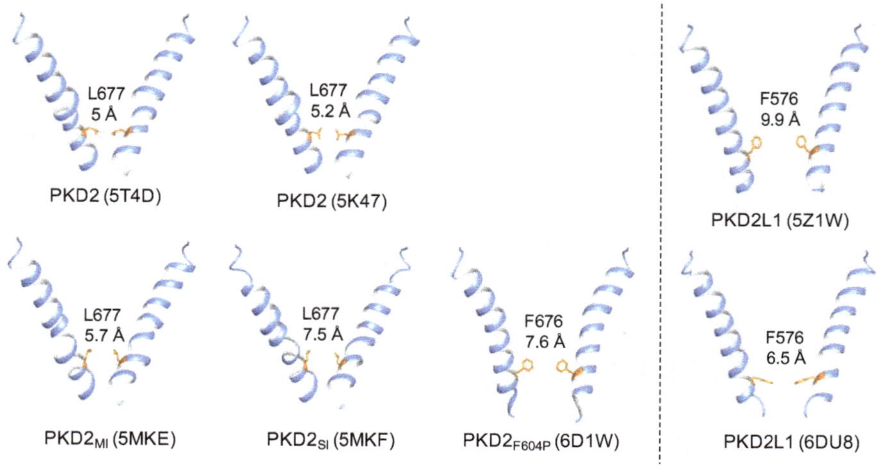

FIGURE 5.7 S6 conformation of published human PKD2 and PKD2L1 structures. The residue forming the constriction was shown and highlighted in orange. The pore diameter determined by the constriction-forming residue was also shown. The PDB accession number for each structure was included in parentheses. $PKD2_{MI}$: PKD2 structure bound with multiple calcium ions. $PKD2_{SI}$: PKD2 structure bound with single calcium ion. $PKD2_{F604P}$: structure of PKD2 F604P mutant.

L677 faces the pore and forms a constriction with a diameter of 5–6 Å, which is consistent with our functional data that L677N mutation resulted in a constitutive channel activity. The concept of L677 acting as a gate residue is further supported by the structure of PKD2 activating mutant F604P in which a combined twisting and splaying motion of L677 residue induced the pore opening to 7.6 Å (Figure 5.7). Previous studies showed that intracellular Ca^{2+} increases PKD2 open probability at low concentrations while inhibiting channel opening at elevated levels.[52] This bell-shaped regulation is now well explained by the distinct orientations of L677 induced by Ca^{2+}, which give rise to a large pore opening (7.5 Å) at low Ca^{2+} concentration (Figure 5.7, see $PKD2_{SI}$, PKD2 structure bound with single Ca^{2+} ion), but a small pore opening (5.7 Å) at elevated Ca^{2+} concentrations (Figure 5.7, see $PKD2_{MI}$, PKD2 structure bound with multiple Ca^{2+} ions). Therefore, different conformations of the gate residue L677 may lead to different pore sizes of PKD2, thus reflecting distinct states of PKD2 channels.

The two reported PKD2L1 structures bear a similar S6 conformation and pore size with the PKD2 F604P mutant and are thus presumably to reside in an open state.[24,25] However, these structural data do not seem to display double-residue gate organization revealed by our functional data. Again, we think significantly different experimental conditions may be a major underlying reason for the discrepancy.

5.4 SINGLE- VS. DOUBLE-RESIDUE GATE IN PKD2 AND PKD2L1

Although PKD2 and PKD2L1 have identical amino acid sequences in the cytoplasmic part of S6 in which gate residues are found, our data showed that PKD2L1 possesses

```
hPKD2L1 540 YFVTYVFFVFFVLLNMFLAIIN 561
  hPKD2 660 YFTTFVFFMFFILLNMFLAIIN 681
P2-hPKD2 660 YFTTFVFFMFFILLNMFLAIIN 681
```

FIGURE 5.8 Comparison of gate residues in PKD2 and PKD2L1. Shown are the amino acid sequences of the cytoplasmic part of S6 in human PKD2L1 and PKD2. Gate residues are marked yellow, and the box indicates the LIAM motif.

a double-residue gate formed by L557/A558, while PKD2 has a single-residue (conventional) gate formed by L677 (corresponding to PKD2L1 L557). PKD2L1, but not PKD2, has detectable channel activity in the absence of an agonist,[45,48] suggesting that PKD2L1 may have a larger pore than PKD2 in the resting state. Our structural data on PKD2-activated mutant F604P showed a larger pore size than the WT channel and the presence of π- to α-helical transition in the S6 helix from WT PKD2 to mutant F604P, which accompanies S6 twisting. We wondered whether this S6 twisting is sufficiently significant so that mutant F604P would possess a double-residue gate. Indeed, our functional study showed that in PKD2 mutant F604P, both F676 and A677 exhibit characteristics of being gate residues (Figure 5.8). Thus, compared with WT PKD2, mutant F604P is more similar to PKD2L1 in terms of the pore size and possession of a double-residue gate. Of note, based on similar approaches, we also found single-residue gates in TRPV4, -M8, and -C4 and double-residue gate in TRPV5 and -V6.[46]

5.5 DETERMINATION OF PKD2/PKD2L1 SURFACE EXPRESSION WITH BIOTINYLATION ASSAY

5.5.1 Materials

1. Phosphate-buffered saline (PBS) solution: 137 mM NaCl, 2.7 mM KCl, 5 mM Na_2HPO_4, and 1.8 mM KH_2PO_4, pH 8.0
2. EZ-link Sulfo-NHS-Lc-Biotin (Pierce, cat no. 21335)
3. Quenching buffer: 192 mM glycine and 25 mM Tris-HCl in PBS solution
4. CelLytic M lysis buffer (Sigma, cat no. C2978)
5. Streptavidin agarose beads (Pierce, cat no. 20347)
6. NP-40 buffer: 150 mM NaCl, 1% NP-40 and 50 mM Tris, pH 8.0

5.5.2 Methods

1. Transfer 10 oocytes expressing PKD2 or PKD2L1 into a well in a 24-well plate. Wash 3 times with 1 mL ice-cold PBS (pH 8.0) with gentle shaking for 30 s.
2. Incubate the oocytes with 1 mL 0.5 mg/mL Sulfo-NHS-SS-Biotin in PBS solution (pH 8.0) for 30 min in cold room with gentle shaking in the dark.
3. Remove the biotin solution and wash the oocytes using 1 mL ice-cold quenching buffer with gentle shaking for 30 s to stop the reaction.

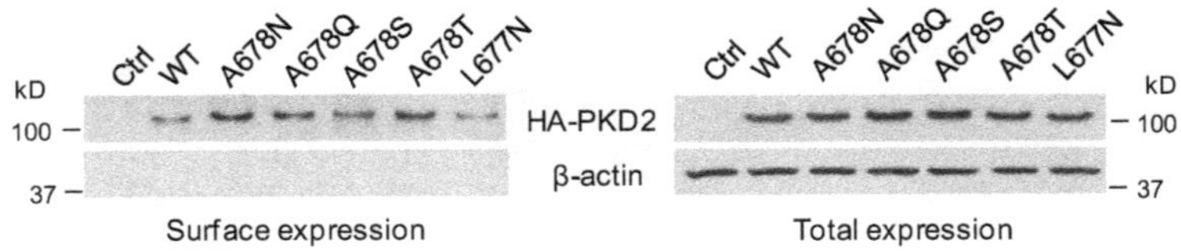

FIGURE 5.9 Representative Western blot of surface expression and total expression of HA-tagged human PKD2 WT or indicated mutants overexpressed in oocytes. Ctrl, H_2O-injected oocytes. β-actin protein served as an internal control to assess cytosolic protein contamination. The relative protein levels can be determined by measuring the band intensities with ImageJ software. (Adapted from Zheng, W et al. *Nat. Commun.* 9, 2302, 2018.)

4. Remove the quenching buffer and wash the oocytes three times with ice-cold PBS (pH 8.0).
5. Transfer the oocytes into a 1.5 mL Eppendorf tube and gently remove the PBS solution as much as possible.
6. Add 500 μL CelLytic M lysis buffer supplemented with protease inhibitor cocktail and pipette up and down repeatedly to completely break the oocytes. Do not vortex.
7. Spin the cell lysate at 15,000 × g for 15 min at 4°C.
8. Transfer the supernatant into a fresh 1.5 mL Eppendorf tube. Do not disturb the pellet and the top yolk layer. An aliquot of 60 μL lysate was taken for total protein measurement.
9. Add 50 μL 50% streptavidin agarose beads and incubate the mixture at cold room overnight with slight agitation.
10. Spin the beads at 2000 × g for 2 min at 4°C and remove the supernatant.
11. Wash the beads using NP-40 buffer with agitation for 5 min at room temperature for a total of 5 times.
12. Remove the NP-40 buffer and resuspend the beads in 80 μL SDS-PAGE sample buffer. The biotinylated PKD2 or PKD2L1 protein is denatured at 65°C for 5 min and subjected to Western blot (Figure 5.9).

5.6 DETERMINATION OF PKD2/PKD2L1 SURFACE EXPRESSION WITH WHOLE-MOUNT OOCYTE IMMUNOFLUORESCENCE ASSAY

5.6.1 MATERIALS

1. Phosphate-buffered saline (PBS) solution: 137 mM NaCl, 2.7 mM KCl, 5 mM Na_2HPO_4, and 1.8 mM KH_2PO_4, pH 7.4.
2. Paraformaldehyde (PFA), triton X-100, bovine serum albumin (BSA), skim milk.
3. Anti-Flag (cat no. 14793) or -HA (cat no. 3724) primary antibodies from Cell Signalling Technology and secondary donkey anti-rabbit lgG conjugated with AlexaFluor 488 from Jackson ImmunoResearch Laboratories.
4. Vectashield antifade mounting medium (Vector Laboratories, cat no. H-1000).
5. SecureSear imaging spacer (Sigma, cat no. GBL654008–100EA).

5.6.2 Methods

1. Transfer 5 oocytes expressing HA-tagged PKD2 or Flag-tagged PKD2L1 into a well in a 24-well plate. Wash three times with 500 μLPBS buffer.
2. Remove the PBS from the last wash and fix the oocytes with 500 μL 4% PFA in PBS for 15 min at room temperature.
3. Quench the unreacted PFA by washing the oocytes 3 times in 500 μL 50 mM NH_4Cl for 5 min each time.
4. Wash the oocytes twice with 500 μL PBS.
5. Permeabilize the oocytes with 0.1% Triton X-100 in PBS for 4 min at room temperature.
6. Wash the permeabilized oocytes 3 times with 500 μL PBS.
7. Block the permeabilized oocytes 30 min with 3% BSA in PBST (0.1% Tween-20 in PBS), followed by another 30 min blocking with 3% skim milk in PBST.
8. Incubate the oocytes with a primary antibody (1:200 dilution in PBST containing 3% skim milk) overnight at 4°C.
9. The next day, wash the oocytes in PBST 3 times for 10 min each time.
10. Incubate the oocytes with a secondary AlexaFluor 488-conjugated antibody (1:1000 dilution in PBST containing 3% skim milk) for 30 min at room temperature.
11. Wash the oocytes 3 times in PBST for 10 min each time.
12. Mount the oocytes with Vectashield mounting medium on slides using 5 SecureSeal imaging spacers: put the oocytes into spacer, remove water with a pipette, and add one drop of Vectashield. Then use a coverslip to seal the oocytes in the spacer and paint nail polish around the edges to seal. Do not squeeze the oocytes too much or they will burst. Let nail polish dry on slides in the dark and store the slides in the dark.
13. Examine the oocytes on an AIVI spinning disc confocal microscope (Figure 5.10).

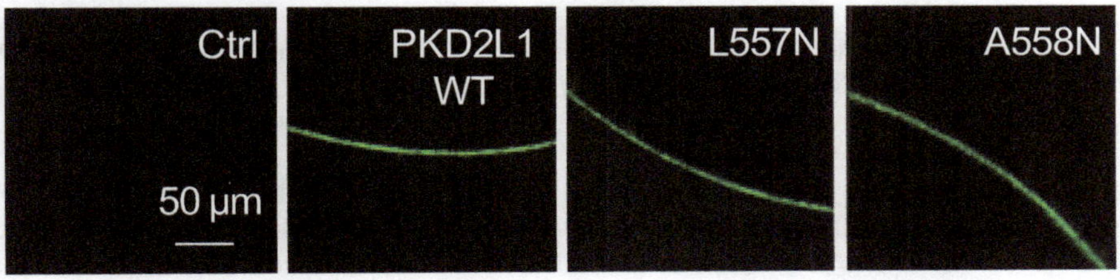

FIGURE 5.10 Representative whole-mount immunofluorescence showing the oocyte surface expression of Flag-tagged human PKD2L1 WT, mutant L557N or A558N. Ctrl, H_2O-injected oocytes. The relative surface expression levels can be assessed by quantifying the fluorescence intensities using Volocity 6.2 software. (Adapted from Zheng, W et al. *Nat. Commun.* 9, 2302, 2018.)

5.7 CONCLUDING REMARKS

PKD2 has been identified to be mutated in about 15% of ADPKD, and its implication in the disease pathogenesis has been extensively studied. However, how it functions as an ion channel remains elusive, largely due to lack of a reliable, well-recognized function readout. In this chapter, based on hydrophobic gating theory, we have described a strategy, via hydrophilic substitutions and electrophysiology, that allowed us to identify and characterize the pore gate residues in PKD2 and PKD2L1 and to open or close the channels in the absence of an agonist, which provided a valuable tool to study PKD2 and PKD2L1 channel function. We identified a series of GOF and LOF gate mutants with increased and decreased channel activity to different extents; this provided good candidates for generating a GOF animal model through knockin of a point mutation to study the channel function of PKD2 and PKD2L1 under *in vivo* and physiological conditions. The same strategy should also be applicable to study other TRP channels with unknown agonist and undetectable current (e.g., TRPC1).

ACKNOWLEDGMENTS

This work was supported by the Natural Sciences and Engineering Research Council of Canada (NSERC; to X.-Z.C.), the National Natural Science Foundation of China (81570648, to X.-Z.C.), the National Institute of Diabetes and Digestive and Kidney Diseases (R01DK104924, to J.-B.P.), W.Z. was supported by Alberta Innovates—Doctoral Graduate Student Scholarship and fellowships from James Hudson Brown—Alexander B. Coxe and the Kavli Institute for Neuroscience. We thank Erhu Cao and David Bulkley for data collection at the electron microscope core at the University of Utah, imaging processing, 3D reconstruction, and model building for the PKD2 F604P mutant. We also thank Yifan Chen for help on data collection at UCSF.

REFERENCES

1. Montell, C. The TRP superfamily of cation channels. *Sci. STKE.* **2005**, re3, 2005.
2. Clapham, D. E. TRP channels as cellular sensors. *Nature.* **426**, 517–524, 2003.
3. Venkatachalam, K. and Montell, C. TRP channels. *Annu. Rev. Biochem.* **76**, 387–417, 2007.
4. Madej, M. G. and Ziegler, C. M. Dawning of a new era in TRP channel structural biology by cryo-electron microscopy. *Pflugers Arch.* **470**, 213–225, 2018.
5. Harris, P. C. and Torres, V. E. Polycystic kidney disease. *Annu. Rev. Med.* **60**, 321–337, 2009.
6. Wilson, P. D. Polycystic kidney disease. *N. Engl. J. Med.* **350**, 151–164, 2004.
7. Tsiokas, L., Kim, E., Arnould, T., Sukhatme, V. P., and Walz, G. Homo- and heterodimeric interactions between the gene products of PKD1 and PKD2. *Proc. Natl. Acad. Sci. USA.* **94**, 6965–6970, 1997.
8. Qian, F. et al. PKD1 interacts with PKD2 through a probable coiled-coil domain. *Nat. Genet.* **16**, 179–183, 1997.
9. Su, Q. et al. Structure of the human PKD1/PKD2 complex. *Science.* **361**, 2018.
10. Torres, V. E. and Harris, P. C. Autosomal dominant polycystic kidney disease: The last 3 years. *Kidney Int.* **76**, 149–168, 2009.

11. Koulen, P. et al. Polycystin-2 is an intracellular calcium release channel. *Nat. Cell Biol.* **4**, 191–197, 2002.
12. Arif, P. M. et al. Function and regulation of TRPP2 ion channel revealed by a gain-of-function mutant. *Proc. Natl. Acad. Sci. USA*. **113**, E2363–E2372, 2016.
13. Liu, X. et al. Polycystin-2 is an essential ion channel subunit in the primary cilium of the renal collecting duct epithelium. *Elife*. **7**, 2018.
14. Nauli, S. M. et al. Polycystins 1 and 2 mediate mechanosensation in the primary cilium of kidney cells. *Nat. Genet.* **33**, 129–137, 2003.
15. Hanaoka, K. et al. Co-assembly of polycystin-1 and -2 produces unique cation-permeable currents. *Nature*. **408**, 990–994, 2000.
16. Delling, M. et al. Primary cilia are not calcium-responsive mechanosensors. *Nature*. **531**, 656–660, 2016.
17. Kim, S. et al. The polycystin complex mediates Wnt/Ca(2+) signalling. *Nat. Cell Biol.* **18**, 752–764, 2016.
18. Wilkes, M. et al. Molecular insights into lipid-assisted Ca^{2+} regulation of the TRP channel Polycystin-2. *Nat. Struct. Mol. Biol.* **24**, 123–130, 2017.
19. Grieben, M. et al. Structure of the polycystic kidney disease TRP channel Polycystin-2 (PC2). *Nat. Struct. Mol. Biol.* **24**, 114–122, 2016.
20. Shen, P. S. et al. The structure of the polycystic kidney disease channel PKD2 in lipid nanodiscs. *Cell*. **167**, 763–773, 2016.
21. DeCaen, P. G., Liu, X., Abiria, S., and Clapham, D. E. Atypical calcium regulation of the PKD2-L1 polycystin ion channel. *Elife*. **5**, 2016.
22. Chen, X. Z. et al. Polycystin-L is a calcium-regulated cation channel permeable to calcium ions. *Nature*. **401**, 383–386, 1999.
23. Shimizu, T., Janssens, A., Voets, T., and Nilius, B. Regulation of the murine TRPP3 channel by voltage, pH, and changes in cell volume. *Pflugers Arch.* **457**, 795–807, 2009.
24. Hulse, R. E., Li, Z., Huang, R. K., Zhang, J., and Clapham, D. E. Cryo-EM structure of the polycystin 2-l1 ion channel. *Elife*. **7**, 2018.
25. Su, Q. et al. Cryo-EM structure of the polycystic kidney disease-like channel PKD2L1. *Nat. Commun.* **9**, 1192, 2018.
26. DeCaen, P. G., Delling, M., Vien, T. N., and Clapham, D. E. Direct recording and molecular identification of the calcium channel of primary cilia. *Nature*. **504**, 315–318, 2013.
27. Delling, M., DeCaen, P. G., Doerner, J. F., Febvay, S., and Clapham, D. E. Primary cilia are specialized calcium signalling organelles. *Nature*. **504**, 311–314, 2013.
28. Huang, A. L. et al. The cells and logic for mammalian sour taste detection. *Nature*. **442**, 934–938, 2006.
29. Ishimaru, Y. et al. Transient receptor potential family members PKD1L3 and PKD2L1 form a candidate sour taste receptor. *Proc. Natl. Acad. Sci. USA*. **103**, 12569–12574, 2006.
30. Lopezjimenez, N. D. et al. Two members of the TRPP family of ion channels, Pkd1l3 and Pkd2l1, are co-expressed in a subset of taste receptor cells. *J. Neurochem.* **98**, 68–77, 2006.
31. Inada, H. et al. Off-response property of an acid-activated cation channel complex PKD1L3-PKD2L1. *EMBO Rep.* **9**, 690–697, 2008.
32. Hussein, S. et al. Acid-induced off-response of PKD2L1 channel in Xenopus oocytes and its regulation by Ca^{2+}. *Sci. Rep.* **5**, 15752, 2015.
33. Horio, N. et al. Sour taste responses in mice lacking PKD channels. *PLOS ONE*. **6**, e20007, 2011.
34. Ishimaru, Y. Molecular mechanisms underlying the reception and transmission of sour taste information. *Biosci. Biotechnol. Biochem.* **79**, 171–176, 2015.
35. Rasaiah, J. C., Garde, S., and Hummer, G. Water in nonpolar confinement: From nanotubes to proteins and beyond. *Annu. Rev. Phys. Chem.* **59**, 713–740, 2008.

36. Beckstein, O. and Sansom, M. S. Liquid-vapor oscillations of water in hydrophobic nanopores. *Proc. Natl. Acad. Sci. USA*. **100**, 7063–7068, 2003.
37. Hummer, G., Rasaiah, J. C., and Noworyta, J. P. Water conduction through the hydrophobic channel of a carbon nanotube. *Nature*. **414**, 188–190, 2001.
38. Beckstein, O. and Sansom, M. S. The influence of geometry, surface character, and flexibility on the permeation of ions and water through biological pores. *Phys. Biol.* **1**, 42–52, 2004.
39. Aryal, P., Sansom, M. S., and Tucker, S. J. Hydrophobic gating in ion channels. *J. Mol. Biol.* **427**, 121–130, 2015.
40. Wang, W. et al. The structure of an open form of an E. coli mechanosensitive channel at 3.45 A resolution. *Science*. **321**, 1179–1183, 2008.
41. Bass, R. B., Strop, P., Barclay, M., and Rees, D. C. Crystal structure of Escherichia coli MscS, a voltage-modulated and mechanosensitive channel. *Science*. **298**, 1582–1587, 2002.
42. Aryal, P., bd-Wahab, F., Bucci, G., Sansom, M. S., and Tucker, S. J. A hydrophobic barrier deep within the inner pore of the TWIK-1 K2P potassium channel. *Nat. Commun.* **5**, 4377, 2014.
43. Yoshimura, K., Batiza, A., Schroeder, M., Blount, P., and Kung, C. Hydrophilicity of a single residue within MscL correlates with increased channel mechanosensitivity. *Biophys. J.* **77**, 1960–1972, 1999.
44. Ou, X., Blount, P., Hoffman, R. J., and Kung, C. One face of a transmembrane helix is crucial in mechanosensitive channel gating. *Proc. Natl. Acad. Sci. USA*. **95**, 11471–11475, 1998.
45. Zheng, W. et al. Hydrophobic pore gates regulate ion permeation in polycystic kidney disease 2 and 2L1 channels. *Nat. Commun.* **9**, 2302, 2018.
46. Zheng, W. et al. Identification and characterization of hydrophobic gate residues in TRP channels. *FASEB J.* **32**, 639–653, 2018.
47. Zheng, W. et al. A novel PKD2L1 C-terminal domain critical for trimerization and channel function. *Sci. Rep.* **5**, 9460, 2015.
48. Zheng, W. et al. Regulation of TRPP3 channel function by N-terminal domain palmitoylation and phosphorylation. *J. Biol. Chem.* **291**, 25678–25691, 2016.
49. Zheng, W. et al. Direct binding between Pre-S1 and TRP-like domains in TRPP channels mediates gating and functional regulation by PIP2. *Cell Rep.* **22**, 1560–1573, 2018.
50. Jo, S., Lim, J. B., Klauda, J. B., and Im, W. CHARMM-GUI membrane builder for mixed bilayers and its application to yeast membranes. *Biophys. J.* **97**, 50–58, 2009.
51. Wang, L., Holmes, R. P., and Peng, J. B. Molecular modeling of the structural and dynamical changes in calcium channel TRPV5 induced by the African-specific A563T variation. *Biochemistry*. **55**, 1254–1264, 2016.
52. Cai, Y. et al. Calcium dependence of polycystin-2 channel activity is modulated by phosphorylation at Ser812. *J. Biol. Chem.* **279**, 19987–19995, 2004.

6 Analyzing the GPCR Function of Polycystin-1

Stephen C. Parnell, Robin L. Maser, Brenda S. Magenheimer, and James P. Calvet

CONTENTS

6.1 INTRODUCTION

Cyst growth in autosomal dominant polycystic kidney disease (ADPKD), caused by inherited mutations in either *PKD1* or *PKD2*, leads to massive kidney enlargement and ultimately to renal failure. Clinical ascertainment and mutation screening have shown that mutation of either gene results in essentially the same disease, although ADPKD arising from *PKD1* mutations (~73%–85%) is more prevalent than *PKD2* disease (~15%–27%) and leads to earlier end-stage renal disease.[1–4] *PKD1* and *PKD2* encode the transmembrane proteins polycystin-1 (PC1) and polycystin-2 (PC2). PC1 is a large (4303 aa) protein with a >3000 aa N-terminal extracellular region, 11 membrane-spanning segments, and a smaller C-terminal cytosolic domain of about 200 aa.[3,5] Initial analysis of the PC1 protein sequence predicted structural features that were consistent with it being a membrane receptor, and more recent advances have revealed that PC1 is related to the adhesion-class of G protein-coupled receptors (GPCRs).[3,6] PC2 (TRPP2) is a member of the transient receptor potential (TRP) family of membrane channels, has nonspecific cation conductance and acts as a Ca^{2+}-regulated channel.[7–9] PC2 is also an endoplasmic reticulum Ca^{2+} release channel which functions in an IP3 receptor-dependent fashion.[10] Early experiments demonstrated that the C-terminal cytosolic regions (C-tails) of PC1 and PC2 directly interact via coiled-coil domains.[11,12] Observations that PC1 and PC2 are coexpressed in many embryonic and adult tissues and organs and the fact that the same disease phenotype results from mutations in either gene led to the interpretation that they likely function together in most tissues. Recent evidence suggests that PC1 and PC2 adopt a 1:3 stoichiometry to form a heteromeric cation channel.[13–15]

The predicted structures of PC1 and PC2 led to early suggestions that they form a membrane receptor-ion channel complex.[13,14] Supporting evidence came from studies in which coexpression of PC1 and PC2 in transfected cells was shown to generate a Ca^{2+} signal[16] and where ciliary PC1-PC2 mediated fluid-flow stimulated transient elevations in intracellular Ca^{2+} in a ryanodine receptor-dependent fashion.[17] Such observations led to the concept that the polycystin complex responds to ligand-mediated[18] or mechanosensory [17] stimuli to regulate channel activity and initiate signal transduction. Early work from our group was the first to describe the PC1 protein as an atypical GPCR.[19,20] Expression of the C-terminal cytosolic tail of PC1 was shown to stimulate a number of signaling pathways in transfected cells, leading to the activation of promoter-reporters such as AP-1 and NFAT.[20–23] A membrane-proximal region of the PC1 C-tail was found to contain a heterotrimeric G-protein binding and activation domain[19] that initiates signaling by activating Gi/o, Gq/11 and

G12/13.[20] PC1 has been shown to activate Gi/o and release Gβγ subunits to modulate ion channel activity in neuronal cells.[24] As such, while it appears that PC1-PC2 is a signaling-responsive Ca^{2+} channel, the biochemical and cellular mechanisms of the complex are not well understood.[25–27]

A number of other observations have suggested that heterotrimeric G proteins play a role in polycystin-1-mediated signaling, as reviewed by Hama and Park.[28] Polycystin-1 has been shown to bind and stabilize RGS7 (regulator of G protein signaling 7),[29] a member of a family of proteins that control G-protein-dependent signaling by accelerating the hydrolysis of GTP bound to Gα subunits of certain heterotrimeric G proteins. RGS7 has also been identified as a possible genetic modifier of the bpk allele, which causes PKD in a mouse model that is similar to human autosomal recessive PKD.[30] G-protein accessory proteins have been shown to have a role in regulating these processes as well, including the activator of G-protein signaling (AGS) proteins, specifically AGS3 or G-protein signaling modulator-1 (GPSM1), which was shown to activate Gβγ-dependent PC1-PC2 channel activity.[31] The PC1 C-tail has also been shown to interact with Gα12,[32,33] suggesting a mechanism by which the levels of PC1 can regulate Gα12/JNK-activity, leading to a direct involvement of Gα12 in the development of kidney cysts in *Pkd1*-deficient mice.[34] In *Xenopus* pronephric development, it has been shown that PC1 recruits GαS, and that GαS knockdown causes a PKD phenotype by abnormally increasing the levels of Gβγ subunits.[35]

Many lines of evidence suggest that an essential function of PC1 is to regulate signaling via heterotrimeric G proteins. As such, this chapter will review the methods and procedures that have been used by our group to demonstrate a G-protein signaling role for PC1. Detailed methods from several articles[19–21,36] that were used to establish these principles are presented below so that the reader can extend these studies to facilitate future research to explore the regulation of G-protein activation and the downstream effectors of PC1-activated G-protein subunits.

6.2 BINDING AND ACTIVATION OF HETEROTRIMERIC G-PROTEINS

The potential role of PC1 to act as a ligand-activated membrane receptor led to the suggestion that the C-terminal cytosolic domain may directly interact with cellular signaling proteins such as heterotrimeric G proteins. To test whether or not the PC1 C-tail might be involved in G-protein signaling, *in vitro* protein binding assays (GST pull-down and co-immunoprecipitation) can be employed in which bacterially expressed PC1 fusion constructs carrying different regions of the C-terminal cytosolic domain are used to determine if they stably bind G proteins. These experiments have shown that the membrane-proximal half of the C-terminal cytosolic domain of PC1 contains a binding domain for heterotrimeric G proteins and that a 20-amino acid sequence contained within this binding region can activate purified G proteins *in vitro*. These were the first experiments to demonstrate that PC1 might function by directly activating heterotrimeric G-protein signal transduction.[19]

6.2.1 G-Protein Activation Peptide

Inspection of the C-terminal cytosolic domain of PC1 reveals a highly conserved 20-amino acid, G-protein activation peptide (RRLRLWMGFSKVKEFRHKVR) conforming to the consensus motif BB...BBxB or BB...BBxxB (B = R, K, or H) present in a number of single- and multi-spanning G-protein-coupled receptors.[37] Evidence indicated that short peptides possessing these characteristics can stimulate guanine nucleotide exchange by G proteins. To determine if this highly basic 20-amino acid PC1 sequence has this capability, the peptide sequence was tested in an *in vitro* G-protein activation assay.[19]

6.2.1.1 GTPase Assay

1. Synthesize the 20-aa, G-protein activation peptide using Fmoc chemistry.
2. Cleave the peptide from the column, deprotect, then precipitate and wash twice using cold ether, and lyophilize.
3. Analyze the 20-aa peptide by C18 reverse-phase HPLC; identify the HPLC peak containing the peptide by amino acid analysis and mass spectrometry, and purify it by C18 reverse-phase HPLC.
4. Assay by steady-state GTP exchange and hydrolysis as described previously.[38,39] The assay uses 10 nM bovine brain heterotrimeric G proteins incubated for 20 min at 37°C in the presence or absence of 20 μM Mg^{2+}.
5. As a control, create Mg^{2+}-free conditions by adding 10 mM EDTA to the reaction.

The 20-aa peptide can stimulate an ~3-fold increase in GTP exchange and subsequent GTP hydrolysis as assayed by release of Pi, using purified bovine brain heterotrimeric Gi/Go. The stimulation is concentration dependent, with an EC50 at a peptide concentration of 0.2 μM and a peak activity at 0.3 μM. Control peptides show very little or no stimulation.[37–39] Measuring GTPase activity in the presence and absence of 10 mM EDTA demonstrates that activation is Mg^{2+} dependent. Under conditions of peak peptide concentration, GTPase activity is stimulated in the presence of Mg^{2+}, but not in the absence of Mg^{2+}. Thus, in these experiments it is shown that PC1 contains a C-tail peptide sequence capable of activating nucleotide exchange on Gα subunits in a receptor-like manner.

6.2.2 Binding of G Proteins to the C-Terminal Cytosolic Domain of Polycystin-1

To test the binding of heterotrimeric G proteins to the C-terminal cytosolic domain of polycystin-1, GST-fusion proteins containing various portions of the cytosolic domain are constructed for bacterial expression and purification. The methods are outlined below and are from Parnell et al.[19] The polycystin-1 portions of these fusion constructs are shown in Figure 6.1 and in the work in Parnell et al.[19] The longest is comprised of 222 amino acids, starting with leucine (L) and ending with a C-terminal threonine (T), hence PC1-LT or PC1-LT222. Others are N-terminally truncated or both N- and C-terminally truncated.

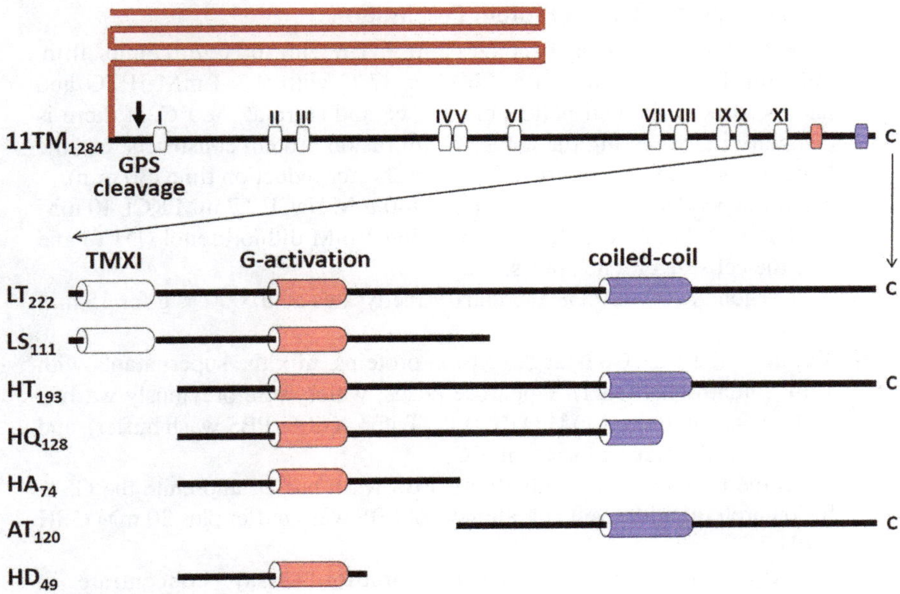

FIGURE 6.1 Polycystin-1 C-terminal fragment constructs. (*Upper*) The PC1–11TM1284 construct is from the 3′ region of a mouse Pkd1 cDNA containing all 11 transmembrane domains (I–XI) and the C-tail (aa 3010–4293). The N-terminal portion of PC1 is shown in red. The PC1–11TM1284 construct includes the GPS cleavage site. (*Lower*) The PC1 C-tail constructs include the C-terminal 222 aa of PC1 (PC1-LT222), a truncated version (PC1-LS111), the C-terminal 193 aa of PC1 (PC1-HT193), a truncated version (PC1-HQ128), a further truncated version (PC1-HA74), the C-terminal 120 aa of PC1 (PC1-AT120), and a membrane-proximal construct of 49 aa (PC1-HD49). The C-tail constructs showing the last PC1 transmembrane domain (TMXI), the 20 aa G-protein activation peptide (G-activation), and the coiled-coil region are drawn to approximate scale. The C-terminal 120 aa of PC1 (PC1-AT120) lacks the heterotrimeric G-protein activation domain but contains the coiled-coil region. It does not bind heterotrimeric G proteins. The membrane-proximal PC1-LS111 and PC1-HA74 constructs contain the heterotrimeric G-protein activation domain but lack the coiled-coil region. These latter two constructs are active in heterotrimeric G-protein signaling. The PC1-HD49 construct is too short to stably bind G proteins. The constructs are subcloned downstream of GST for bacterial expression or downstream of sIg (CD5 signal peptide, CH2-CH3 domains of human IgG, and the CD7 transmembrane domain) for mammalian expression. (Adapted from Parnell SC et al. 1998. *Biochem Biophys Res Commun* 251(2):625–631; Parnell SC et al. 2002. *J Biol Chem* 277(22):19566–19572. Puri S et al. 2004. *J Biol Chem* 279(53):55455–55464.)

6.2.2.1 Polycystin-1 Fusion Protein and Control Constructs

Mouse PC1 GST-fusion proteins are generated by cloning fragments of a cDNA representing the 3′ end of Pkd1 mRNA into pGEX-4T-1 for expression in *E. coli* (DH5α or BL21). A control C-terminal polycystin-2 GST-fusion protein representing the last 280 amino acids of polycystin-2 is made by cloning an RT-PCR product, using HeLa-cell RNA. GST is produced from the pGEX-4T-1 vector lacking Pkd1 DNA.

6.2.2.2 Polycystin-1 Fusion Protein Preparation

1. Grow bacteria-expressing fusion proteins in LB with 100 μg/mL ampicillin, inducing for 30 min to 4 h at 20°C or 37°C with 0.5–1 mM IPTG and harvest; freeze the cell pellets on dry ice and store at −80°C. If there is difficulty in expressing the longer C-tail fusion protein constructs, use the lower induction temperature (20°C) and shorter induction time (30 min).
2. Thaw the pellets at 4°C in 50 mL PBS (140 mM NaCl, 2.7 mM KCl, 10 mM Na_2HPO_4, 1.8 mM KH_2HPO_4) containing 1 mM dithiothreitol (DTT) and lyse the cells in a French press.
3. Add Triton X-100 to 0.1% and clarify the lysates at 10,000 × g for 15 min at 4°C.
4. To purify the PC1 GST-tagged fusion proteins, mix the supernatants with 1 mL glutathione (GSH)-Sepharose beads, which were previously washed with PBS containing 1 mM DTT, 0.1% Triton X-100 (PBS wash buffer), and incubate with rotation for 4 h at 4°C.
5. Wash the beads 4 times with 10 mL PBS wash buffer, and elute the GST-fusion proteins with five 1 mL aliquots of PBS wash buffer plus 20 mM GSH at pH 7.4.
6. Identify peak fractions using the Bradford assay, concentrate in Microcon-30 kDa microconcentrators, and store at −20°C.

6.2.2.3 G-Protein Preparation

1. Purify bovine brain heterotrimeric G protein as previously described.[38,40]
2. Prepare rat brain lysates[41] by isolating 10 frozen rat brains (11.4 μg) and thawing them in 20 mL TED (20 mM Tris, pH 8, 1 mM EDTA, 1 mM DTT, 25 KIU aprotinin) containing 3 mM $MgCl_2$ and 10% sucrose.
3. After thawing, add an equal volume of the same buffer and homogenize the brains in a Polytron for ~1 min at 4°C.
4. Add an additional 160 mL of TED plus 3 mM $MgCl_2$ and centrifuge the homogenate at 6400 × g for 30 min at 4°C.
5. Recover and disperse the pellet with a glass rod in 67 mL of TED and centrifuge at 12,900 × g for 1 h at 4°C.
6. Disperse the resulting pellet in 67 mL TED plus 0.1 M NaCl, 0.1% sodium cholate, and centrifuge at 12,900 × g for 1 h at 4°C.
7. Disperse the pellet in 20 mL TED and stir on ice while adding an additional 20 mL TED plus 2% sodium cholate over a 10-min period.
8. After stirring for 1 h, centrifuge the lysate at 143,000 × g for 1 h at 4°C and recover the clear supernatant, divide into 1 mL aliquots, and freeze at −80°C.

6.2.2.4 GST Pull-Down Assays

1. For pull-down assays,[39] prepare GSH-Sepharose beads by washing three times in PBS containing 1 mM DTT and three times in CHAPS buffer (50 mM HEPES, pH 7.4, 1 mM EDTA, 1 mM DTT, 0.5% CHAPS).
2. Bind equimolar amounts of full-length GST or GST-fusion proteins to 15–20 μL beads by incubating with intermittent agitation for 30 min at 4°C.

3. Mix 30–80 μL purified bovine brain heterotrimeric Gi/Go or 100 μL rat brain lysate with the protein-bound beads in a total volume of 0.5–1.5 mL CHAPS buffer and rotate overnight at 4°C.
4. Wash the beads 6 times in CHAPS buffer plus 120 mM NaCl, mix with Laemmli loading buffer, boil for 3 min, load on a 4%/12.5% SDS-polyacrylamide gel, and electrophorese.
5. For Western blotting (see below), equilibrate gels for 30 min with transfer buffer (25 mM Tris, 192 mM glycine, 20% methanol) plus 0.01% SDS and transfer to nitrocellulose for 1 h at 60 V.

6.2.2.5 Co-Immunoprecipitation Assays

Polycystin-1 antibody

1. Synthesize an 11-amino acid peptide (RVSLWPNNKVH) from the C-terminus of mouse polycystin-1 as a multiple antigen peptide (peptide 18).
2. Produce rabbit antiserum (Cocalico Biologicals) and obtain PC1-specific antibodies (A18) by affinity purification[42] using the PC1-LT222 GST-fusion protein cross-linked to GSH-Sepharose beads.
3. Elute bound antibody from the beads with 0.1 M glycine, pH 2.5.
4. Neutralize the solution with 2 M Tris base, and determine the concentration by A280.
5. Add BSA to a final concentration of 1 mg/mL; aliquot the purified antibody (A18) and store at −80°C.

Co-Immunoprecipitation

1. For co-immunoprecipitation, mix the GST-fusion protein and G proteins at a molar ratio of 1:4 (100 ng or 250 ng of the PC1-HT193 GST-fusion protein and 325 ng or 1.3 μg purified Gi/Go in a 10 or 25 μL reaction volume) in binding buffer (5 mM HEPES, pH 7.4, 0.1 mM EDTA, 0.1 mM DTT, 0.05% CHAPS, 2 μg/mL aprotinin, 2 μM leupeptin, 0.2 mM PMSF) for 1 h at room temperature.
2. Then combine the proteins with 2× immunoprecipitation buffer (300 mM NaCl, 20 mM Tris, pH 7.4, 2 mM EDTA, 2 mM EGTA, 0.4 mM PMSF, 2% Triton X-100, 1% NP-40), 25 or 50 μL of purified A18 antibody, and water in a final volume of 0.5 mL, and incubate the mixture overnight at 4°C.
3. To recover the immune complexes, add 30–40 μL of anti-rabbit IgG-agarose and incubate the mixtures with rotation for 30 min at 4°C.
4. Centrifuge the mixtures in a microcentrifuge for 4 min at 4°C, and wash the pellets 3 times in immunoprecipitation buffer for 10 min each time.
5. Resuspend the pellets in electrophoresis sample buffer, boil for 5 min, and centrifuge for 5 min.
6. Electrophorese the supernatants and transfer the proteins to Immobilon-P PVDF membranes (Millipore) in transfer buffer plus 0.05% SDS at 50 V for 30 min followed by 100 V for 1 h.

6.2.2.6 Western Blots with Anti-G-Protein Antibodies

1. Block the membranes in TBS/T buffer (10 mM Tris, pH 7.5, 0.9% NaCl, 0.1% Tween-20) plus 5% nonfat dry milk, rinse 3 times in TBS/T, and incubate with primary antibody (rabbit anti-Giα1/Giα2, rabbit anti-Giα3/Goα, or rabbit anti-Gβ at 1:1,000; Calbiochem) in TBS/T for 1.5 h.
2. After washing 3 times for 5 min in TBS/T, incubate the blots with AP-conjugated goat anti-rabbit IgG (Sigma) at 1:10,000 for 30 min, wash 4 times for 10 min in TBS/T, equilibrate 5 min in AP buffer (100 mM Tris, pH 9.5, 100 mM NaCl, 5 mM $MgCl_2$), and image.

6.2.2.7 Western Blots with Anti-GST Antibody

1. Block the membranes in PBS/T buffer (PBS, 0.3% Tween-20) plus 10% nonfat dry milk and incubate with primary antibody (goat anti-GST at 1:1,000; Pharmacia Biotech) in PBS/T plus 10% milk for 1 h.
2. After rinsing twice with PBS/T and washing twice for 10 min with PBS/T, incubate the blots with AP-conjugated rabbit anti-goat IgG at 1:30,000 in PBS/T plus 10% milk for 1 h, rinse twice with PBS/T, wash twice for 10 min with PBS/T, equilibrate 5 min in AP buffer, and image.

6.2.2.8 Western Blots Following Co-Immunoprecipitation

1. Block nonspecific protein binding with 2% normal goat serum in TBS/T buffer, and incubate the blots with a mixture of the rabbit anti-Giα1/Giα2 and rabbit anti-Giα3/Goα at 1:1000 in TBS/T plus 2% goat serum for 2 h.
2. After washing in TBS/T with 0.05% NP-40, incubate the blots with AP-conjugated goat anti-rabbit IgG at 1:30,000 for 1.5 h, wash, and image.

6.2.3 The Minimal G-Protein Binding Domain

As shown in Parnell et al.,[19] the full-length PC1-LT222 fusion protein can bind and pull-down Gα subunits as detected with anti-Gα antibodies specific to Giα1/Giα2 and Giα3/Goα. Neither GST alone nor a PC2 GST-fusion protein containing the C-terminal cytosolic domain of polycystin-2 binds G proteins. Since bovine brain or rat brain heterotrimeric G proteins were used for the pull-down assay, there would have been Gβγ subunits present, which should also be in the PC1-bound material. As expected, PC1 was able to pull down Gβγ subunits in contrast to GST alone or polycystin-2, consistent with PC1 binding to heterotrimers.[19]

A number of truncated C-terminal constructs[19] were made and tested for binding activity with purified bovine brain heterotrimeric G proteins. The highest amount of binding is seen with the full-length C-tail PC1-LT222 construct. Removal of 94 amino acids from the N- and C-termini of LT222, the latter truncating the coiled-coil region, results in somewhat decreased binding for the HQ128 construct. Removal of only 29 amino acids from the N-terminus of LT222 also leads to somewhat decreased binding, as seen for the HT193 construct. Further C-terminal truncation, which completely removes the coiled-coil region giving the HA74 protein, further reduces binding. Removal of an additional 25 amino acids (HD49) virtually eliminates all stable binding.

The coiled-coil region in the distal portion of the C-tail interacts with PC2 and potentially other proteins. To determine if this region is capable of binding G proteins, an AT120 construct was made and tested. As was shown in Parnell et al.,[19] this region of the C-tail domain was not effective in stably binding G proteins. The ability of the C-terminal cytosolic domain of PC1 to stably bind G proteins was confirmed by co-immunoprecipitation using the PC1-HT193 GST-fusion protein and an affinity-purified anti-polycystin-1 antibody, A18, which were used to co-immunoprecipitate bovine brain G proteins, as assayed by blotting with a mixture of anti-Gα antibodies. Together these experiments suggest that the proximal C-tail region binds heterotrimeric G proteins, and that there is a minimal G-protein binding region of 74 amino acids (HA74) which contains a 20-aa G-protein activation peptide.

6.3 HETEROTRIMERIC G-PROTEIN SIGNALING TO JNK AND AP-1

The C-terminal cytosolic C-tail of PC1 has been shown to be involved in protein interactions and initiating signal transduction. The first biochemical analyses supporting this idea utilized transient transfection of the C-tail of PC1, which showed that polycystin-1 modulates Wnt signaling via stabilization of β-catenin[18,29] and activates the transcription factor AP-1 via c-Jun N-terminal kinase (JNK) and protein kinase C.[23]

To determine whether heterotrimeric G proteins mediate this PC1-initiated signaling, the effects of expression of the Gβγ sequestering constructs, dominant-negative (DN) Gαi2 and β-adrenergic receptor kinase C-terminal tail (βARK-ct), were tested. The evidence has demonstrated that PC1 activates JNK signaling via Gβγ subunits. It has also been shown that DN p115RhoGEF inhibits AP-1 activation and that wildtype (WT) Gα subunits effectively augment polycystin-1 signaling to AP-1. The following methods describe the protocols that were used to demonstrate these findings. Further rationale can be found in Parnell et al.[20]

6.3.1 DNA Constructs

1. The sIg-PKD-MN6 (MN6) construct contains the C-terminal cytosolic tail of human PC1 (amino acids 4077–4241) fused to the membrane targeting cassette sIg.7 (obtained from Dr. G. Walz[23]). The sIg.7 cassette consists of the CD5 signal peptide, the CH2-CH3 domains of human IgG, and the CD7 transmembrane domain. MN6 terminates in the coiled-coil region of polycystin-1. For a control, insert sIg.7 alone in pcDM12 (a derivative of pcDM8; Addgene) (sIg-0–12; control for sIg-PKD-MN6).
2. Make the sIg-PKD222 construct from a mouse Pkd1 cDNA (in our case, obtained from Dr. G. Germino), which can be subcloned downstream of sIg.7 in the Invitrogen vector pcDNA1.1/Amp such that it encodes the C-terminal 222 amino acids of polycystin-1.[19] For a control, insert sIg.7 alone in pcDNA1.1/Amp (sIg-0–1.1; control for sIg-PKD222).
3. Make the sIg-PKD1284 polycystin-1 fusion construct from the 3′ portion of a mouse Pkd1 cDNA clone[43] encoding amino acid residues 3010–4293,

which can be subcloned downstream of the CD5 signal sequence and the CH2-CH3 IgG domains in pcDNA1.1/Amp. For a control, introduce a stop codon in place of PC1 amino acid 3092 to create the construct sIg-stop.
4. The HA-JNK1β construct; βARKct and vector control pRK5; DN and WT Gαi2; GST-Jun-(1–79); Myc-tagged DN p115RhoGEF (Lsc-RGS, amino acids 1–283[44]) in the Stratagene vector pCMV3-Tag3 and out-of-frame control plasmid are described in Parnell et al.[20]
5. EE-tagged Gαi1, Gαi2, Gαi3, Gαq, Gα12, and Gα13 in pcDNA3.1 can be obtained from the Guthrie cDNA Resource Center (https://www.cdna.org/files/catalog.pdf).
6. pFR-Luc, pFA2-cJun, pAP-1-Luc, pFC-MEKK, and pBlueScript (pBS) can be obtained from Agilent, and the pRL-null vector from Promega.

6.3.2 Exogenous JNK Immune Complex Kinase Assay: Effects of Gβγ Sequestering

The following experiments examine the effects of PC1 signaling on activation of the JNK pathway[23] by assaying activation of transfected HA-JNK1β. Activation of HA-JNK1β is assayed by its ability to phosphorylate a substrate peptide GST-Jun-(1–79) *in vitro*. Because JNK is known to be activated by Gβγ subunits,[45,46] HEK293T cells are cotransfected with cDNAs encoding a C-terminal PC1 fusion protein, together with HA-JNK1β, and a Gβγ-sequestering construct, either the β-adrenergic receptor kinase C-terminal tail (βARK-ct),[47] DN Gαi2, or WT Gαi2. Inhibition of HA-JNK1β activity by a Gβγ-sequestering construct provides evidence that JNK activation is mediated by heterotrimeric G-protein activation.

1. Maintain human embryonic kidney (HEK) 293T cells (ATCC) in DMEM (1x MOD) with L-glutamine and 1 g/L glucose (Corning Cellgro) with 500 U/L penicillin and 0.5 mg/L streptomycin (Sigma) and 10% fetal calf serum at 37°C in 5% CO_2. Plate cells at 4×10^5 cells per T25 flask and grow for 2 days prior to transfection.
2. Cotransfect cells for 16 h using a modified calcium phosphate protocol as described previously.[48] Transfect a PC1 construct (2 μg human sIg-PKD-MN6 C-tail construct or 3 μg mouse 11-TM sIg-PKD1284 construct, or a corresponding amount of their respective controls lacking PC1 sequence, sIg-0–12 or sIg-stop) plus 2 μg HA-JNK1β reporter construct and a Gβγ sequestering construct, βARK-ct, DN Gαi2, or WT Gαi2 as follows:
   i. sIg-PKD-MN6 or sIg-0–12 plus 0, 1, 2, or 3 μg βARK-ct brought to 3 μg with pRK5 for a total of 7 μg DNA.
   ii. sIg-PKD-MN6 or sIg-0–12 plus 0, 2, 4, or 6 μg DN Gαi2 or WT Gαi2 brought to 6 μg with pRK5 for a total of 10 μg DNA.
   iii. sIg-PKD1284 or sIg-stop plus 0, 1, 3, or 5 μg βARK-ct brought to 5 μg with pRK5 for a total of 10 μg DNA.
   iv. sIg-PKD1284 or sIg-stop plus 0, 2, 4, or 6 μg DN Gαi2 or WT Gαi2 brought 6 μg with pRK5 for a total of 11 μg DNA.

3. Following the addition of DNA precipitates, incubate cells at 37°C in 5% CO_2 for 4 h, then replace the medium with serum-free growth medium or with medium containing 0.5% serum.
4. To test the involvement of the Gi family, pertussis toxin (200 ng/mL) can be added to cells either 1 h before or 4 h after adding the DNA precipitate.
5. At 16 h after replacement of the medium, wash cells in PBS and scrape into 0.5 mL per T25 flask of Tris/Triton lysis buffer (TLB) (20 mM Tris, pH 7.4, 137 mM NaCl, 25 mM β-glycerol phosphate, 2 mM EDTA, 1 mM Na_3VO_4, 2 mM $Na_2P_2O_7$, 1% Triton X-100, 10% glycerol) plus 1 mM phenylmethylsulfonyl fluoride, 5 μg/mL leupeptin, 5 μg/mL aprotinin, 2 mM benzamidine, and 0.5 mM DTT (TLB+).
6. Vortex scraped cells for 20 s in lysis buffer at 4°C and incubate on ice for 20 min, and then clarify the lysates by centrifugation (14,000 × g) for 10 min. Place supernatants at −80°C until use. Determine protein concentrations using the Bio-Rad Detergent Compatible Assay Kit or a comparable kit.
7. Wash protein A/G+-agarose beads (50 μL per reaction; Santa Cruz Biotechnology) twice in TLB and once in TLB+ and suspend in 0.5 mL of TLB+ per 6 reactions plus 0.4 μg per reaction of HA-probe Antibody (F-7) (Santa Cruz Biotechnology). Use equal concentrations of protein for the assays.
8. For assaying HA-JNK1β kinase activity, rotate the beads at 4°C for 45 min and spin down briefly in a microcentrifuge; decant the supernatant, and resuspend the beads in aliquots of 100 μL per reaction. Add 200 μg of cellular lysate and TLB+ to a final volume of 0.6 mL; rotate the reactions at 4°C for 2 h and then wash 3 times in TLB and twice in kinase buffer (25 mM HEPES, pH 7.4, 25 mM β-glycerol phosphate, 25 mM $MgCl_2$, 0.1 mM Na_3VO_4, 0.5 mM DTT).
9. Resuspend the beads in 31 μL of kinase buffer containing 1 mM ATP, 8 μg of GST-Jun-(1–79) substrate, and [γ-^{32}P] ATP (3000 Ci/mmol; PerkinElmer Life Sciences) and incubate in a room temperature water bath for 5 min. Terminate reactions by adding 20 μL of 2x Laemmli loading buffer and place on dry ice until loading.
10. Boil reactions for 5 min and then load and electrophorese; detect fusion protein as described previously[49]; ^{32}P incorporation is quantified using a Molecular Dynamics PhosphorImager SI.
11. Data are expressed as fold stimulation relative to the value for sIg-0–12 or sIg-stop without inhibitors, which are set at one.

6.3.3 Endogenous JNK Activation Assay: Effects of Gβγ Sequestering

Activation of JNK can also be assayed by examining the activity of a transfected promoter-reporter construct that responds directly to endogenous JNK activation of c-Jun: the c-Jun activation domain/GAL4 DNA binding domain construct (pFA2-cJun) and the luciferase reporter gene under the control of a GAL4 promoter (pFR-Luc).

1. Plate HEK293T cells at ~7.5×10^5 cells per well on a 6-well plate, grow for 1 day, and transfect for 24–26 h.
2. To determine the level of activation of endogenous JNK, cotransfect cells with the c-Jun activation domain/GAL4 DNA binding domain construct (pFA2-cJun) and the luciferase reporter gene under the control of a GAL4 promoter (pFR-Luc) together with sIg-PKD222 or sIg-0–12 and the Gβγ inhibitors βARK-ct, DN Gαi2, WT Gαi2, or the pRK5 vector alone.
3. Transfect each 3-well triplicate sample with a total of 7 μg of DNA, which includes 2 μg of the Gβγ inhibitor constructs or pRK5, 2 μg of sIg-PKD222 or sIg-0–12, 1 μg of pFR-Luc, 50 ng of pFA2-cJun, 10 ng of pRL-null, and pBS as the filler DNA.
4. Following transfection, lyse cells and determine JNK activation by assaying firefly luciferase activity using the PathDetect c-Jun Trans-Reporting System (Agilent) by following the manufacturer's instructions; normalize the values to Renilla luciferase activity. It is important to normalize values with a control construct that does not respond to PC1, such as a promoterless pCIS-CK negative control plasmid which should show no activity. Data are expressed as relative luciferase units (RLUs).

6.3.4 Endogenous AP-1 Activation Assay: Augmentation and Inhibition

Activation of JNK can also be assayed by examining the activity of a transfected AP-1 promoter-reporter construct that responds to endogenous JNK activation of c-Jun, which in turn activates the AP-1 promoter. Activation of endogenous AP-1 is assayed with the PathDetect Cis-Reporting System, which utilizes a 7x AP-1 reporter.

1. Plate HEK293T cells 24 h prior to transfection at ~7.5×10^5 cells per well in 6-well plates.
2. *For Gα augmentation*: A total of 7 μg of DNA is used for calcium phosphate precipitation, with 3 μg of PC1 or control construct, 1 μg of AP-1 reporter, 5 ng of RL-null, 100 or 500 ng of Gα construct, and pBS as the filler. The following Gα constructs are used: Gαi1, Gαi2, Gαi3, Gαq, Gα12, Gα13.
3. *For Rho inhibition*: A total of 7 μg of DNA is used for calcium phosphate precipitation, with 3 μg of control or PC1 construct, 1 μg of reporter, 5 ng of RL-null, 1 μg of DN p115RhoGEF (DN p115) or control plasmid, and pBS as the filler.
4. Change the medium, lacking serum, 3.5–4 h after transfection is begun, and incubate cells for 24–26 h longer.
5. Lyse the cells in Passive Lysis Buffer (Promega) and assay 20 μL of lysate with the Dual-Luciferase Reporter Assay System (Promega) on a luminometer.
6. Confirm expression levels of the Gα constructs and p115 by Western blotting.

6.3.5 Western Blotting

1. Antibodies for βARK-ct (anti-GRK2), Gαi3, Gαq, Gα12, Gα13, p115RhoGEF (anti-9E10 Myc tag), and JNK (anti-JNK1) are available from Santa Cruz Biotechnology.

2. Antibodies for $G\alpha i1/G\alpha i2$ are available from Calbiochem; and EE-tagged $G\alpha$ subunits from Babco.
3. Perform Western blots with Immobilon-P membranes (Millipore) according to the specifications for the antibodies; perform anti-human IgG Western blots as described previously.[19]

6.3.6 Summary of Signaling to JNK and AP-1

These experiments show that PC1 activates JNK and AP-1 signaling via $G\alpha$ and $G\beta\gamma$ subunits of heterotrimeric G proteins and that the C-tail can couple with Gi, Gq, and G12 family proteins. Polycystin-1 contains sequence elements and structural features found in GPCRs belonging to the subfamily of large orphan receptors of the adhesion GPCR class.[50] Polycystin-1 has an unusual structure for a GPCR, as it has 11 transmembrane domains[5] rather than 7 and, thus, would have to be classified as an atypical GPCR. GPCRs can signal through several pathways that lead to the activation of JNK and AP-1.[45,51] Pathways that activate Rac and/or Cdc42 and JNK via heterotrimeric G proteins have been described, including those activated by $G\alpha 12$ and $G\alpha 13$ subunits,[52,53] $G\beta\gamma$ subunits,[45,46] $G\alpha i$ subunits,[54] and Gq.[55] In addition, AP-1 can be activated by JNK-independent pathways involving Gq and G12/G13 activation of Rho, the p38 MAPKs, and BMK1/ERK5.[56,57] Work by others[23] showed that the PC1 C-terminal cytosolic domain can activate JNK via the small G proteins Rac-1 and Cdc42. Our evidence showed that PC1-mediated signaling to JNK and AP-1 is regulated by heterotrimeric G-protein subunits, possibly through several pathways.[20]

6.4 HETEROTRIMERIC G-PROTEIN SIGNALING TO NFAT

Regulation of intracellular Ca^{2+} mobilization has been associated with the functions of PC1 and PC2. We have demonstrated that PC1 can activate the calcineurin/NFAT (nuclear factor of activated T cells) signaling pathway through $G\alpha q$-mediated activation of phospholipase C (PLC). We determined that PC1 activates a pathway that leads to calcineurin activation and translocation of dephosphorylated NFAT to the nucleus. The following methods were used to demonstrate the role of the PC1 C-tail in activating G-protein signaling to PLC, inositol trisphosphate and ryanodine receptors, Ca^{2+} and calcineurin, and to the dephosphorylation of NFAT and its nuclear translocation and stimulation of an NFAT promoter.

6.4.1 Transfection Analyses for Examining G-Protein Activation

6.4.1.1 PC1 Activation of NFAT by $G\alpha q$

To identify potential signaling pathways from polycystin-1 that can modulate intracellular Ca^{2+} mechanisms, we coexpressed various PC1 C-tail deletion constructs with an NFAT promoter-luciferase reporter construct having four composite NFAT/AP-1 binding sites from the human interleukin-2 promoter.[58] NFAT is a Ca^{2+}-regulated transcription factor. The full-length PC1 C-tail construct (PC1-LT222) was shown to be capable of producing significant activation of the NFAT reporter.[21] Using the various truncated C-tail constructs established that PC1-induced NFAT activation is dependent on an intact G-protein binding and activation region, and does not require the coiled-coil region. Coexpression of Gαq or Gα12 expression vectors with the PC1 cDNA resulted in further stimulation of PC1-LT222-induced NFAT activation.

The procedures for demonstrating the ability of various PC1 C-tail constructs and Gα subunits to activate an NFAT promoter-reporter are provided here and outlined below. HEK293T cells are transiently cotransfected with an NFAT-responsive promoter-firefly luciferase reporter construct (100 ng/well/6-well plate) and a control Renilla luciferase construct (1.5 ng/well) together with 500 ng/well of either the control sIg-0 or one of various PC1 C-tail deletion constructs (e.g., PC1-LT222, PC1-HT193, PC1-AT120, or PC1-LS111). The final DNA amount is brought to 3 μg/well using pBS DNA. The cells are harvested and lysed 24 h posttransfection using Promega 1+ Passive Lysis Buffer. Luciferase activity is then measured using the Dual-Luciferase Reporter Assay System (Promega). The Renilla luciferase values for the wells within an individual 3-well experiment are used to correct firefly luciferase values. It is important to normalize values with a control construct that does not respond to PC1, such as the pCis-CK plasmid, which should not have activity. Anti-human IgG Western blots demonstrate expression levels of the various fusion proteins.

For augmentation of PC1 C-tail-mediated NFAT activation by Gαq, HEK293T cells are transiently cotransfected as above with NFAT luciferase, Renilla luciferase, and 200 ng/well of either sIg-0 or the PC1 C-tail construct PC1-LT222 together with 50 ng/well Gαq or pcDNA3.1 (vector control for Gαq) and are harvested 24 h posttransfection.

6.4.1.2 Inhibition of PC1 C-Tail-Mediated NFAT Activation

We have been able to show that NFAT activation by the PC1 C-tail involves PLC activation, Ca^{2+} release from internal stores, and Ca^{2+} entry. Numerous studies have demonstrated that ligand-dependent activation of Gq-coupled receptors results in the activation of PLC,[59,60] giving rise to inositol trisphosphate (IP3) and diacylglycerol,[61] which in turn stimulate Ca^{2+} mobilization from intracellular stores and Ca^{2+} entry from the extracellular space. To address whether PC1-mediated NFAT activation involves PLC activation, one can make use of a known inhibitor of PLC, U73122. We found that addition of U73122 at 4 h before harvesting cells caused a significant decrease in PC1-mediated NFAT activation. We also tested the inhibitor of IP3 receptors, xestospongin, which significantly decreased PC1-LT222-mediated NFAT activation. The treatment of HEK293T cells with 2-amino-phenylborate (2-APB), which inhibits both the IP3 receptor and the ryanodine receptor, almost completely abolished PC1-mediated NFAT activation. These results suggest that the initial events resulting in PC1-mediated NFAT activation may be, at least in part, dependent on Gq signaling to PLC followed by Ca^{2+} release from intracellular stores.

NFAT activation has been shown to require sustained increases in intracellular Ca^{2+} that are dependent on Ca^{2+} entry via store-operated calcium release-activated Ca^{2+} channels.[62] To determine whether the PC1-mediated NFAT activation is dependent on Ca^{2+} entry, we treated PC1-LT222 and NFAT luciferase-transfected HEK293T cells with gadolinium hydrochloride (GAD) at 4 h before harvesting. GAD treatment completely abolished PC1-mediated NFAT activation. Similar results were obtained when PC1-transfected cells were treated with the extracellular Ca^{2+} chelator, EGTA. These results suggest that NFAT activation by PC1-activated G-protein signaling is dependent both on Ca^{2+} release from internal stores and on Ca^{2+} entry from the extracellular pool.

To test whether PC1-mediated elevation in intracellular Ca^{2+} can activate calcineurin, we used the calcineurin inhibitor cyclosporine A (CSA). HEK293T cells transfected with PC1-LT222 and the NFAT reporter construct were treated with CSA. CSA treatment was found to completely inhibit the PC1-mediated NFAT activation. As a control, we found that PC1-mediated AP-1 activation is not inhibited by CSA treatment. The inhibition by CSA suggests that the PC1-mediated activation of the composite NFAT/AP-1 promoter is strongly dependent on calcineurin activation and therefore is an NFAT-dependent process initiated by heterotrimeric G-protein signaling. The details of these methods are as follows.

For the inhibition studies, cells were transiently transfected as described below with 200 ng/well of either sIg-0 or PC1-LT222. At 4 h before harvesting, the cells were treated with either vehicle or inhibitor and were harvested 24 h posttransfection. Anti-human IgG Western blots were carried out to demonstrate expression levels of the fusion proteins. The inhibitors used are:

1. U73122 (50 μM) is used to inhibit PLC.
2. Xestospongin (10 μM) is used to inhibit the IP3 receptor.
3. 2-APB (50 μM) is used to inhibit both IP3 and ryanodine receptors.
4. GAD (20 μM) is used to block Ca^{2+} channels to inhibit Ca^{2+} entry.
5. CSA (100 ng/mL) is used to inhibit the calcineurin/NFAT pathway.

6.4.1.3 Increased NFAT Activation with LiCl

PC1 has been shown to inhibit GSK-3β, leading to the stabilization of β-catenin and activation of β-catenin target genes.[29] The distal C-tail region of PC1, which contains the coiled-coil domain, was shown to support this activity. This region of the PC1 C-tail corresponds to the PC1-AT120 construct used in our experiments. GSK-3β is known to phosphorylate nuclear NFAT, thus causing it to translocate back to the cytoplasm.[63] Thus, it is possible that there are two activities in the PC1 C-tail that can enhance NFAT activity, one mediated by the G-protein activation domain in the proximal C-tail, leading to a sustained elevation in intracellular Ca^{2+}, and the other mediated by the coiled-coil in the distal C-tail leading to an inhibition of GSK-3β and thus to the nuclear retention of NFAT. Consistent with this finding, the PC1-LT222 construct, which contains both domains, gives rise to a higher level of NFAT activation than does the PC1-LS111 construct, which lacks the GSK-3β inhibitory domain. If this difference is due to the inability of PC1-LS111 to inhibit GSK-3β, it should be possible to increase PC1-LS111 mediated NFAT activity to the same level achieved by PC1-LT222 by inhibiting GSK-3β with LiCl.[64] In fact, NFAT activity was shown to be significantly lower with PC1-LS111 than with PC1-LT222 in the absence of LiCl but was increased to the same level in the presence of LiCl. For these experiments, HEK293T cells were transiently cotransfected with NFAT luciferase, Renilla luciferase, and 500 ng/well PC1 deletion constructs, PC1-LT222 or PC1-LS111. At 16 h prior to harvesting, the cells were treated with either vehicle or LiCl (25 mM), and the cells were harvested 24 h posttransfection. The details of these reagents and methods are as follows.

6.4.2 Cell Lines and DNA Constructs

1. Maintain HEK293T cells in Dulbecco's Modified Eagle's Medium (DMEM) containing 4.5% glucose and L-glutamine supplemented with 10% heat-inactivated fetal bovine serum (HyClone) and penicillin/streptomycin.
2. Maintain M-1 mouse cortical collecting duct cells[65] in DMEM/F12 medium supplemented with 5% fetal bovine serum.
3. Mouse PC1 C-tail constructs are described above (Figure 6.1).
4. HA-tagged and Myc-tagged PC2 constructs[66] were obtained from Dr. L. Tsiokas (University of Oklahoma Health Sciences Center). As a control for HA-PC2, introduce a stop codon giving rise to a construct encoding aa 1–379 (PC2-stop).
5. EE-tagged Gαq in pcDNA3.1 was obtained from Guthrie cDNA Resource Center, as described above.
6. Cis-acting 4× pNFAT-luciferase and pBlueScript are used; the pNFAT-luciferase construct has a TATA box and four 30-bp repeats containing the composite ARRE-2 site from the human interleukin-2 promoter region:
   5′-GGAGGAAAAACTGTTTCATACAGAAGGCGT-3′.
7. The corresponding NFAT (GGAAAA) and AP-1 (TGTTTCA) elements are underlined.[67] The pRL-null construct was from Promega.

6.4.3 NFAT Luciferase Assay

1. Plate HEK293T cells at a density of $\sim 7.5 \times 10^5$ cells/well of a 6-well plastic plate in DMEM plus 10% heat-inactivated fetal bovine serum. 24 h after plating, transiently transfect the cells with the Ca^{2+} phosphate precipitation method. A total of 3 μg of plasmid DNA is used to transfect each well, which contains 50–500 ng of PC1 DNA or control construct, 100 ng of the ARRE-2 NFAT/AP-1 promoter-reporter construct (firefly luciferase), 1.5–5 ng of RL-null, pBlueScript as filler DNA, and 50 ng of EE-tagged Gαq.
2. At 6 h posttransfection, replace the medium with serum-free DMEM and incubate the cultures for an additional 20 h.
3. Add inhibitors 4 h before harvesting.
4. Lyse the cells in Passive Lysis Buffer (Promega); use 20 μL of cell lysate with the Dual-Luciferase Assay System.
5. Analyze data by one-way ANOVA using GraphPad software (GraphPad Software).

6.4.4 Western Blot Analysis

1. Boil HEK293T cell lysates in the presence of 2× sample buffer, fractionate by SDS-PAGE, and transfer to Immobilon-P membranes (Millipore).

2. Perform Western blotting using anti-HA antibody (Roche Applied Science) and anti-PC1 C-terminal peptide (antibody A19).[68] Perform anti-human IgG Western blots as described above and in earlier articles.[5,20,68]
3. Use secondary antibodies conjugated to alkaline phosphatase to detect the immobilized antibodies by chemiluminescence with CDP-Star substrate (Amersham Biosciences) according to the manufacturer's instructions.

6.4.5 Measurement of Intracellular Ca^{2+}

PC1 C-tail activation of heterotrimeric G proteins causes sustained Ca^{2+} increases and calcineurin-dependent NFAT activation. This activation and nuclear translocation of NFAT is known to depend on the activity of the Ca^{2+}/calmodulin-dependent protein phosphatase, calcineurin, which requires increases in intracellular Ca^{2+} levels for NFAT activation.[62] To test whether PC1-mediated, G-protein signaling can lead to a sustained elevation in intracellular Ca^{2+}, HEK293T cells are cotransfected with PC1-LT222 or the control construct, sIg-0, and a green fluorescent protein (GFP) expression construct to identify transfected cells. After 24 h, the cells are loaded with Fura-2/AM and both basal and caffeine-stimulated intracellular Ca^{2+} levels are determined. Both steady-state intracellular Ca^{2+} levels (76.3 +/− 17.8 nM (+/−S.E., $n = 7$) for sIg-0 versus 137.1 +/− 19.5 nM (+/−S.E., $n = 8$) for PC1-LT ($p < 0.05$)) and peak caffeine-induced Ca^{2+} release were found to be higher in PC1-LT222-transfected cells. Thus, it appears that the PC1-LT222 construct can give rise to significant long-term elevations in basal intracellular Ca^{2+}. The details of these methods are as follows:

1. Plate HEK293T cells at ~5×10^4 cells/well/6-well plate on type I collagen-coated glass coverslips 24 h prior to cotransfection with 450 ng of PC1-LT222 or sIg-0 and 22.5 ng of a cytomegalovirus enhanced-GFP construct (per 3 wells) (Invitrogen) to identify transfected cells, plus 9 µg of pBlueScript DNA.
2. At 4 h posttransfection, replace the medium with DMEM plus 0.5% heat-inactivated serum and incubate the cultures for an additional 20 h.
3. Load cells with 5 µm Fura-2/AM in DMEM/F12 equilibrated with 5% CO_2, 95% air for 30 min at 37°C.
4. Rinse the cells with a HCO_3-Ringer's solution containing 2 mM $CaCl_2$, and mount the coverslips in a thermally controlled chamber on the stage of a Nikon inverted microscope equipped with a monochromator. The chamber is continuously perfused with Ringer's solution equilibrated with 5% CO_2, 95% air at 37°C.
5. Identify transfected cells by viewing GFP fluorescence with a fluorescein isothiocyanate (FITC) filter set (490-nm excitation and 535-nm emission).
6. Make baseline Fura-2 measurements with dual excitation wavelengths of 340 and 380 nm. The measurement of emitted light at 510 nm is restricted to GFP-expressing cells using an adjustable iris in front of a digital photomultiplier detection system (Photon Technology International). Felix 32 analysis software controls the monochromator and data acquisition to generate the 340/380 fluorescence ratio (F_{340}/F_{380}).

7. After a steady state (F_{340}/F_{380}) is established, add 10 mM caffeine to release Ca^{2+} from ryanodine-sensitive stores.
8. At the end of each experiment, permeabilize the cells with 2 μM ionomycin in Ringer's solution containing 2 mM Ca^{2+} to determine the maximum (F_{340}/F_{380}) ratio (R_{max}) and then add 10 mM EGTA to determine the minimum ratio (R_{min}). Measure background correction for the glass coverslips and GFP fluorescence in several cells expressing GFP at the same intensity and subtract from the experimental values.
9. F_{340}/F_{380} ratios are converted to $[Ca^{2+}]$ using the equation $[Ca^{2+}] = K_d \times ((R - R_{min})/R_{max} - R)) \times (S_{f380}/S_{b380})$, where the dissociation constant (K_d) of Fura-2 for Ca^{2+} is 224 nM, R_{max} and R_{min} are F_{340}/F_{380} ratios for Ca^{2+}-saturating and Ca^{2+}-free conditions, and S_{f380} and S_{b380} are fluorescence signals at 380 nm for free Ca^{2+} and bound Ca^{2+}, respectively.[69]
10. A significant difference in intracellular $[Ca^{2+}]$ between cells transfected with sIg-0 and PC1-LT222 is determined using a parametric Student's unpaired *t*-test. Values are represented as the mean ±S.E.

6.4.6 NFAT Nuclear Translocation Assay

As a further assay for PC1-mediated G-protein signaling, the dephosphorylation and nuclear translocation of NFAT can be examined. NFAT requires dephosphorylation by calcineurin for its nuclear translocation.[62] To determine whether NFAT is dephosphorylated following transfection of PC1-LT222, we assayed the relative amounts of phosphorylated and dephosphorylated forms of a cotransfected HA-tagged NFATc1-GFP fusion protein by Western blot analysis using an anti-HA antibody to determine their relative electrophoretic mobilities.[70] The predominant form of HA-NFATc1-GFP in sIg-0 control-transfected cells was the slower migrating phosphorylated band representing cytosolic NFAT. The treatment of these cells with the Ca^{2+} ionophore A23187 shifted most of the HA-NFATc1-GFP to the faster migrating dephosphorylated form found in the nucleus. Cotransfection with PC1-LT222 also resulted in a shift to the faster migrating dephosphorylated form, which was prevented by CSA treatment. Intracellular localization of HA-NFATc1-GFP was assessed by visualization of GFP fluorescence. Cotransfection with sIg-0 resulted in predominantly cytoplasmic localization of HA-NFATc1-GFP, whereas cotransfection with PC1-LT222 resulted in significant nuclear localization of HA-NFATc1-GFP.[71] Cotransfection with HA-NFATc1-GFP and PC1–11TM, which contains all of the 11 transmembrane domains as well as the C-tail, also resulted in nuclear translocation of HA-NFATc1-GFP. The 11TM construct was also effective in activating the NFAT luciferase reporter. The methods are as follows:

Dephosphorylation and nuclear translocation of NFAT by PC1-LT222 and PC1–11TM mediated G-protein signaling are demonstrated by transfecting HEK293T cells with an HA-NFATc1-GFP expression vector[70] in the presence of either PC1-LT222 or sIg-0. Cells were treated with 10 μM calcium ionophore, A23187, or with 100 ng/ml CSA. The cells were lysed with 1× passive lysis buffer, and Western blotting was performed using an anti-HA antibody. For NFAT nuclear localization, HEK293T cells were cotransfected with 100 ng/well HA-NFAT1c-GFP expression vector and

200 ng of either PC1-LT222 or PC1–11TM and their respective controls, either sIg-0 or sIg-stop. At 20 h posttransfection, the cells were washed and fixed and the nuclei were counterstained with DAPI. The following outlines the methods:

1. Seed HEK293T cells in Lab-TekII chamber slides (Nunc) and cotransfect with 100 ng of an HA-NFATc1-GFP expression vector[70] and 200 ng of PC1-LT222 or PC1–11TM or their respective controls.
2. At 20 h posttransfection, wash the cells three times with PBS and fix for 10 min with freshly prepared 4% paraformaldehyde at room temperature.
3. Then wash the cells with three changes of PBS for 5 min each and counterstain with DAPI for an additional 5 min.
4. Rewash the slides with PBS, air dry, mount with Antifade (Molecular Probes), and examine with a Nikon fluorescence microscope equipped with a Spot 32 camera.
5. Merge HA-NFATc1-GFP and DAPI images to distinguish cytosolic and nuclear NFAT.
6. To confirm the fluorescence data, determine the phosphorylation states of cytosolic and nuclear HA-NFATc1-GFP by Western blotting using an anti-HA antibody.

6.5 CONCLUDING REMARKS

As shown in the Mayo Clinic ADPKD Mutation Database (http://pkdb.mayo.edu) most variants in the PKD1 gene are truncating loss-of-function mutations, including large deletions, and splicing, frameshift, and nonsense mutations, which would be expected to significantly alter the expression level of the PC1 protein. However, there are also a number of PKD1 C-terminal tail single amino acid changes associated with ADPKD that are thought to be likely disease-causing mutations. These include a dense cluster of variants within the G-protein binding and activation region of the C-tail (Figure 6.2) that would be predicted to affect G-protein signaling specifically, without necessarily affecting other properties of the polycystin-1 protein such as biogenesis or stability. The presence of these ADPKD mutations in the G-protein binding and activation region argues strongly that an essential function of PC1 is to bind and activate G proteins.

One such mutation in the PC1 C-tail is a 3-base-pair deletion found in an ADPKD patient causing the deletion (ΔL) of a single conserved leucine residue within the polycystin-1 C-terminal tail.[72] We have shown that this ΔL mutation is one of several single amino acid mutations that significantly affects G-protein signaling, as demonstrated in AP-1 transient transfection assays as described in Section 6.3.[20] Confirmation of the importance of ΔL was obtained by generating a knockin mouse model (Pkd1ΔL), which showed that this mutation causes a full-blown PKD phenotype.[36] Homozygous Pkd1ΔL/ΔL mice were found to have an embryo-lethal phenotype, similar to that seen with mouse models carrying truncating, loss-of-function mutations. Combination of Pkd1ΔL with a floxed conditional Pkd1 deletion allele during embryonic development caused a severe cystic phenotype in newborns. As such, Pkd1ΔL acts like a severe truncating allele even though normal levels of

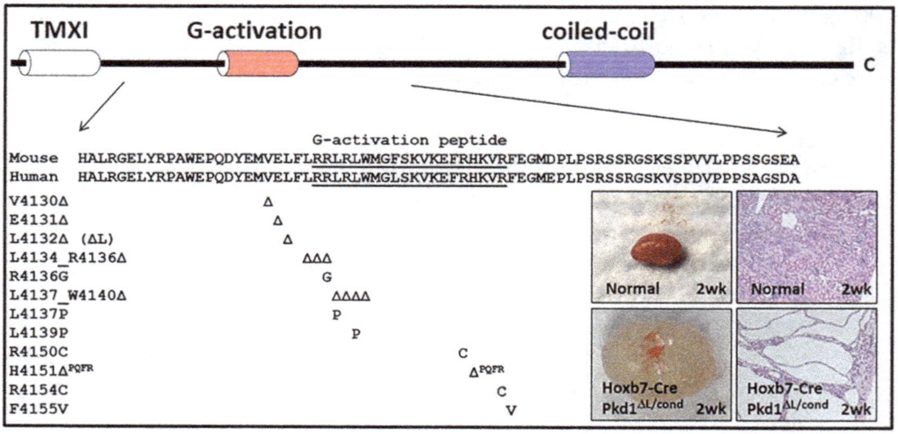

FIGURE 6.2 Alignment of the PC1-HA74 sequences from mouse and human PC1 spanning the 20-aa G-protein activation peptide (G-activation peptide) (underlined). The PC1 C-terminal tail is shown from the last transmembrane domain (TMXI) to the C-terminus (C), and the relative positions of the G protein activation peptide and coiled-coil region. Missense and small deletion or insertion mutations in the Mayo Clinic ADPKD Mutation Database (http://pkdb.mayo.edu) that are in PC1-HA74 are listed, including the human L4132Δ (ΔL) mutation that was used to make the L4122Δ knockin mouse model.[36] As these mutations are all considered likely to be disease-causing, they strongly argue that the PC1-HA74 heterotrimeric G protein binding and activation region provides a critical function. (*Inset*) Kidneys from 2-week-old normal (*top*) or Hoxb7-Cre *Pkd1*$^{\Delta L/cond}$ (*bottom*) mice, showing gross morphology (*left*) or histology (*right*). The severe polycystic phenotype (bottom) is caused by a loss of PC1 heterotrimeric G protein signaling by the ΔL mutation. (Figure adapted from Parnell SC et al. 2018. *Hum Mol Genet* 27(19):3313–3324.)

full-length PC1ΔL protein are produced, cleaved, and form mature, glycosylated protein. In addition, the PC1ΔL protein was able to form immunoprecipitable complexes with PC2, but was unable to support PC2 channel activity when the two proteins were cotransfected in CHO cells. Thus, these experiments argue that the G-protein binding and activation region of the PC1 C-tail is critical to PC1 function by regulating PC2 channel activity, and that loss of PC1-regulated heterotrimeric G protein activation is the fundamental cause of ADPKD.

REFERENCES

1. Cornec-Le Gall E, Audrezet MP, Le Meur Y, Chen JM, and Ferec C (2014) Genetics and pathogenesis of autosomal dominant polycystic kidney disease: 20 years on. *Hum Mutat* 35(12):1393–1406.
2. Harris PC and Torres VE (2014) Genetic mechanisms and signaling pathways in autosomal dominant polycystic kidney disease. *J Clin Invest* 124(6):2315–2324.
3. Paul BM and Vanden Heuvel GB (2014) Kidney: Polycystic kidney disease. *Wiley Interdiscip Rev Dev Biol* 3(6):465–487.
4. Lanktree MB et al. (2018) Prevalence estimates of polycystic kidney and liver disease by population sequencing. *J Am Soc Nephrol* 29(10):2593–2600.

5. Nims N, Vassmer D, and Maser RL (2003) Transmembrane domain analysis of polycystin-1, the product of the polycystic kidney disease-1 (PKD1) gene: Evidence for 11 membrane-spanning domains. *Biochemistry* 42(44):13035–13048.
6. Promel S, Langenhan T, and Arac D (2013) Matching structure with function: The GAIN domain of adhesion-GPCR and PKD1-like proteins. *Trends Pharmacol Sci* 34(8):470–478.
7. Semmo M, Kottgen M, and Hofherr A (2014) The TRPP subfamily and polycystin-1 proteins. *Handb Exp Pharmacol* 222:675–711.
8. Cai Y et al. (2004) Calcium dependence of polycystin-2 channel activity is modulated by phosphorylation at Ser812. *J Biol Chem* 279(19):19987–19995.
9. Celic AS et al. (2012) Calcium-induced conformational changes in C-terminal tail of polycystin-2 are necessary for channel gating. *J Biol Chem* 287(21):17232–17240.
10. Koulen P et al. (2002) Polycystin-2 is an intracellular calcium release channel. *Nat Cell Biol* 4(3):191–197.
11. Qian F et al. (1997) PKD1 interacts with PKD2 through a probable coiled-coil domain. *Nat Genet* 16(2):179–183.
12. Tsiokas L, Kim E, Arnould T, Sukhatme VP, and Walz G (1997) Homo- and heterodimeric interactions between the gene products of PKD1 and PKD2. *Proc Natl Acad Sci USA* 94(13):6965–6970.
13. Yu Y et al. (2009) Structural and molecular basis of the assembly of the TRPP2/PKD1 complex. *Proc Natl Acad Sci USA* 106(28):11558–11563.
14. Zhu J et al. (2011) Structural model of the TRPP2/PKD1 C-terminal coiled-coil complex produced by a combined computational and experimental approach. *Proc Natl Acad Sci USA* 108(25):10133–10138.
15. Su Q et al. (2018) Structure of the human PKD1-PKD2 complex. *Science* 361(6406).
16. Hanaoka K et al. (2000) Co-assembly of polycystin-1 and -2 produces unique cation-permeable currents. *Nature* 408(6815):990–994.
17. Nauli SM et al. (2003) Polycystins 1 and 2 mediate mechanosensation in the primary cilium of kidney cells. *Nat Genet* 33(2):129–137.
18. Kim S et al. (2016) The polycystin complex mediates Wnt/Ca(2+) signalling. *Nat Cell Biol* 18(7):752–764.
19. Parnell SC et al. (1998) The polycystic kidney disease-1 protein, polycystin-1, binds and activates heterotrimeric G-proteins *in vitro*. *Biochem Biophys Res Commun* 251(2):625–631.
20. Parnell SC et al. (2002) Polycystin-1 activation of c-Jun N-terminal kinase and AP-1 is mediated by heterotrimeric G proteins. *J Biol Chem* 277(22):19566–19572.
21. Puri S et al. (2004) Polycystin-1 activates the calcineurin/NFAT (nuclear factor of activated T-cells) signaling pathway. *J Biol Chem* 279(53):55455–55464.
22. Le NH et al. (2004) Aberrant polycystin-1 expression results in modification of activator protein-1 activity, whereas Wnt signaling remains unaffected. *J Biol Chem* 279(26):27472–27481.
23. Arnould T et al. (1998) The polycystic kidney disease 1 gene product mediates protein kinase C alpha-dependent and c-Jun N-terminal kinase-dependent activation of the transcription factor AP-1. *J Biol Chem* 273(11):6013–6018.
24. Delmas P et al. (2002) Constitutive activation of G-proteins by polycystin-1 is antagonized by polycystin-2. *J Biol Chem* 277(13):11276–11283.
25. DeCaen PG, Delling M, Vien TN, and Clapham DE (2013) Direct recording and molecular identification of the calcium channel of primary cilia. *Nature* 504(7479):315–318.
26. Delling M, DeCaen PG, Doerner JF, Febvay S, and Clapham DE (2013) Primary cilia are specialized calcium signalling organelles. *Nature* 504(7479):311–314.
27. Jin X et al. (2014) Cilioplasm is a cellular compartment for calcium signaling in response to mechanical and chemical stimuli. *Cell Mol Life Sci* 71(11):2165–2178.

28. Hama T and Park F (2016) Heterotrimeric G protein signaling in polycystic kidney disease. *Physiol Genomics* 48(7):429–445.
29. Kim E et al. (1999) Interaction between RGS7 and polycystin. *Proc Natl Acad Sci USA* 96(11):6371–6376.
30. Guay-Woodford LM, Wright CJ, Walz G, and Churchill GA (2000) Quantitative trait loci modulate renal cystic disease severity in the mouse bpk model. *J Am Soc Nephrol* 11(7):1253–1260.
31. Kwon M et al. (2012) G-protein signaling modulator 1 deficiency accelerates cystic disease in an orthologous mouse model of autosomal dominant polycystic kidney disease. *Proc Natl Acad Sci USA* 109(52):21462–21467.
32. Yu W et al. (2010) Polycystin-1 protein level determines activity of the Galpha12/JNK apoptosis pathway. *J Biol Chem* 285(14):10243–10251.
33. Yu W et al. (2011) Identification of polycystin-1 and Galpha12 binding regions necessary for regulation of apoptosis. *Cell Signal* 23(1):213–221.
34. Wu Y et al. (2016) Galpha12 is required for renal cystogenesis induced by Pkd1 inactivation. *J Cell Sci* 129(19):3675–3684.
35. Zhang B, Tran U, and Wessely O (2018) Polycystin 1 loss of function is directly linked to an imbalance in G-protein signaling in the kidney. *Development* 145(6).
36. Parnell SC et al. (2018) A mutation affecting polycystin-1 mediated heterotrimeric G-protein signaling causes PKD. *Hum Mol Genet* 27(19):3313–3324.
37. Nishimoto I et al. (1993) Alzheimer amyloid protein precursor complexes with brain GTP-binding protein G(o). *Nature* 362(6415):75–79.
38. Okamoto T et al. (1990) A simple structure encodes G protein-activating function of the IGF-II/mannose 6-phosphate receptor. *Cell* 62(4):709–717.
39. Smine A et al. (1998) Regulation of brain G-protein go by Alzheimer's disease gene presenilin-1. *J Biol Chem* 273(26):16281–16288.
40. Sternweis PC and Robishaw JD (1984) Isolation of two proteins with high affinity for guanine nucleotides from membranes of bovine brain. *J Biol Chem* 259(22):13806–13813.
41. Katada T, Oinuma M, and Ui M (1986) Mechanisms for inhibition of the catalytic activity of adenylate cyclase by the guanine nucleotide-binding proteins serving as the substrate of islet-activating protein, pertussis toxin. *J Biol Chem* 261(11):5215–5221.
42. Bar-Peled M and Raikhel NV (1996) A method for isolation and purification of specific antibodies to a protein fused to the GST. *Anal Biochem* 241(1):140–142.
43. Lohning C, Nowicka U, and Frischauf AM (1997) The mouse homolog of PKD1: Sequence analysis and alternative splicing. *Mamm Genome* 8(5):307–311.
44. Rumenapp U et al. (2001) The M3 muscarinic acetylcholine receptor expressed in HEK-293 cells signals to phospholipase D via G12 but not Gq-type G proteins: Regulators of G proteins as tools to dissect pertussis toxin-resistant G proteins in receptor-effector coupling. *J Biol Chem* 276(4):2474–2479.
45. Coso OA, Teramoto H, Simonds WF, and Gutkind JS (1996) Signaling from G protein-coupled receptors to c-Jun kinase involves beta gamma subunits of heterotrimeric G proteins acting on a Ras and Rac1-dependent pathway. *J Biol Chem* 271(8):3963–3966.
46. Yamauchi J, Kaziro Y, and Itoh H (1999) Differential regulation of mitogen-activated protein kinase kinase 4 (MKK4) and 7 (MKK7) by signaling from G protein beta gamma subunit in human embryonal kidney 293 cells. *J Biol Chem* 274(4):1957–1965.
47. Koch WJ, Hawes BE, Inglese J, Luttrell LM, and Lefkowitz RJ (1994) Cellular expression of the carboxyl terminus of a G protein-coupled receptor kinase attenuates G beta gamma-mediated signaling. *J Biol Chem* 269(8):6193–6197.
48. Rogers JA, Read RD, Li J, Peters KL, and Smithgall TE (1996) Autophosphorylation of the Fes tyrosine kinase. Evidence for an intermolecular mechanism involving two kinase domain tyrosine residues. *J Biol Chem* 271(29):17519–17525.

49. Parnell SC, Magenheimer BS, Maser RL, and Calvet JP (1999) Identification of the major site of *in vitro* PKA phosphorylation in the polycystin-1 C-terminal cytosolic domain. *Biochem Biophys Res Commun* 259(3):539–543.
50. Sugita S, Ichtchenko K, Khvotchev M, and Sudhof TC (1998) alpha-Latrotoxin receptor CIRL/latrophilin 1 (CL1) defines an unusual family of ubiquitous G-protein-linked receptors. *G-protein coupling not required for triggering exocytosis. J Biol Chem* 273(49):32715–32724.
51. Coso OA et al. (1995) Transforming G protein-coupled receptors potently activate JNK (SAPK). Evidence for a divergence from the tyrosine kinase signaling pathway. *J Biol Chem* 270(10):5620–5624.
52. Collins LR, Minden A, Karin M, and Brown JH (1996) Galpha12 stimulates c-Jun NH2-terminal kinase through the small G proteins Ras and Rac. *J Biol Chem* 271(29):17349–17353.
53. Voyno-Yasenetskaya TA, Faure MP, Ahn NG, and Bourne HR (1996) Galpha12 and Galpha13 regulate extracellular signal-regulated kinase and c-Jun kinase pathways by different mechanisms in COS-7 cells. *J Biol Chem* 271(35):21081–21087.
54. Yamauchi J, Kawano T, Nagao M, Kaziro Y, and Itoh H (2000) G(i)-dependent activation of c-Jun N-terminal kinase in human embryonal kidney 293 cells. *J Biol Chem* 275(11):7633–7640.
55. Levi NL et al. (1998) Stimulation of Jun N-terminal kinase (JNK) by gonadotropin-releasing hormone in pituitary alpha T3–1 cell line is mediated by protein kinase C, c-Src, and CDC42. *Mol Endocrinol* 12(6):815–824.
56. Marinissen MJ, Chiariello M, Pallante M, and Gutkind JS (1999) A network of mitogen-activated protein kinases links G protein-coupled receptors to the c-Jun promoter: A role for c-Jun NH2-terminal kinase, p38s, and extracellular signal-regulated kinase 5. *Mol Cell Biol* 19(6):4289–4301.
57. Marinissen MJ, Chiariello M, and Gutkind JS (2001) Regulation of gene expression by the small GTPase Rho through the ERK6 (p38 gamma) MAP kinase pathway. *Genes Dev* 15(5):535–553.
58. Rao A, Luo C, and Hogan PG (1997) Transcription factors of the NFAT family: Regulation and function. *Annu Rev Immunol* 15:707–747.
59. Lee CH, Park D, Wu D, Rhee SG, and Simon MI (1992) Members of the Gq alpha subunit gene family activate phospholipase C beta isozymes. *J Biol Chem* 267(23):16044–16047.
60. Wu DQ, Lee CH, Rhee SG, and Simon MI (1992) Activation of phospholipase C by the alpha subunits of the Gq and G11 proteins in transfected Cos-7 cells. *J Biol Chem* 267(3):1811–1817.
61. Liu B and Wu D (2004) Analysis of G protein-mediated activation of phospholipase C in cultured cells. *Methods Mol Biol* 237:99–102.
62. Hogan PG, Chen L, Nardone J, and Rao A (2003) Transcriptional regulation by calcium, calcineurin, and NFAT. *Genes Dev* 17(18):2205–2232.
63. Haq S et al. (2000) Glycogen synthase kinase-3beta is a negative regulator of cardiomyocyte hypertrophy. *J Cell Biol* 151(1):117–130.
64. Frame S and Cohen P (2001) GSK3 takes centre stage more than 20 years after its discovery. *Biochem J* 359(Pt 1):1–16.
65. Stoos BA, Naray-Fejes-Toth A, Carretero OA, Ito S, and Fejes-Toth G (1991) Characterization of a mouse cortical collecting duct cell line. *Kidney Int* 39(6):1168–1175.
66. Rundle DR, Gorbsky G, and Tsiokas L (2004) PKD2 interacts and co-localizes with mDia1 to mitotic spindles of dividing cells: Role of mDia1 IN PKD2 localization to mitotic spindles. *J Biol Chem* 279(28):29728–29739.
67. Macian F, Lopez-Rodriguez C, and Rao A (2001) Partners in transcription: NFAT and AP-1. *Oncogene* 20(19):2476–2489.

68. Sutters M et al. (2001) Polycystin-1 transforms the cAMP growth-responsive phenotype of M-1 cells. *Kidney Int* 60(2):484–494.
69. Grynkiewicz G, Poenie M, and Tsien RY (1985) A new generation of Ca^{2+} indicators with greatly improved fluorescence properties. *J Biol Chem* 260(6):3440–3450.
70. Aramburu J et al. (1998) Selective inhibition of NFAT activation by a peptide spanning the calcineurin targeting site of NFAT. *Mol Cell* 1(5):627–637.
71. Kim SJ, Ding W, Albrecht B, Green PL, and Lairmore MD (2003) A conserved calcineurin-binding motif in human T lymphotropic virus type 1 p12I functions to modulate nuclear factor of activated T cell activation. *J Biol Chem* 278(18):15550–15557.
72. Afzal AR et al. (1999) Novel mutations in the 3 region of the polycystic kidney disease 1 (PKD1) gene. *Hum Genet* 105(6):648–653.

7 Methods to Study the Vasculature in ADPKD

Patricia Outeda and Terry Watnick

CONTENTS

7.1 DEFECTIVE VASCULATURE IN ADPKD

Autosomal dominant polycystic kidney disease (ADPKD) is the most common monogenic form of end-stage kidney disease with over 12 million people worldwide affected.[1] Almost all cases of ADPKD are caused by mutations in either *PKD1* or *PKD2*, encoding two integral membrane proteins, polycystin-1 (PC1) and polycystin-2 (PC2), respectively.[2–4] PC1 is a large non-kinase membrane receptor, while PC2 is a calcium-permeable, transient receptor potential (TRP)-like channel.[5] PC1 and PC2

TABLE 7.1
Other Vascular Complications Associated with PKD

Vascular Complications	Frequency	References
Cardiac valve abnormalities	25%	22
Pericardial effusion	35%	23
Extracranial aneurysms (ascending aorta; popliteal, splenic, and coronary arteries)	N/A	24–26

interact via their carboxy-termini[6] and tetragonal opening for polycystins (TOP) domains[7,8] and participate in a common signaling pathway that has yet to be fully elucidated.[6,10] Although the hallmark of the disease is the development of renal cysts, ADPKD is a systemic disorder with numerous extra-renal complications.[1] Of these, the vascular complications are the most feared because they are a significant cause of excess morbidity and mortality prior to the onset of renal failure.[11,12] ADPKD has been linked to dissections in almost every large artery including the ascending aorta, splenic artery, coronary arteries, iliac arteries, and internal carotid arteries.[13,14] Intracranial aneurysms (ICAs), however, are the most frequently fatal vascular complication of ADPKD. Several observational studies have demonstrated that ADPKD is associated with an ~5-fold increase in the prevalence of ICA: ~9%–12% versus 1%–2% in the general population.[15–21] Numerous other vascular complications have been reported in the literature and are summarized in Table 7.1.

There are experimental data that support a functional role for polycystins in the vasculature. Genetic reporter studies in mice show that *Pkd1* is highly expressed throughout the cardiovascular system including all major branches of the aorta, the Circle of Willis (where ICAs form), and the aortic outflow tract.[27] At the cellular level, both polycystins can be detected on Western blots prepared from vascular smooth muscle cells (VSMCs) and endothelial cells (ECs).[28–31] Furthermore, *Pkd1* and *Pkd2* knockout mice die during mid-gestation with a dramatic vascular phenotype. At E14.5, *Pkd1* and *Pkd2* null embryos have edema (see Figure 7.1) and focal hemorrhages and a defect in vascularization of the fetal placenta.[32] These embryonic phenotypes are likely related to a functional role for polycystins in endothelial cells.[32,33] Further studies show that selective inactivation of *Pkd1* or *Pkd2* in endothelial cells results in a subset of the "vascular" phenotypes previously identified in null animals including polyhydramnios and ~30%–40% perinatal lethality, both likely due to defective placental vasculature. Interestingly, deletion of *Pkd1* or *Pkd2* in endothelial cells did not result in edema, which is a ubiquitous finding in null animals and has been attributed to an endothelial cell defect. Additional data from both murine and zebrafish models have demonstrated that edema in Pkd1/2-null mice is due to abnormal lymphatic morphogenesis. At a cellular level, *Pkd1*- and *Pkd2*-deficient lymphatic endothelial cells (LECs) have defects in oriented cell migration with a failure to establish front-rear polarity.

Although there is little known specifically about polycystin signaling pathways in endothelial cells, there is an extensive literature dealing with general aspects of polycystin signaling.[34,35] The current working model places PC1 and PC2 at the nexus of a calcium-dependent, signal-transduction pathway. Since the two proteins physically interact, it is

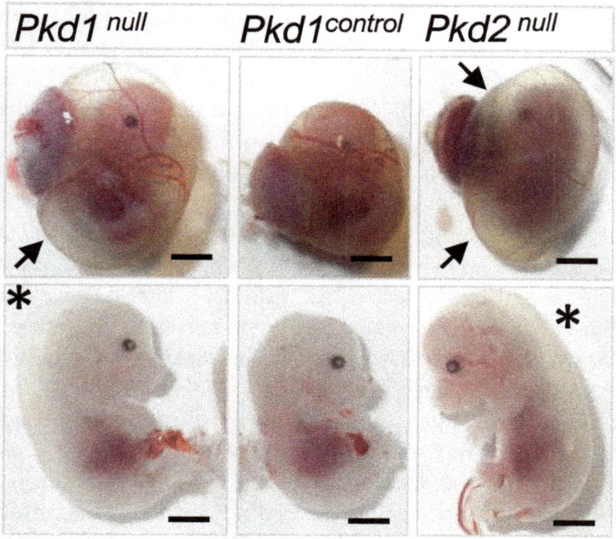

FIGURE 7.1 Defective vascular phenotype in Pkd1 and Pkd2 null embryos. Representative images of *Pkd1* and *Pkd2* null embryos and controls harvested at E14.5. Arrows indicate polyhydramnios and asterisks indicate edema. Scale bar 1 mm.

thought that PC1 acts as a cell surface receptor that regulates the channel activity of PC2 in response to information derived from the extracellular milieu. The ligand(s) for PC1 have yet to be isolated but multiple studies have found that the polycystin complex is located in the primary cilia of renal epithelial cells where it may sense and be activated by mechanical stimuli such as flow.[36–38] Recently, investigators have demonstrated that PC1 and PC2 can be co-immunoprecipitated from endothelial cells as well.[31,39,40] In addition, these proteins are both detected in the primary cilia of endothelial cells that line major vessels (aorta and mesenteric arteries) as well as endothelial cells grown in culture.[39,40] Endothelial cells lacking either PC1 or PC2 are defective in their response to fluid shear stress as evidenced by a lack of cytosolic Ca^{2+} increase and subsequent failure to produce nitric oxide (NO).[39,40] Interestingly *Pkd*-null endothelial cells still display normal responses to mechanical stimuli and to agonist stimulation, such as acetylcholine. The authors of these studies conclude that hypertension in ADPKD could be caused by a failure to produce NO in response to fluid sheer stress. One inconsistency in this hypothesis, however, is that heterozygous vessels in humans with ADPKD presumably contain one-half the wildtype complement of polycystin, yet $Pkd2^{+/-}$ endothelial cells appear to respond normally to fluid shear stress.

In humans, Wang et al. demonstrated that small resistance vessels isolated from patients with ADPKD do have a defect in acetylcholine (which functions as an activator of nitric oxide synthase [NOS]) induced endothelium-dependent relaxation, even in the absence of hypertension or chronic renal insufficiency.[41,42] Moreover, constitutive NOS activity was reduced in these vessels, and the plasma concentration of NO along with the rate of renal NO excretion was diminished in individuals with ADPKD. All of these derangements were noted to be more severe in ADPKD patients with hypertension compared with those without it. Similar results were obtained in aortas from (non-hypertensive) mice that were heterozygous for a *Pkd1* null allele.[43] Taken together, the data are consistent with the idea that ADPKD leads to reduced levels of NO but whether this is the direct result of defective flow sensing or due to abnormalities in other signaling pathways remains unclear.

In summary, there remains little known about the function of polycystins in the vasculature. The phenotypes observed in *Pkd1/2* null mice are primarily developmental and due to endothelial cells defects. There are no *Pkd1/2* mutant mice that fully recapitulate the human adult vascular phenotype. None of the mouse models described to date spontaneously develop ICA, although some do develop aortic disease (see *Pkd1*^{nl} above). Mice homozygous for a hypomorphic *Pkd1* allele (*Pkd1*^{nl}) with an intronic neomycin cassette that results in significantly reduced PC1 expression develop dissecting aortic aneurysms.[44,45] The vascular phenotype, which was first observed in a mixed genetic background, however disappeared when the allele was made congenic on a C57Bl/6 background (*D. Peters, personal communication*). When *Pkd1* is deleted in vascular smooth muscle cells the mice survive and develop mild elastic fiber fragmentation in the ascending aorta at 6 months of age. In this chapter we focus primarily on the methods and assays that can be used to study the biology of polycystins in endothelial cells, since this appears to be the major cell type involved in the dramatic vascular phenotype observed in *Pkd1/2* null mice.

7.2 MOUSE MODELS TO STUDY THE VASCULATURE IN PKD

7.2.1 Pkd1 and Pkd2 Knockout Mouse Models and Hypomorphic Alleles

There are numerous genetically engineered mouse models with mutations in *Pkd1* or *Pkd2* that have been reported in the literature[27–55], and several are available to the research community through Jackson Labs (*Pkd1*: stock no. 010671, *Pkd2*: stock no. 017292). In all cases, mice with homozygous null mutations in *Pkd1* or *Pkd2* die during embryonic development with polyhydramnios, edema, and hemorrhage. These models are ideal for studying embryonic vascular phenotypes associated with angiogenesis and can serve as a source of tissues for the assays described below as well as immunohistochemical studies and Western blot analysis of target signaling pathways.

7.2.2 Pkd1 and Pkd2 Conditional Mouse Models and Cre Driver Lines

Pkd conditional lines harboring LoxP sites in different regions of the *Pkd1* and *Pkd2* genes (floxed alleles) allow deletion of these genes in a temporal and cell-type specific fashion.[32,50] *Pkd1/2* conditional alleles are available through Jackson laboratories (*Pkd1*: stock no. 010671, *Pkd2*: stock no. 017292) or various core centers (http://www.baltimorepkdcenter.org). The use of conditional gene inactivation has several advantages including the ability to study adult phenotypes by circumventing embryonic lethality, which is observed in global *Pkd1/2* knockouts and to define the cell autonomous function of these genes. An important caveat is that Cre drivers may not be 100% efficient. Therefore, the use of a reporter line, such as the mT/mG or confetti mouse, indicating the degree of deletion, that can be bred into the conditional background may be useful.[56,57] A detailed list of Cre recombinase lines that can be used to delete *Pkd* genes in endothelial cells and vascular smooth muscle cells can be found in several recent reviews.[58,59] In Table 7.2, we summarize the most commonly used Cre transgenic mouse lines that can be applied to study the vasculature in ADPKD.[31–33,60] Of note, deletion of *Pkd1/2* in either endothelial cells or vascular smooth muscle cells results in a viable animal.

7.3 *IN VIVO* ASSAYS TO STUDY THE VASCULATURE IN PKD

7.3.1 The Mouse Embryonic Hindbrain

The mouse embryonic hindbrain is an excellent and easily adaptable model to investigate angiogenesis *in vivo* (Figure 7.2). The hindbrain is highly vascularized during embryonic development, and the stereotypical distribution of its vascular network is easily visualized by whole-mount staining using endothelial cell markers.[79,80] The vascularization of the murine embryonic hindbrain starts at E9.5 (days postfertilization) with blood vessels sprouting into neuroectodermal tissue and then growing radially into the hindbrain parenchyma. Sprouting endothelial cells sense and are guided by VEGF-A, which is secreted by neural progenitors.[81–84] The main advantage of the mouse hindbrain is that it allows quantification of critical parameters such as the number of vessels, branch points, or filopodia formation during endothelial cell migration.[80,85] Since vascularization starts early during development, one can study the vasculature of mice with lethal embryonic phenotypes.[49,55] A major disadvantage is that it does not allow *ex vivo* genetic or pharmacological manipulation like other models, for example, the retinal explants.[86] Detailed protocols on how to dissect and stain the embryonic hindbrain have been published and can be easily standardized and applied to study angiogenesis in *PKD* mutant mice.[80,87] Here we provide a general protocol for hindbrain dissection and staining that can be easily performed in *Pkd*-null embryos to investigate and characterize the vascular development.

Materials

Pkd1-or *Pkd2*-null embryos harvested at E10.5–11.5; rat anti-mouse PECAM monoclonal antibody or CD31 (clone MEC13.3; BD Pharmingen, cat. no. 550274); phosphate-buffered saline tablets or PBS Dulbecco's phosphate-buffered saline (10×) (Gibco, cat. no. 10010–023); biotinylated lectin from *Bandeiraea simplicifolia* BS-I isolectin β4 or IB4 (Sigma-Aldrich, cat. no. L2140); AlexaFluor goat anti-rat antibodies and AlexaFluor streptavidin antibodies (1:500); Triton X-100 (Sigma Ultra, cat. no. T9284); heat-inactivated fetal bovine serum or FBS (Gemini's BenchMark, cat. no. 100–106); Vectashield mounting medium with DAPI (Vector Laboratories; cat. no. H-1200); blocking buffer: 2% BSA, 3% FBS, and 0.5% Triton-X100 in PBS; and phosphate buffered saline with Tween 20 (PBST): 0.1% (v/v) Tween 20 in 1× PBS. Equipment: Stereomicroscope, Dumont tweezers (#5) forceps, scissor, and razor.

Method

Dissect the uterus from the pregnant dam at the required gestational age (usually between E10.5 and E11.5) using scissors and tweezers and place the uterus with

TABLE 7.2
Most Common Murine Mouse Models to Study the Vasculature in PKD

	Vascular Endothelial Cells
Constitutively Active	
Flkl Cre[67]	The murine *Vegfr2* (or *flk1*) gene promoter drives Cre recombinase expression in ECs in embryos from E11.5 or E13.5 onward.
Tie1 Cre[63]	The *Tie1* gene promoter drives Cre recombinase expression in most embryonic ECs from E8–9 onward and hematopoietic cells. Not expressed in all ECs in the adult vasculature.
Tie2 Cre[65]	The *Tie2* gene promoter drives Cre recombinase expression in ECs in embryonic and adult vasculature. Also expressed in a monocyte subpopulation and mesenchymal cells.
VE-Cadherin-Cre[60]	The VE-cadherin gene promoter drives Cre recombinase expression in ECs in E7.5 in the yolk sac and embryos. Also expressed in ECs in the adult vasculature, lymphatic system and some hematopoietic cells.
Spatial and Temporal Inducible ERT2 (drives Cre recombinase expression after tamoxifen [4-OHT] administration)	
Cdh5-Cre[69]	The *Cdh5* gene promoter (or VE-cadherin) drives Cre recombinase expression after tamoxifen (4-OHT) administration.
Pdgfb-Cre[62]	The platelet-derived growth factor B (*Pdgfb*) gene promoter drives Cre recombinase expression in most of ECs in postnatal vascular beds but not in adult organs after tamoxifen (4-OHT) administration.
	Lymphatic Endothelial Cells
Constitutively Active	
Lyve1-Cre[68]	The *Lyve1* gene promoter drives Cre recombinase expression in LECs some endothelial cells and hematopoietic cells.
Spatial and Temporal Inducible ERT2 (drives Cre recombinase expression after tamoxifen [4-OHT] administration)	
Prox1-Cre ER^{T2} [61, 70]	The *Prox1* gene promoter drives expression of Cre recombinase in LECs and some other Prox1 expressing cells (liver, heart) after tamoxifen (4-OHT) administration but not in ECs except in venous valves.
Sox18-CreER^{T2} [72–75]	The *Sox18* gene promoter drives expression of Cre recombinase in LECs and also in a low percentage of cells in E15.5 kidneys and ovaries.
	Vascular Smooth Muscle Cells
Constitutively Active	
Tagln-Cre^{KI} [76]	The *Tagln* gene promoter drives Cre recombinase expression in vascular smooth muscle cells in the aorta, hepatic and pulmonary arteries but not in all the organs.
Spatial and Temporal Inducible ERT2 (drives Cre recombinase expression after tamoxifen [4-OHT] administration)	
Myh11-CreER^{T2} [77]	The *Myh* gene promoter drives Cre recombinase expression in vascular smooth muscle cells in multiple organs after tamoxifen (4-OHT) administration.
Acta2-Cre-ER^{T2} [78]	The *Acta2* gene promoter drives Cre recombinase expression in vascular smooth muscle cells after tamoxifen (4-OHT) administration.

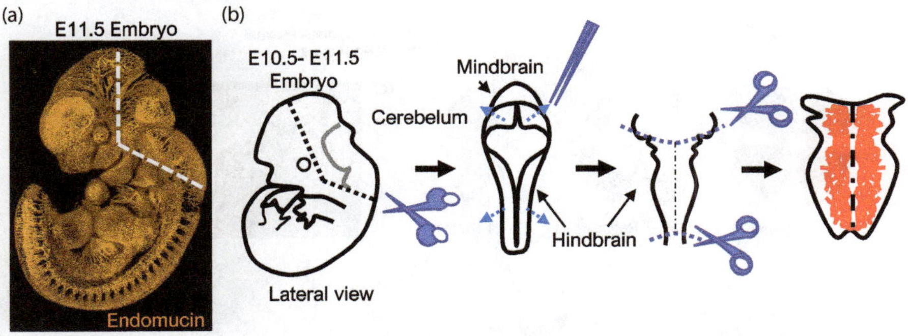

FIGURE 7.2 The Mouse embryonic hindbrain model. (a) Immunofluorescence staining of a wildtype embryo at E11.5 using anti-endomucin (orange) shows vascular vessels distribution during embryonic development. (b) Schematic representation showing the area—white dotted line in (a) and black dotted line in (b)—of the head that need to be dissected to isolate the hindbrain. After removal of midbrain and spinal cord the tissue is fixed, permeabilized and stained to visualize the vessels using antibodies of choice.

intact embryos in a Petri dish with cold PBS. Remove the embryos from the uterine horns with forceps and place them into a dish with clean cold PBS. Under a stereomicroscope, remove the front of the head and face to extract the hindbrain at the level of the fourth ventricle (as illustrated in Figure 7.2). Disconnect the midbrain and spinal cord and transfer the isolated hindbrains to an Eppendorf tube containing cold 4% formaldehyde to fix for 2 h at 4°C with gentle agitation.

Note: several hindbrains with the same genotype can be pooled into one Eppendorf tube.

Wash the tissue 3 times using PBS at room temperature and then incubate the hindbrains in blocking buffer (see Materials section) for 30 min to 1 h. Then, add the primary antibody, usually PECAM1 (1:200) or biotinylated IB4 (1:200). Incubate in primary antibody at 4°C for 48–72 h with gentle rotation. Thoroughly wash the hindbrains 5 times for 30 min each in PBT at room temperature and incubate overnight at 4°C with gentle rolling containing secondary fluorescent antibodies diluted at 1:500 in blocking solution. Wash the hindbrains in room temperature with PBT five times for 30 min each. For imaging, design a chamber using several layers of black electrical tape on a microscope slide and transfer each hindbrain to flat mount in mounting medium. Cover the chamber with a glass coverslip and image the vessels using a confocal laser scanning microscope. The parameters that can be quantified using this method include vessel density or pericyte density and vessel diameter, number of branch points or number of tip cells and number of filopodia per tip cell.

7.3.2 The Postnatal Retina

The vascularization of the mouse retina is one of the most valuable models for studying angiogenesis *in vivo* (Figure 7.3). The importance of this model is demonstrated in numerous studies that used the postnatal retina to decipher key mechanisms regulating

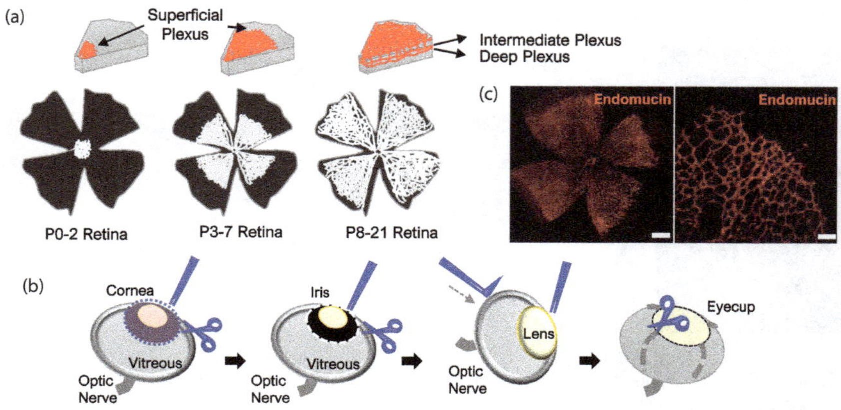

FIGURE 7.3 The postnatal retina model. (a) Schematic representation of retinal vascularization in postnatal mice from P0 to P21. Vascularization of the retina starts quickly after birth (P0) with a radial outgrowth of vessels in the superficial layer of the retina progressing towards the edges. At P7, vessels sprout vertically into deeper layers of the retina to finally form the deep and intermediate plexus, which is fully mature by the end of the third postnatal week (P21). (b) Schematic representation of retina dissection. After dissection of the globe of the eye, cut the edge of the cornea and the iris all around using a forceps and scissor, as indicated. Remove and discard the cornea and the iris. Press the posterior half of the globe gently against the Petri dish to force the lens out and remove it using forceps. Cut the eyecup perpendicular to the corneo-sclera using a sharp scissor at least 4 times at an even distance to form the "petals" and fix and permeabilize to stain using specific antibodies. (c) Retinal whole mounts from wildtype P7 pups (C57BL/6) stained with endomucin (red). Scale bar 1 mm and 100 μm, respectively.

angiogenesis, such as the role of VEGF or the Notch signaling pathway.[88–91] Retinal vascularization in mice takes place during the first week after birth when a radial outgrowth of vessels around the optic nerve at P1 quickly progresses toward the retinal edges in a highly regulated temporal and spatial manner that can be easily observed (see Figure 7.3 illustrating the postnatal vascularization of the retina).[92] During the second and third week after birth, vessels undergo vertical sprouting to vascularize both outer and inner plexiform layers. By the end of the third postnatal week, the retina is entirely vascularized, and the vascular network is fully mature.[93–96] The postnatal retina can be used to study the vasculature in PKD by crossing *Pkd1* or *Pkd2* conditional mice with an appropriate Cre driver that will delete these genes in endothelial cells, largely bypassing embryonic lethality.

Materials

Conditionally inactivated *Pkd* mice (see Section 2.2.2), PBS Dulbecco's phosphate-buffered saline (10×) (Gibco, cat. no. 10010–023), formaldehyde (PFA) (Fisher Scientific, cat. no. F/1501/PB15), Triton X-100 (Sigma Ultra, cat. no. T9284), BSA (Sigma, cat. no. 7906), fluorescein isothiocyanate (FITC)-conjugated lectin from *Bandeiraea simplicifolia* isolectin B4 or IB4 (Sigma, cat. no. L9831), rat anti-mouse endomucin (clone V.7C7, Santa Cruz, cat. no. sc65495), Vectashield mounting medium with DAPI (Vector Laboratories; cat. no. H-1200), Pasteur pipettes, stereomicroscope, glass slides, coverslips.

Method

Euthanize pups at P4-P7 using approved procedures and remove the ocular globe by making several incisions around the eye and severing the extraocular muscles (as shown in Figure 7.3). Transfer the ocular globe to a 12–24-well plate using a plastic transfer pipette containing 0.5–1 mL of cold PBS. Under a stereomicroscope remove the retina by inserting blunt-end forceps in the subretinal space. Fix the retinas at room temperature using 4% PFA for 2 h and wash them at least 3 times using cold PBS. For whole-mount staining block and permeabilize retinas using 2% BSA, 3% FBS and 0.5% Triton X-100 overnight at 4°C. To visualize the vascular network, incubate the retina for 24 h using FITC-conjugated isolectin B4 or endomucin. Wash 3–5 times using PBS and 0.2% Tween-20 for at least 30 min each. Flat mount each retina on a glass slide using mounting media containing DAPI and image vascular network or retinal tip cell filopodia using confocal microscopy.

7.3.3 The Subcutaneous Matrigel Plug Assay

The subcutaneous plug Matrigel assay was originally described by Passaniti et al. in 1992 and is now extensively used to evaluate proangiogenic and antiangiogenic molecules using animal models such as mouse, zebrafish, or chicken (Figure 7.4).[97–102] Briefly, in mice a predetermined volume of Matrigel, supplemented with antiangiogenic or proangiogenic factors, is subcutaneously injected into the ventral region or close to the dorsal midline of a mouse, where it solidifies to form a plug (as illustrated in Figure 7.4). The age of the animal as well as location of injection influences the degree of vascularization. For example, older animals (from 12–24 months of age) produce more vessels than younger ones, and ventral injections in the groin area are more rapidly vascularized in comparison with dorsal injection.

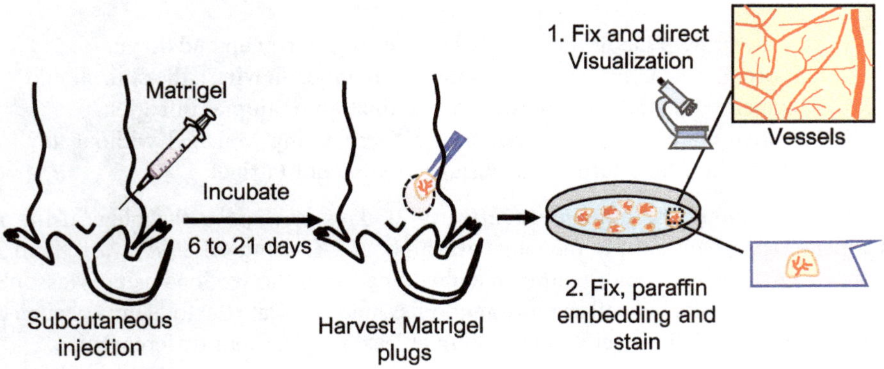

FIGURE 7.4 Subcutaneous Matrigel assay. At day 0, inject Matrigel, subcutaneously, ventrally, in the groin area. Matrigel will solidify after reaching body temperature to form a plug that will be vascularized over time. Allow the vessels to grow and euthanize the mice after 6 to 21 days to remove the vascularized plugs. Fix and visualize Matrigel plugs directly or embed in paraffin and stain (using endothelial cells specific markers). Perform the analysis by counting the number of vessels per area.

Once the Matrigel is placed subcutaneously it will be intensely vascularized over a period of time (usually from 6 to 21 days) mimicking *in vivo* angiogenesis. The vascularized areas can be analyzed by embedding the plugs in paraffin and sectioning and then staining with (1) specific antibodies (such as isolectin B4 or PECAM) to highlight the new formed vessels or (2) Masson's trichrome staining. This assay is reasonably easy to perform and inexpensive but can lack reproducibility because of the challenges of creating identical subcutaneous plugs.[103,104] To maximize reproducibility, we recommend keeping variables such as gender, age, and the injected area uniform among experiments. This method requires adult animals and therefore conditional *Pkd1/2* mouse lines with cell specific inactivation of these genes should be used for subcutaneous plug injection. Several related protocols have been adapted with the goal of generating more reproducible three-dimensional (3D) plugs and these are described in more detail in the literature.[105–111]

Materials

Conditionally inactivated *Pkd* mice (see Section 7.2.2), growth factor–reduced Matrigel (GFR-Matrigel) (BD Biosciences, cat. no. 354230), 25-gauge needles (Monoject), PBS Dulbecco's phosphate-buffered saline (10×) (Gibco, cat. no. 10010–023), VEGF (Peprotech, cat. no. 450–32), bFGF (Peprotech, cat. no. 450–33), formaldehyde (Fisher Scientific, cat. no. F/1501/PB15), Triton X-100 (Sigma Ultra, cat. no. T9284), Vectashield mounting medium with DAPI (Vector Laboratories; cat. no. H-1200).

Method

Thaw GFR-Matrigel on ice and add proangiogenic factors to the suspension (VEGF 30 ng/mL or bFGF 150 ng/mL). Load syringes with 0.5–1 mL of the Matrigel growth factor suspension and inject it subcutaneously into the ventral area of each mouse or close to the dorsal midline under sterile conditions.

Notes:

1. Mix the growth factor thoroughly by pipetting slowly up and down.
2. After removing hair (using hair-removal cream or shaving), the skin should be wiped with ethanol to prevent contamination or animal infection.
3. Negative controls should be included by injecting Matrigel without any supplement. Make sure to inject same volume of Matrigel.

Animals should be euthanized after 6–21 days to remove the plugs using a scalpel or sharp scissor. Fix the plugs overnight at 4°C using 4% PFA, wash them 3 times using cold PBS before paraffin embedding, stain the sections using Masson's trichrome or endothelial cell specific antibodies, and calculate the total area occupied by endothelial cells in several fields and in at least plugs of four different mice.

7.3.4 The Aortic Ring Assay

The rat aortic ring assay was developed by Nicosia and Otinetti in 1990 and has been adapted for use in mice and used in countless studies (Figure 7.5).[112–115] The aortic ring assay allows the analysis of cellular proliferation, migration, tube formation,

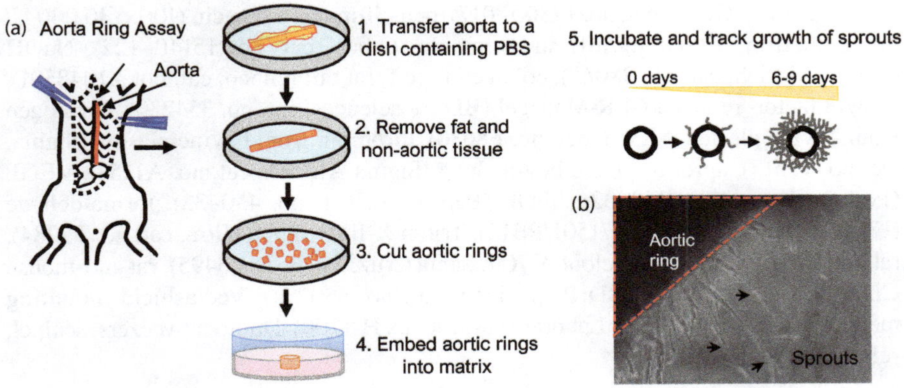

FIGURE 7.5 Aortic ring assay. (a) Schematic showing the location of the mouse aorta. Dissect the thoracic aorta from euthanized mice. Using forceps and scissors separate the spine and the aorta. Cut the aorta from the diaphragm to the heart and place it in PBS in a 10 cm^2 dish. Remove the surrounding non-aortic tissue using Dumont tweezers under the stereoscope and transfer the aorta to a clean 10 cm^2 dish on ice. Section the aorta into small rings using a scalpel and transfer the rings to a prechilled well plate. Carefully add the matrix on top of the aortic ring and when polymerized add media. Incubate the plates at 37° C and acquire images every day or every other day from day 0 to days 6–9 to track the growth of the sprouts. (b) Example of sprouts (black arrows) from an aortic ring (red dotted line) harvested from a ~1 month C57BL6 mouse after 4 days of incubation using collagen.

branching morphogenesis, perivascular cell recruitment and microvessel remodeling. It is relatively inexpensive and fast and allows the investigator to simultaneously study several rings derived from the same mouse under multiple conditions.[116] Aortic rings can be transfected using siRNA or transduced with lentivirus to rescue phenotypes or to aggravate defective angiogenesis.[116–118] One limitation of the methodology is that explants can only be kept in culture for a limited time.[112] There are some variables that should be considered and kept constant between experiments such as the age of the animal, the width of the ring, and the media added to the explant. Briefly, the method requires (1) dissection of the thoracic aorta, (2) preparation of the rings, (3) embedding of the rings in a matrix (usually Matrigel, fibrin, or collagen), (4) treatment with antiangiogenic/angiogenic factors, and (5) microscope imaging and microvessel sprout quantification (as illustrated in Figure 7.5). Here we provide a general protocol for the mouse aorta ring assay. Detailed protocols with slight adaptations can be found in the literature.[114,116,119]

Materials

Conditionally inactivated *Pkd* mice (see Section 7.2.2), PBS Dulbecco's phosphate-buffered saline (10×) (Gibco, cat. no. 10010–023), Opti-MEM (Gibco, cat. no. 31985070), heat-inactivated fetal bovine serum or FBS (Gemini's BenchMark, cat. no. 100–106), Dulbecco's Modified Eagle's Medium (DMEM, Gibco BRL), BSA (Sigma, cat. no. 7906), Dulbecco's Modified Eagle's Medium, high glucose + L-glutamine + pyroxidine hydroxide + 110 mg/L sodium pyruvate without sodium bicarbonate

(DMEM [10×], Gibco, cat. no. 12800–017), penicillin-streptomycin (100×; 10,000 U/mL penicillin, 10,000 μg/mL streptomycin; Gibco, cat. no. 15140–122), NaOH (Sigma-Aldrich, cat. no. S8045), collagen type I, rat tail (Gibco, cat. no. A1048301), growth factor–reduced GFR-Matrigel (BD Biosciences, cat. no. 354230), fibrinogen from bovine plasma (Sigma, cat. no. F8630), thrombin from bovine plasma (Sigma, cat. no. T7513), aprotinin from bovine lung (Sigma-Aldrich, cat. no. A1153), VEGF (Peprotech, cat. no. 450–32), bFGF (Peprotech, cat. no. 450–33), formaldehyde (Fisher Scientific, cat. no. F/1501/PB15), Triton X-100 (Sigma Ultra, cat. no. T9284), rat anti-mouse endomucin (clone V.7C7, Santa Cruz, cat. no. sc65495), rat anti-mouse CD31 (Clone MEC13.3, BD Pharmigen, cat. no. 550274), Vectashield mounting medium with DAPI (Vector Laboratories; cat. no. H-1200), Dumont tweezers, scalpel, scissors, razor.

Method

Dissect aortas from euthanized animals under sterile conditions, transfer them to a tissue culture dish containing cold PBS, and remove the surrounding fat and non-aortic tissue using a stereomicroscope (as illustrated in Figure 7.5). Flush any blood from the vessel lumen using a 27G needle and cut small rings (~0.5 mm wide).

Notes: It is important to keep the size of the rings constant since differences in width can introduce differences in vessel sprouting.

1. Optional: Rings can be serum-deprived overnight in Opti-MEM at 37°C and 5% CO_2. Embed the rings using Matrigel, fibrin or collagen type I.
2. Optional: Add sterile coverslips to individual wells before the embedding if planning to perform immunofluorescence staining to visualize the sprouts.

Here, we provide a general protocol for the most common matrices used to perform this assay.

Aortic ring embedding

1. *Matrigel embedding*: Thaw GFR-Matrigel on ice, add 20–40 μL per well (in a 96- or 24-well plate) and place the plate in an incubator at 37°C to allow the matrix to polymerize. Transfer one ring per well to the top of the layer using a pipette tip (cut the tip with a sterile scalpel) or a forceps with the luminal axis perpendicular to the bottom of the well and add a second layer of Matrigel on top of the ring (20–40 μL). Allow the Matrigel polymerization at 37°C in the incubator for 10–15 min and add 100–500 μL of culture media in each well.
2. *Fibrin embedding*: Prepare fibrin gel (3 mg/mL) by adding fibrinogen into 20 mL of Opti-MEM under sterile conditions. Aliquot the fibrinogen in small volumes (0.2–0.5 mL) and keep it on ice. Place individual aortic rings into the wells (6–10 at a time) and polymerize the fibrinogen by adding 0.2–0.5 U of thrombin to each aliquot (1 U/mL). Cover each ring using 30 μL of polymerizing fibrin gel and incubate at room temperature for about 10 min.

Note: To prevent degradation of fibrin gels by proteases, add aprotinin (10 µg mL−1) to the fibrinogen solution.

3. *Collagen Type I embedding*: Prepare collagen using DMEM to a final concentration of 1 mg/mL, and keep it on ice until use. *Note: Adjust the pH using 5N NaOH.* Place one ring per well in a 96–48 well plate (6–10 aortic rings at a time), add 50–100 µL to each well of the collagen solution, and transfer the plate to an incubator with 5% CO_2 at 37°C for 1 h.

Add 150–500 µL of the explant culture medium (Opti-MEM with 2.5% FBS and 30 ng/ L VEGF) and proceed to image time 0 using an inverted phase-contrast microscope. Change medium every other day and repeat image acquisition for further analysis.

Immunostaining and sprout growth quantification: Immunostaining (preferably with fluorescent antibodies) of aortic rings is required to distinguish microvasculature sprouting from growth of other cell types (i.e., fibroblasts). To do this, remove the explant culture medium and wash each well gently with PBS at least 3 times. Fix the explants with 4% PFA for at least 1 h and wash with cold PBS at least 3 times. Block and permeabilize the explant using 2% BSA and 0.5% Triton X-100 for 1 h. Prepare antibodies using blocking buffer to stain endothelial cells—PECAM1 or CD31 (1:200), endomucin (1:50), or pericytes (α-smooth muscle actin, α-SMA [1:200], or B1:S1 lectin-FITC [0.1 mg/mL])—and incubate with the explants, overnight, at 4°C. On the next day, wash the explants 3 times with PBS and 0.2% Tween-20 for 30 min each wash and add 50–100 µL of blocking buffer with Alexa Fluor secondary antibodies and incubate for 2–3 h in the dark at room temperature. Wash 3 times, 30 minutes each, with PBS and 0.2% Tween-20 for 30, and add DAPI-containing mounting medium. The explants can now be visualized from plates or they can be transferred to a slide for confocal microscopy. The time course for angiogenic sprouting is variable and depends primarily on the model used, the embedding matrix, treatment, and age of the animals from which the aortic ring is derived (angiogenic response is higher in young animals of ~1 month of age than old animals of ~10 months of age). In general, exponential growth is observed during the first few days after culture and reaches a peak between 6 and 9 days before slowly regressing.

Note: Standardization using an initial experiment to determine the kinetics of sprouting/regression using specific conditions should be performed.

The most commonly used parameters to assay sprouting are (1) the number of microvessels emerging from the ring, (2) sprout length, (3) branch points, (4) the area covered by the sprouts, and (4) the supporting cell coverage. For detailed methods on how to quantify the vascularized area in the aortic ring assay, consult the literature.[114,116,120–122]

7.4 CELL-BASED SYSTEMS FOR STUDYING THE VASCULATURE IN PKD

Cell-based systems are complementary to *Pkd* mutant mice for dissecting endothelial cell-signaling pathways. In this section, we describe some of the standard endothelial cell assays that can be applied to study polycystin signaling in this cell type. These protocols can be adapted for use with primary endothelial cells isolated from genetically modified mice or from endothelial cell lines with genetically introduced *Pkd1/2* mutations.

7.4.1 Endothelial Cell Lines for Use in PKD Research

The list of commercially available endothelial cell lines is lengthy and includes cells derived from a variety of organisms (e.g., human, rat, rabbit, mouse bovine) and vessel subtypes (e.g., arteries, veins, lymphatic vessels, microvasculature). In Table 7.3 we list a subset of the most commonly used murine and human endothelial cell lines.

These cell lines, along with individualized protocols for culturing (growth medium, seeding density, etc.), can be obtained from general culture collection centers, such as the American Type Culture Collection (ATCC—https://www.atcc.org/). Gene-editing strategies such as CRISPR or TALENs can then be used to introduce mutations in either *PKD1* or *PKD2*. Appropriate targeting sequences (see Table 7.4) have been tested and validated for use in IMCD (inner medullary–collecting duct) or hESC (human embryonic stem cells) cells[123–125] but can also be used to engineer mutations in immortalized endothelial cell lines such as EA.hy926. In the case of primary cells such as HUVECs, gene editing is not practical since these cells can only be used for a restricted number of passages (usually 4–6) but siRNA can be used to efficiently knockdown either *PKD1* or *PKD2*.

Independent of the method selected for gene silencing, efficacy should be confirmed. In the case of siRNA, gene silencing should be validated by qRT-PCR and/or Western blot to confirm decreased mRNA levels and reduced protein expression, respectively. The ideal controls for siRNA experiments are cells transfected with

TABLE 7.3
Endothelial Cell Lines Commonly Used in Cell-Based Assays

Cell Name	Organism	Origin	Type
		Cell Lines	
Svec4–10	Mouse	Axillary lymph node	Lymphatic
C166	Mouse	Yolk sac	Endothelial
EA.hy926	Human	Hybrid	Endothelial
HMEC-1	Human	Foreskin	Endothelial
		Primary Cells	
HDLEC	Human	Juvenile foreskin	Lymphatic
HUVEC	Human	Umbilical cord	Endothelial
HPAEC	Human	Pulmonary artery	Endothelial
HAEC	Human	Aorta	Endothelial
HDMVEC	Human	Skin	Endothelial

TABLE 7.4
Targeting Sequences to Knockout or Knockdown Expression of *Pkd1/2*

Origin	Type	Sequence	References
		Gene Targeted: Pkd1	
CRISPR	*H. sapiens*	*5′-GTGGGTGCGAGCTTCCCCCCGGG-3′*	124
	M. musculus	*5′-TCTGGGGCAGGCCGCACCTC-3′*	
		5′-GTCGCACCGCAGACGGGCCA-3′	
		5′-AGAGCTCGCCACGCAAGGCG-3′	
		5′-ACCCCAGGACTATGAGATGG-3′	
siRNA	*H. sapiens*	*5′-TTGTAGACACAGAACTCCTCG-3′*	33,72
		5′-AATGTCTTGCCAAAGACGGAC-3′	
		5′-TCCTGACCGTGCTGGCATCTA-3′	132
		5′-CGTGGTCTTCAATGTCATTTA-3′	
	M. musculus	*5′-AAGCCAATGAGGTCACCAG-3′*	131
		5′-AACGCAGCAGTAATCTGCT-3′	
		5′-TTCTCTCCAGGAACACTGG-3′	
		Gene Targeted: Pkd2	
CRISPR	*H. sapiens*	*5′-GCGTGGAGCCGCGATAACCCCGG-3′*	123
	M. musculus	*5′-CGGCTGCGGCTGCACGCGTCTGG-3′*	125
siRNA	*H. sapiens*	*5′-ATTTCTCTCGG TTAGTGCTGC-3′*	33
		5′-ATTAGCTTCCTC AATCTCTGC-3′	
	M. musculus	*5′-TTTCCAATATCTCTTCCAC-3′*	131
TALEN	*M. musculus*	*5′-CAGTAAATGCAGAGAGGA-3′*	124
		5′-ACATGTAGAGTCAGTCCA-3′	
		5′-CTCTCATGCTTGATCACC-3′	
		5′-TCTGCTTTTCATAGGTTA-3′	

scrambled nucleotides. Since there is always a possibility of off-target effects, we recommend using at least two siRNAs for each gene to be targeted.[126]

In the case of gene editing via CRISPR or TALENs, the gene mutation should be verified by sequencing different clones to confirm the presence of biallelic mutations. In addition, Western blot should be used to document the absence of protein if truncating mutations are introduced. Wildtype clones that are generated at the same time but lack mutations and have normal expression of the targeted protein can be used as controls. TALEN and CRISPR methods may have fewer off-target effects compared with siRNA but nonetheless remains a potential source of artifact.[127–130] In addition, if independent clones are isolated, there is a possibility of variation that is independent of the gene of interest. Therefore, to ensure reproducibility, we recommend using a minimum of three independent clonal isolates for both mutant and control cell lines.

In Table 7.4, we listed some of the validated targeting sequences that have been published in the literature. Detailed protocols on how to generate these cells can be found in the literature[33,72,123–125,131,132] and can be easily adapted to endothelial cells.

7.4.2 Embryonic Endothelial Cell Isolation

Another approach for generating *Pkd1/Pkd2* mutant endothelial cells is to isolate them from genetically engineered mice, either germline knockouts or conditional mice. The advantage of primary cell isolation is that it avoids clonal artifacts as well as potential off-target effects that can be associated with the use of CRISPR, TALENs, or siRNA. The disadvantages of primary cell isolation include a relatively laborious procedure that yields cells that can only be used for a limited number of passages (usually 2–6). In addition, if endothelial cells are isolated from embryos, the yield is relatively small. For these reasons, it may be necessary to repeat cell isolation several times, which can result in variability due to differences in the purity of cell preparations. Cell preparations are never 100% pure because of contamination with other cell types, usually fibroblasts. In addition, primary cell cultures usually require the use of more expensive growth media with a richer mixture of micronutrients, hormones, and growth factors.

Since *Pkd*-null animals do not survive to birth, the isolation of primary endothelial cells can be performed using *Pkd1/2* mutant and control embryos at E12.5–E14.5 days. Endothelial cell isolation relies on preparing extensively digested tissues that are incubated with magnetic beads coated with specific cell markers such as PECAM1 (or CD31) for blood endothelial cells or Lyve1 for lymphatic endothelial cells.[133] In this method, a population of endothelial cells that likely contains several subtypes of endothelial cells is isolated from the whole embryo. As an alternative, microvascular endothelial cells can be isolated from individual tissues (e.g., lung, brain, or kidney) derived from adult mice with floxed *Pkd* genes. The resulting endothelial cells can then be transduced with either an adenoviral or lentiviral Cre recombinase (or empty vector) to generate PKD mutant endothelial cells along with controls.

We have used these methods to isolate blood and lymphatic endothelial cells from *Pkd* null embryos, and the yield of cells was enough to perform biochemical studies, RNA isolation for RNA-sequencing, or primary culture for several passages.

Materials

Prepare both the murine endothelial cell growth media (mECGM) and antibody-coated magnetic beads the day before the animals are to be harvested (see Section Preparation of magnetic beads for blood endothelial cell CD31$^+$ isolation). PBS Dulbecco's phosphate-buffered saline (10×) (Gibco, cat. no. 10010–023), BSA (Sigma, cat. no. 7906). *Preparation of mECGM*: Supplement Dulbecco's Modified Eagle's Medium (DMEM, Gibco BRL) with 20% heat inactivated fetal bovine serum or FBS (Gemini's BenchMark, cat. no. 100–106), 2 mM L-glutamine (Gibco), 1 mM sodium pyruvate (Invitrogen), 20 mM HEPES (Gibco), 1% non-essential amino acids (Gibco), 150 μg/mL EC growth supplement (Sigma, Poole, UK) and 12 U/mL heparin (CP Pharmaceuticals, Wrexham, UK).

1. *Preparation of magnetic beads for blood endothelial cell CD31$^+$ isolation*: For 5 embryos, transfer 75 μL of sheep anti-rat IgG Dynabeads (Invitrogen, cat. no. 11035) to a clean Eppendorf tube containing 1 mL of blocking buffer (PBS supplemented with 0.2% BSA and 2 mM EDTA) and resuspend. Place

the tube in a magnet (DynaMag Magnet, ThermoFisher Scientific) for 2 min and discard the supernatant. Repeat washes at least 3 times. After the last wash, add 0.5 mL of blocking buffer, resuspend the beads, and add 25 μL of purified rat anti-mouse CD31 (Clone MEC13.3, BD Pharmigen, cat. no. 550274). Rotate overnight or at least for 1 h at 4°C.

2. *Preparation of magnetic beads for lymphatic endothelial cell Lyve⁺ isolation*: For 5 embryos, transfer 75 μL of sheep anti-rat IgG Dynabeads (Invitrogen, cat. no. 11035) to a clean Eppendorf tube containing 1 mL of blocking buffer (PBS supplemented with 0.2% BSA and 2 mM EDTA) and add rat anti-mouse F4/80 (Invitrogen) and rat anti-mouse CD45 (BD Pharmingen) (1:100) to deplete macrophages and other hematopoietic cells. In a separate Eppendorf tube, add 75 μL of goat anti-rabbit IgG Microbeads, wash them according to manufacturer's instructions, and add 5 μL of anti-Lyve1 antibody (AngioBio) per embryo. Rotate overnight (or at least for 1 h) at 4°C.

Method

Euthanize timed pregnancies at E14.5 under sterile conditions using IACUC-approved methods. After spraying the abdominal area with 70% ethanol, section the abdominal cavity to expose the uterus and then collect and place in a clean Petri dish with cold 1× PBS. Excise the uterine horn to expose the embryonic sacs and release embryos carefully from the yolk sac; separate each embryo from its placenta by cutting the umbilical cord. Transfer the embryos to a clean and sterile Petri dish containing cold PBS and decapitate each embryo. At this point, the tail can be used for genotyping using standard primers and protocols. Transfer each embryo individually to a clean Petri dish and chop the embryonic tissue using a clean, sterile razor. Add 1.5–2 mL of collagenase buffer (2 mg/mL Collagenase type II and 5–10 U of DNase I per 1 mL in DMEM) previously warmed to 37°C, and transfer each embryo to a 15 mL Falcon tube. Place the tubes with minced embryos in a water bath at 37°C and rock gently for 30–40 min. Remove the samples from the water bath every 10 min and mechanically dissociate the cells with a P1000 pipette up and down. Add 5 mL of plain cold DMEM to each Falcon tube, and then resuspend the cells and separate individual cells from nondigested pieces of tissue using a 40 μm strainer. Centrifuge each 15 mL Falcon tube containing individual embryos for 5 min at 1200 rpm. Then remove the supernatant and gently resuspend the cell pellet in 500 μL to 1 mL of blocking buffer (0.2% BSA in PBS), transfer to an Eppendorf tube, and rotate at 4°C for 10–15 min to inhibit nonspecific binding.

1. *For blood endothelial cell isolation*: Wash the anti-CD31-coated (see Section Preparation of magnetic beads for blood endothelial cell CD31⁺ isolation) magnetic sheep anti-rat IgG Dynabeads beads with blocking buffer, placing the Eppendorf tube on a magnet for 2 min, and then remove the supernatant. Repeat the wash step at least 3 times. After the final wash, resuspend the CD31-conjugated beads in blocking buffer (50 μL per sample). Add 50 μL of the antibody-conjugated beads to each Eppendorf tube containing a cell suspension and incubate, rotating the bead-cell suspensions for 30 min at

4°C. Place the Eppendorf tubes with the cells and magnetic beads on the magnet for 2 min, remove the supernatant, and add 1 mL of blocking buffer. Remove the tubes from the magnet and rotate them for at least 2 min to ensure that the cells are suspended. Place the tubes in the magnet for 2 min and remove supernatant. Wash 4–5 more times with blocking buffer.

After the final wash, resuspend the cells using endothelial growth culture media and plate them in gelatin-coated (2% gelatin type B from bovine skin, Sigma) 6-well plates (one sample per well).

Note: Alternatively, instead of using magnetic beads, cells can also be sorted using fluorescence-activated cell sorting (FACS). After tissue disaggregation, cells can be stained using fluorophore-conjugated antibodies and separated or sorted using flow cytometer.

2. *For lymphatic endothelial cell isolation*: Since the number of lymphatic endothelial cells per embryo is small, we recommend pooling several embryos (3–4 embryos if possible). *CD31*$^{+}$ *cells* can also be isolated simultaneously using the sequence of beads incubations described below.[133] To deplete macrophages and hematopoietic cells, we recommend incubating each sample with 50 μL of F4/80 and CD45-coated magnetic beads, rotating at 4°C for 15–20 min. Place the Eppendorf tubes with the cells and magnetic beads on the magnet for 2 min, collect the supernatant, and transfer to a new clean Eppendorf tube. Discard *F4/80+* and *CD45+ cells* that are bound to the F4/80 and CD45-coated magnetic beads. Next, add anti-Lyvel-coated magnetic beads to the supernatant (*F4/80−* and *CD45− cells*) and incubate at 4°C for 20–30 min, rotating continuously. Place the Eppendorf tubes with the cells and magnetic beads on the magnet for 2 min and then collect the supernatant (containing non-LEC endothelial cells) and transfer it to a new clean Eppendorf tube for blood endothelial cell isolation if this is desired. If the plan is to only isolate LECs, then the supernatant can be discarded. Wash the *Lyvel*$^{+}$ *cells* attached to the magnet by adding 1 mL of blocking buffer. Remove the tubes from the magnet and rotate them for at least 2 min to ensure that the cells are in suspension. Place the tubes in the magnet for an additional 2 min and remove supernatant. Repeat at least 4–5 more washes with blocking buffer to ensure specific binding of *Lyvel*$^{+}$ *cells*. If *CD31*$^{+}$ *cells* are to be collected from the supernatant, then add 50 μL of anti-CD31 IgG Dynabeads. Incubate at 4°C for 20–30 min, rotating, and perform the same steps as before to wash *CD31+ cells* binding to anti-CD31 IgG Dynabeads.

   Note: The same cell-isolation procedures can be used to isolate primary LEC or endothelial cells from small pups or adult tissues.

The degree of purity for each cell culture should be estimated by growing an aliquot of the isolated cells on a glass coverslip coated with 2% gelatin or fibronectin and performing immunostaining using, (1) For LECs, rabbit anti-mouse Prox1 (AngioBio) or goat anti-mouse VEGFR3 (R&D Systems) or, 2) for blood, endothelial rat anti-mouse CD31 (clone MEC13.3; BD Pharmingen, cat. no. 550274), goat anti-mouse VEGFR2 (R&D Systems), or goat anti-rat Neuropilin-1 (R&D Systems).

7.5 *IN VITRO* ASSAYS TO EVALUATE ENDOTHELIAL CELL PROLIFERATION

7.5.1 The MTT Cell Proliferation Assay

The tetrazolium salts MTT—3-(4, 5-dimethylthiazolyl-2)-2, 5-diphenyltetrazolium bromide—cell proliferation assay measures the cell proliferation rate in metabolically active cells. The reduction of yellow MTT to insoluble formazan can be quantified by spectrophotometry.[134] Although the MTT assay is widely accepted as a reliable method to examine cell proliferation, results should be confirmed using an alternate measure of cell proliferation since some agents can affect MTT processing without affecting endothelial cell viability.[135]

Materials

MTT (3-(4,5-Dimethyl-2-thiazolyl)-2,5-diphenyl-2H-tetrazolium bromide, (M2128, Sigma, Germany) stock solution (5 mg/mL in PBS), Corning 96 Well TC-Treated Microplates, DMSO (D8418, Sigma), VEGF (Peprotech, cat. no. 450–32), cell growth media.

Method

Trypsinize endothelial cells from a 10 cm^2 cell culture dish and count viable cells using a hemocytometer. Adjust the cell density to a 2×10^4 cells/mL, and then seed the cells in a 96-well cell culture plate using 0.1%–0.5% FBS media supplemented with 20 ng/mL of VEGF. Grow the cells for 24–72 h using normal culture conditions. Add 20 μL of MTT (5 mg/mL) in 200 μL of cell culture medium and incubate the cells for 4 h. Remove 150 μL of supernatant from each well and add 150 μL of DMSO. Measure the absorbance using a microplate reader.

Note: Perform the experiment using three duplicate wells for each sample. A minimum of three independent experiments should be performed and used for analysis.

7.5.2 BrdU-Labeled Cell Proliferation Assay

5-Bromo-2′-deoxyuridine (bromodeoxyuridine [BrdU]) is a thymidine analogue that is incorporated into the DNA of dividing cells during S phase of the cell cycle and can be easily detected by standard immunohistochemical techniques using a primary antibody directed against BrdU.[136,137] BrdU incorporation is therefore an indicator of DNA synthesis and is used as a marker of cell proliferation.

Materials

BrdU (Abcam, ab142567), BSA (Sigma, cat. no. 7906), Vectashield mounting medium with DAPI (Vector Laboratories; cat. no. H-1200), VEGF (Peprotech, cat. no. 450–32), fluorochrome-conjugated anti-BrdU antibody (1:15; Santa Cruz, cat. no. sc-32323).

Method

Grow endothelial cells on coverslips for 24 h using regular growth medium. After 24 h, incubate the cells for an additional 16 h in basal media supplemented with 0.1%–0.5% FBS and 10 ng/mL of BrdU in the presence or absence of recombinant VEGF (20 ng/mL).

Note: The incubation time for BrdU is specific for each cell line. Primary cells require up to 24 h, while rapidly proliferating cell lines need less time. The time required to achieve the optimal signal-to-noise ratio should be optimized.

The incorporation of BrdU into proliferating cells can be detected by immunofluorescence. Wash the cells using PBS and fix them with 2% PFA for 20 min and then permeabilize with 1% Triton X-100 in PBS for 5 min. Denature the DNA by incubating the cells with 2 N HCl for min at 37°C, followed by five rinses in PBS. Incubate cells with 3% bovine serum albumin (BSA) in PBS and then probe with a fluorochrome-conjugated anti-BrdU antibody (1:15) in a humidified chamber at 37°C for 60 min. After several washes with PBS, mount the slides using mounting media and DAPI (4′,6-diamidino-2-phenylindole) to stain the nuclei. Image several fields at 40× using an inverted fluorescence microscope (3–4 different fields from each duplicate). The proliferation index can be calculated as the percentage of BrdU positive cells/total number of cells (DAPI-positive). Each experiment should be performed in triplicate and repeated a minimum of 4 to 5 times.

7.6 *IN VITRO* ASSAYS TO EVALUATE ENDOTHELIAL CELL MIGRATION

7.6.1 Wound-Healing Assay

The wound-healing assay is a simple and economical method broadly used in many disciplines to study *in vitro two-dimensional* cell migration.[138] The assay can be performed in standard well plates (from 12- to 96-well plates) by simply scratching a cell monolayer using a pipette tip as shown in Figure 7.6a.[139,140] The cell-free gap induces directional cell migration, and by capturing images of migrating cells at fixed time intervals, the speed at which the wound is closed can be measured.[141] The investigator can manipulate gene expression, extracellular matrix composition or other variables to investigate the mechanisms involved in cell migration. Appropriate controls should be performed and analyzed in parallel.[142,143] A number of factors should be considered and controlled to maximize reproducibility, including consistently sizing the width/depth of the pipette wound and maintaining a constant cell density between experimental conditions and controls.[144]

Materials

Growth endothelial cell medium and basal medium (adapted to the cells of choice), Corning 96-well plate (Sigma, CLS3595), VEGF or VEGF-C (Peprotech), pipettes, tips, and PBS.

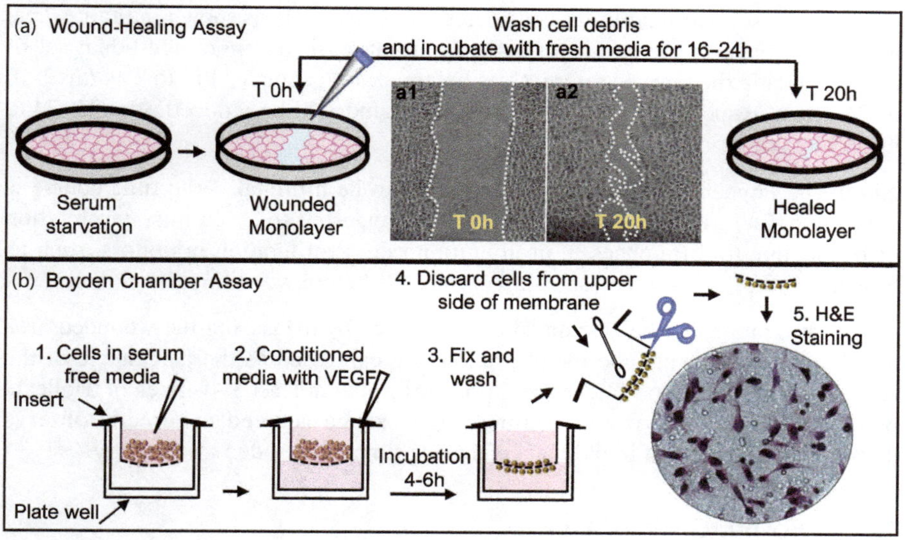

FIGURE 7.6 *In vitro* assays to evaluate endothelial cell migration. (a) Wound-healing assay. Prepare an endothelial cell monolayer. Scratch the monolayer with a pipette tip as indicated and allow the cells to migrate into the wound. The migratory behavior of cells can be measured by the rate at which they cover the wounded area. Panel (a1) shows an image taken just after wounding a monolayer of EA.hy 926 cells time 0 (T0 h). Panel (a2) shows the cells after 20 hours of migration (T20 h). Dotted lines indicate the width of the gap. (b) Boyden Chamber Assay Cells in serum free media are seeded into a cell culture insert and placed inside the well of a cell culture plate. Media containing the chemoattractant is added to the bottom chamber. The cells migrate from the upper chamber to the lower chamber through pores in the membrane and remain attached to the bottom of the filter. Imaging shows H&E staining of cells that have migrated to the underside of the filter.

Method

Seed resuspended endothelial cells in culture medium at 0.5–3×10^5 cells/mL in a 96 well plate.

Note: Cell density is cell-line dependent; we recommend optimizing the conditions for each cell line prior to migration assays.

When cells reach 80% confluence, remove endothelial cell media and add endothelial basal media supplemented with 0.2% FBS (to synchronize the cell cycle). The next morning, gently scratch the monolayer with a 1 μL sterile pipette tip across the center of the well. To ensure that the line is straight, place a sterile ruler on top of the plate as a guide. To mark specific fields for image acquisition, scratch the underside of the plate (2–3 lines) in a specific location near the main wound. After scratching the monolayer gently, wash the well twice with basal media to remove detached cells. Add fresh basal media supplemented with growth factors of interest (e.g., VEGF-A for HUVECs, VEGF-C for LECs, etc.), and place the plates under a phase-contrast microscope leaving the marked reference at the edge of the captured field. This first image capture is considered time zero (T0), as shown in Figure 7.6a. Return the plate to the incubator and capture at least one more additional set of images placing the marked reference as before at, for example, T16 to T24 (after 16 or 24 hours, respectively). The width of the wound will be reduced after 16–24 h, mostly by migration of cells into the wound.

Note: The impact of proliferation is expected to be minimal if the time course is kept to 16–24 h. Longer time points are not suggested to avoid misinterpretation of results due to differences in proliferation rate. Proliferation inhibitors (such as 5-hydroxyurea) can also be used to treat the cells before wounding the monolayer.

Cell migration can be calculated using ImageJ by measuring the wounded area at different time points and the migration rate expressed as the percentage of the reduced area. Each experiment should be performed at least 3–4 times in triplicate or quadruplicate. The wound-healing assay can be adapted to directly observe directional migration of individual cells using time-lapse video microscopy.[141,145–148]

7.6.2 Boyden Chamber Assay

The Boyden chamber assay (also named transwell assay) is commonly used to study directional endothelial cell migration (chemotaxis) by establishing a gradient of inducing/repressing angiogenesis stimuli (Figure 7.6b).[149] Endothelial cells resuspended in media (usually basal endothelial cell media with no additives) are placed in an upper chamber with a permeable membrane coated with an extracellular matrix component (usually gelatin or Matrigel). The same basal media containing a chemoattractant is added to the lower chamber (as shown in Figure 7.6b). The cells, attracted by the stimulus, migrate through the membrane and attach to the other side of the membrane. Cell migration can be measured by staining and counting the number of cells attached to the membrane after a specific period of time. *PKD1*- or *PKD2*-depleted lymphatic endothelial cells migrate less efficiently toward VEGF-C gradients, indicating a defect in directional migration.[33]

Materials

Subconfluent endothelial cells, Corning Transwell polycarbonate membrane cell-culture inserts with 8.0 μm pore (Corning Inc., cat. no. CLS3422–48EA), Dulbecco's Modified Eagle's Medium (DMEM, Gibco BRL), human recombinant VEGF (Peprotech, cat. no. 100–20), human recombinant VEGF-C (Peprotech, cat. no. 100–20C) if using human endothelial cells or lymphatic endothelial cells respectively, BSA (Sigma, cat. no. 7906), Mayer's hematoxylin solution (Sigma-Aldrich, cat. no. MHS32), Eosin Stain Solution 5% (Sigma-Aldrich, cat. no. R03040–74), Permount mounting media (Fisher Scientific, cat. no. SP15–500).

Method

Coat the polycarbonate membranes with 0.1% gelatin for 5 minutes and wash with PBS. Block the membrane using 0.2% BSA. Detach endothelial cells from the plate using 0.25% Trypsin (Gibco) for 1 min and count triplicates of 2×10^4 cells each in 100 μL of basal endothelial cell medium or plain DMEM. Seed the cells (experimental and control) into the upper chamber. Add 500 μL of basal endothelial cell medium or plain DMEM supplemented with 0.5% FBS and the chemoattractant to the lower chamber. It is important to avoid bubbles in the bottom of the well that will interfere with a homogenous distribution of the cells in the membrane.

Note: Negative controls without chemoattractant in the lower chamber should be included.

Incubate plates for 4–6 h at 37°C and 5% CO_2 to allow the cells to migrate. After incubation, remove media from top and bottom chambers and wash the membrane three times with PBS. Fix the cells using 4% PFA for 10 min, and remove the cells that have attached to the membrane in the upper chamber by wiping it with a cotton swab. Wash the upper chamber 3 times with PBS to remove cell debris and stain the cells that had migrated onto the lower membrane surface with hematoxylin and eosin. Using forceps and a sharp razor, cut the membrane off the insert, place it on a microscope slide, and mount using a cover slide and Permount mounting media. Count the number of attached cells using a 10× objective using ImageJ (or FIJI). Data can be represented as the average number of migratory cells per well.

Note: The method can be adapted by replacing the polycarbonate membrane of the insert with polyethylene terephthalate (PET) and fluorescently labelling the cells with DAPI or phalloidin. Migrating cells can be counted using a fluorescent plate reader.[150]

7.7 *IN VITRO* ASSAYS TO EVALUATE ENDOTHELIAL CELL TUBE FORMATION

7.7.1 Capillary-Like Tube Formation Assay

The capillary-like or tube-like tubulogenesis assay is another well known "*in vitro*" method to study angiogenesis (Figure 7.7a). This is a versatile method since it allows the use of different endothelial cells (e.g., HUVECs, HMEC-1, or primary cells isolated from genetically modified animals) and different extracellular matrix components, and it is ideal for a quick screening of factors modulating angiogenesis.[151–155] When plated on a matrix (commonly Matrigel, fibrin, or collagen), endothelial cells quickly reorganize, migrate, and differentiate to interconnect and form a complex network with quantifiable tube-like structures that mimics "*in vivo*" angiogenesis.[156,157] These tube-like structures can be visualized using a phase-contrast or fluorescent microscope (as illustrated in Figure 7.7a). The most common matrix for 3D culture is Matrigel, which is a mixture of proteins secreted by murine Engelbreth-Holm-Swarm sarcoma cells and which resembles the extracellular environment of many tissues.[158,159] Collagen gels are extensively used in angiogenesis assays and fibrin gels have also become more popular.[160,161] It is important to keep variables such as passage number (especially for primary cells), cell density, and time allowed for the capillary network to form consistent across experimental groups, and therefore optimization may be required.

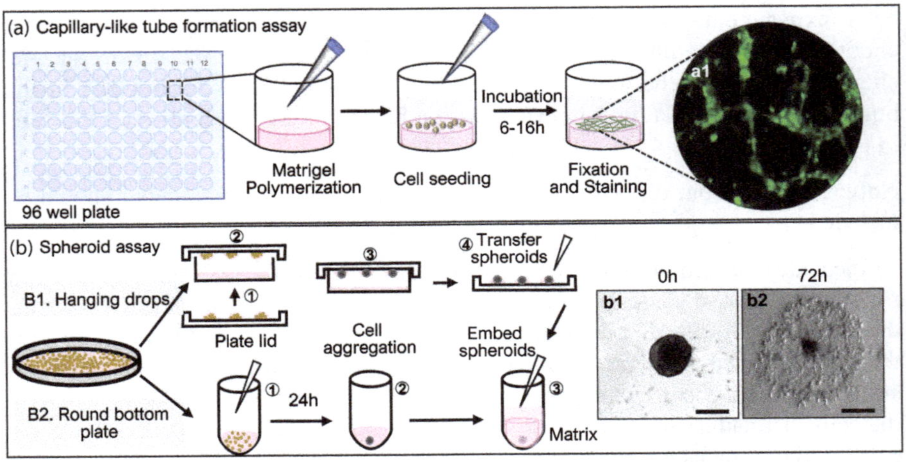

FIGURE 7.7 *In vitro* assays to evaluate endothelial cell tube formation. (a) Workflow for the capillary-like tube formation assay. A small volume of Matrigel (~50 μL) is added to the bottom of the wells in a 96-well plate and allowed to polymerize. Cells are seeded on top of the matrix and incubated for 6–16 h to allow the tubes to be form. In panel (a1), 2×10^4 LECs in 100 μL of VEGF-C enriched media were seeded on Matrigel and allowed to migrate for 10 h. After fixation, cells were stained with phalloidin. (b) Workflow for the spheroid assay using two different methods for generating spheroids. In the hanging drop method, panel (b1), spheroids are generated by cell aggregation in drops of media that hang from the lid of the tissue culture dish (steps 1–2). After 24–72 h of incubation (steps 3–4), the spheroids are transferred to a round-bottom plate and embedded in matrix (step 5). In the second method, panel (B2), spheroids are generated after seeding a mixture of cells, media, and methylcellulose in a non-adherent round-bottom well (step 1) that allows cell aggregation (step 2). Spheroids generated using one of these methods are next embedded in a matrix (step 3) and allowed to sprout and migrate over a period of time (usually 24–72 h). Panel (b1) shows a spheroid generated from EA.hy 926 cells after 24 h with the hanging drop method. The spheroids were embedded in fibrin gels and incubated for 72 hours, shown in panel (b2).

Materials

Endothelial cells, growth factor-reduced BD Matrigel (BD Bioscience, 354230), EBM-2 Basal Medium and EGM-2 SingleQuots Supplements (Lonza, CC-3156 and CC-4176), human recombinant VEGF (Peprotech, cat. no. 100–20), human recombinant VEGF-C (Peprotech, cat. no. 100–20C), Corning 96 Well TC-Treated Microplates (Sigma-Aldrich CLS3595).

Method

1. *Coating the cell tissue culture plate with Matrigel*: Thaw an appropriate volume of growth factor-reduced BD Matrigel at 4°C and coat prechilled plates, adding 50 μL of Matrigel per well. Avoid bubble formation and allow to polymerize at 37°C for 30 min.
2. *Cell preparation*: Seed endothelial cells in a 6-well plate at a density that will reach 80%–90% confluency within 24 h. Serum starve the cells using EBM-2 Basal Medium for 6–7 h or overnight. Trypsinize the cells and

determine the cell density by counting the cells. Resuspend 8×10^4 cells in 400 μL of EBM-2 Basal Medium supplemented with human recombinant VEGF if working with endothelial cells (30–60 ng/mL) or VEGF-C if working with LECs (200 ng/mL). Dispense 100 μL of the cell suspension in each Matrigel-coated well and incubate the plate at 37°C and 5% CO_2 for 6–16 h to allow capillary-like formation. Fix structures with 4% PFA and stain the cells using phalloidin and DAPI. Take photographs of at least three representative areas per well and count the number of tube-like structures, branched points, or tube length using software such as the Angiogenesis Analyzer plugin for Image J.

7.7.2 Spheroid Assay

The spheroid assay is a well-established, versatile, and widely used method to study angiogenesis in a 3D environment (Figure 7.7b). In contrast with the previously described methods, the spheroid assay allows the investigator to study all the steps involved in angiogenic sprouting.[162,9] In this method, endothelial cells are allowed to form aggregates, using the methods described below. When embedded in a matrix, endothelial cell aggregates or spheroids have the capacity to degrade the surrounding matrix and then migrate to form small sprouts, mimicking *in vivo* angiogenesis.[162] This method has been used to decipher key aspects of angiogenesis such as the role of Delta-Notch signaling pathways and to characterize the metabolic profile of vessels sprouts.[53,46,47] This method was also adapted for live cell imaging and was used to characterize the behavior of sprouting endothelial cells that compete for the tip cell position.[46,71]

Materials

Endothelial cells grown using manufacturer's protocol. Methylcellulose (Sigma-Aldrich, M0512), 10× M199 (Gibco BRL, Eggenstein, Germany), trypsin-EDTA (0.25%) (Gibco, cat. no. 25200056), heat-inactivated fetal bovine serum or FBS (Gemini's BenchMark, cat. no. 100–106), Corning Costar Ultra-Low Attachment Multiple Well Plate (Sigma-Aldrich, cat. no. CLS-7007–24EA), collagen I rat protein, tail (ThermoFisher, cat. no. A1048301), human recombinant VEGF (Peprotech, cat. no. 100–20), human recombinant VEGF-C (Peprotech, cat. no. 100–20C).

Method

1. *Generation of endothelial cell spheroids*: There are several factors that should be considered prior the initiation of the experiment. First the number of cells that can form the aggregates or spheroids varies between different cell lines (400–1200) and should be determined prior the initiation of the experiment.

Here, we summarize the two most common methods for generating spheroids:

2. *Hanging drops*: Grow endothelial cells, wash them twice with PBS, and trypsinize the cells to calculate the cell density using a cell hematocytometer. Neutralize the trypsin by adding cell growth medium and wash the cells

using the same medium. We have found that 400–600 HUVECs cells were required to form spheroids, but this should be optimized for each type of endothelial cell. Transfer 7.2×10^4 cells to a sterile multichannel pipette reservoir with a final volume of 3 mL, and pipet 25 μL onto the lid of a 96-well plate, forming small drops as illustrated in Figure 7.7b. Incubate the drops upside down at 37°C and 5% CO_2 for 24 h.

Note: Avoid bubbles when pipetting the drops; additionally, add some medium in the 96 wells to avoid evaporation.

3. *Nonadherent round bottom 96-well plates*: Resuspend 7.2×10^4 cells in a final volume of 3 mL of culture medium containing 0.25% (w/v) methylcellulose; pipet 25 μL of the solution to each well of a 96-well, clear, round-bottom ultra-low attachment microplate, and incubate the plate at 37°C and 5% CO_2 for h to allow cell aggregation.

Embedding the Spheroids

Prepare collagen stock solution by mixing 8 volumes collagen (at 2 mg/mL, 4°C) with 1 volume of 10× M199 medium and ~1 volume of 0.2 N NaOH to adjust the pH to 7.4, and mix with 4 mL ECGM basal medium (without supplements) containing 20% FBS and 0.5% (w/v) methylcellulose. Gently transfer the spheroids using a wide open 1 mL pipette tip or a 10 mL serological pipette to a 15 mL conical tube, spin down the spheroids (200 g for 5 min), remove supernatant, and add the collagen stock solution prepared above. Add 1 mL of the spheroid-collagen solution per well to a 24-well plate and incubate the plate at 37°C and 5% CO_2 for 30 min to allow collagen polymerization. Finally, add 100 μL of basal medium containing VEGF or VEGF-C (if using endothelial cells or lymphatic endothelial cells) and acquire the images representing the initial time point (T0). Next, incubate the plate at 37°C and 5% CO_2 for 24 h to up to 3 days. Stop the assay by adding 4% paraformaldehyde and acquire images of the different spheroids. The readout parameters include the number of sprouts per spheroid, average length of the sprouts, and the cumulative sprout length of all sprouts per spheroid. The experiments should be replicated at least 3–4 times using the same conditions.

7.8 CONCLUDING REMARKS

Vascular complications, including intracranial aneurysms, are a major cause of morbidity and mortality in autosomal dominant polycystic kidney disease (ADPKD). Almost all cases of ADPKD are caused by mutations in either *PKD1* or *PKD2*, encoding two integral membrane proteins, Polycystin-1 (PC1) and Polycystin-2 (PC2), which are expressed in all types of endothelial cells and vascular smooth muscle cells. *Pkd1* and *Pkd2* knockout mice die during mid-gestation with a dramatic vascular phenotype that includes focal hemorrhage, edema, defective vascularization of the fetal placenta and polyhydramnios. Selective inactivation of *Pkd1* or *Pkd2* in endothelial cells results in a subset of the "vascular" phenotypes previously identified in null animals including polyhydramnios and ~30%–40% perinatal lethality.

In addition, studies in murine and zebrafish models have demonstrated that edema in *Pkd1/2* null mice is due to abnormal lymphatic morphogenesis. At the cellular level, endothelial cells lacking expression of polycystins are defective in their response to fluid shear stress, while *Pkd1/2* mutant lymphatic endothelial cells fail to migrate properly. Together the data indicate a functional role for polycystins in the vasculature and specifically in the endothelial cell compartment. In this chapter, we summarized the tools that can be applied to study the role of polycystins in endothelial cells. These include mouse models and assays that use tissues from *Pkd1/2* mutant mice to measure vessel branching. In addition, we describe methods for generating *Pkd1/2* mutant endothelial cells, which can be used in a variety of standard assays to evaluate the effects of polycystin depletion.

ACKNOWLEDGMENTS

This work was supported by grant R01DK09503 to TW and by the NIDDK sponsored Baltimore Polycystic Kidney Disease Research and Clinical Core Center, P30DK090868. We thank members of the Baltimore PKD Center for helpful discussions.

REFERENCES

1. Torres, V.E., Harris, P.C., and Pirson, Y., Autosomal dominant polycystic kidney disease. *Lancet*, 2007. **369**(9569): 1287–1301.
2. The polycystic kidney disease 1 gene encodes a 14kb transcript and lies within a duplicated region on chromosome 16. The European polycystic kidney disease consortium. *Cell*, 1994. **78**(4): 725.
3. Mochizuki, T. et al., *PKD2*, a gene for polycystic kidney disease that encodes an integral membrane protein. *Science*, 1996. **272**(5266): 1339–42.
4. Hughes, J. et al., The polycystic kidney disease 1 (*PKD1*) gene encodes a novel protein with multiple cell recognition domains. *Nat Genet*, 1995. **10**(2): 151–60.
5. Harris, P.C. and Torres, V.E., Polycystic kidney disease. *Annu Rev Med*, 2009. **60**(0): 321–37.
6. Qian, F. et al., PKD1 interacts with PKD2 through a probable coiled-coil domain. *Nat Genet*, 1997. **16**(2): 179–183.
7. Su, Q. et al., Structure of the human PKD1-PKD2 complex. *Science*, 2018. **361**(6406).
8. Grieben, M. et al., Structure of the polycystic kidney disease TRP channel polycystin-2 (PC2). *Nat Struct Mol Biol*, 2017. **24**(2): 114–122.
9. Laib, A.M. et al., Spheroid-based human endothelial cell microvessel formation in vivo. *Nature Protocols*, 2009. **4**: 1202.
10. Gallagher, A.R., Germino, G.G., and Somlo, S., Molecular advances in autosomal dominant polycystic kidney disease. *Adv Chronic Kidney Dis*, 2010. **17**(2): 118–130.
11. Fick, G.M. et al., Causes of death in autosomal dominant polycystic kidney disease. *J Am Soc Nephrol*, 1995. **5**(12): 2048–56.
12. Pirson, Y., Chauveau, D., and Torres, V., Management of cerebral aneurysms in autosomal dominant polycystic kidney disease. *J Am Soc Nephrol*, 2002. **13**(1): 269–76.
13. Bobrie, G. et al., Spontaneous artery dissection: Is it part of the spectrum of autosomal dominant polycystic kidney disease? *Nephrol Dial Transplant*, 1998. **13**(8): 2138–41.
14. Robinson, P.N. et al., The molecular genetics of Marfan syndrome and related disorders. *J Med Genet*, 2006. **43**(10): 769–87.

15. Huston, J. 3rd et al., Value of magnetic resonance angiography for the detection of intracranial aneurysms in autosomal dominant polycystic kidney disease. *J Am Soc Nephrol*, 1993. **3**(12): 1871–7.
16. Ruggieri, P.M. et al., Occult intracranial aneurysms in polycystic kidney disease: Screening with MR angiography. *Radiology*, 1994. **191**(1): 33–9.
17. Irazabal, M.V. et al., Extended follow-up of unruptured intracranial aneurysms detected by presymptomatic screening in patients with autosomal dominant polycystic kidney disease. *Clin J Am Soc Nephrol*, 2011. **6**(6): 1274–85.
18. Graf, S. et al., Intracranial aneurysms and dolichoectasia in autosomal dominant polycystic kidney disease. *Nephrol Dial Transplant*, 2002. **17**(5): 819–23.
19. Xu, H.W. et al., Screening for intracranial aneurysm in 355 patients with autosomal-dominant polycystic kidney disease. *Stroke*, 2011. **42**(1): 204–6.
20. Vlak, M.H. et al., Prevalence of unruptured intracranial aneurysms, with emphasis on sex, age, comorbidity, country, and time period: A systematic review and meta-analysis. *Lancet Neurol*, 2011. **10**(7): 626–36.
21. Rinkel, G.J. et al., Prevalence and risk of rupture of intracranial aneurysms: A systematic review. *Stroke*, 1998. **29**(1): 251–256.
22. Lumiaho, A. et al., Mitral valve prolapse and mitral regurgitation are common in patients with polycystic kidney disease type 1. *Am J Kidney Dis*, 2001. **38**(6): 1208–1216.
23. Qian, Q. et al., Increased occurrence of pericardial effusion in patients with autosomal dominant polycystic kidney disease. *Clin J Am Soc Nephrol*, 2007. **2**(6): 1223–1227.
24. Neves, J.B., Rodrigues, F.B., and Lopes, J.A., Autosomal dominant polycystic kidney disease and coronary artery dissection or aneurysm: A systematic review. *Ren Fail*, 2016. **38**(4): 493–502.
25. Bouleti, C. et al., Risk of ascending aortic aneurysm in patients with autosomal dominant polycystic kidney disease. *Am J Cardiol*, 2019. **123**(3): 482–488.
26. Luciano, R.L. and Dahl, N.K., Extra-renal manifestations of autosomal dominant polycystic kidney disease (ADPKD): Considerations for routine screening and management. *Nephrology Dialysis Transplantation*, 2013. **29**(2): 247–254.
27. Boulter, C. et al., Cardiovascular, skeletal, and renal defects in mice with a targeted disruption of the *Pkd1* gene. *Proc Natl Acad Sci USA*, 2001. **98**(21): 12174–12179.
28. Griffin, M.D. et al., Vascular expression of polycystin. *J Am Soc Nephrol*, 1997. **8**(4): 616–26.
29. Torres, V.E. et al., Vascular expression of polycystin-2. *J Am Soc Nephrol*, 2001. **12**(1): 1–9.
30. Qian, Q. et al., Analysis of the polycystins in aortic vascular smooth muscle cells. *J Am Soc Nephrol*, 2003. **14**(9): 2280–7.
31. Qian, F. et al., Cleavage of polycystin-1 requires the receptor for egg jelly domain and is disrupted by human autosomal-dominant polycystic kidney disease 1-associated mutations. *Proc Natl Acad Sci USA*, 2002. **99**(26): 16981–6.
32. Garcia-Gonzalez, M.A. et al., Pkd1 and Pkd2 are required for normal placental development. *PLOS ONE*, 2010. **5**(9): e12821.
33. Outeda, P. et al., Polycystin signaling is required for directed endothelial cell migration and lymphatic development. *Cell Rep*, 2014. **7**(3): 634–644.
34. Bergmann, C. et al., Polycystic kidney disease. *Nat Rev Dis Primers*, 2018. **4**(1): 50.
35. Harris, P.C. and Torres V.E., Genetic mechanisms and signaling pathways in autosomal dominant polycystic kidney disease. *Journal Clin Invest*, 2014. **124**(6): 2315–2324.
36. Yoder, B.K., Hou, X., and Guay-Woodford, L.M., The polycystic kidney disease proteins, polycystin-1, polycystin-2, polaris, and cystin, are co-localized in renal cilia. *J Am Soc Nephrol*, 2002. **13**(10): 2508–16.
37. Pazour, G.J. et al., Polycystin-2 localizes to kidney cilia and the ciliary level is elevated in orpk mice with polycystic kidney disease. *Curr Biol*, 2002. **12**(11): R378–80.

38. Nauli, S.M. et al., Polycystins 1 and 2 mediate mechanosensation in the primary cilium of kidney cells. *Nat Genet*, 2003. **33**(2): 129-37.
39. Nauli, S.M. et al., Endothelial cilia are fluid shear sensors that regulate calcium signaling and nitric oxide production through polycystin-1. *Circulation*, 2008. **117**(9): 1161–71.
40. AbouAlaiwi, W.A. et al., Ciliary polycystin-2 is a mechanosensitive calcium channel involved in nitric oxide signaling cascades. *Circ Res*, 2009. **104**(7): 860–9.
41. Wang, D., Iversen, J., and Strandgaard, S., Endothelium-dependent relaxation of small resistance vessels is impaired in patients with autosomal dominant polycystic kidney disease. *J Am Soc Nephrol*, 2000. **11**(8): 1371–6.
42. Wang, D. et al., Endothelial dysfunction and reduced nitric oxide in resistance arteries in autosomal-dominant polycystic kidney disease. *Kidney Int*, 2003. **64**(4): 1381–8.
43. Muto, S. et al., Pioglitazone improves the phenotype and molecular defects of a targeted *Pkd1* mutant. *Hum Mol Genet*, 2002. **11**(15): 1731–42.
44. Lantinga-van Leeuwen, I.S. et al., Lowering of *Pkd1* expression is sufficient to cause polycystic kidney disease. *Hum Mol Genet*, 2004. **13**(24): 3069–77.
45. Hassane, S. et al., Pathogenic sequence for dissecting aneurysm formation in a hypomorphic polycystic kidney disease 1 mouse model. *Arterioscler Thromb Vasc Biol*, 2007. **27**(10): 2177–83.
46. De Bock, K. et al., Role of PFKFB3-Driven Glycolysis in Vessel Sprouting. *Cell*, 2013. **154**(3): 651–663.
47. Doebele, C. et al., Members of the microRNA-17–92 cluster exhibit a cell-intrinsic antiangiogenic function in endothelial cells. *Blood*, 2010. **115**(23): 4944–4950.
48. Hopp, K. et al., Functional polycystin-1 dosage governs autosomal dominant polycystic kidney disease severity. *Journal Clin Invest*, 2012. **122**(11): 4257–4273.
49. Lu, W. et al., Perinatal lethality with kidney and pancreas defects in mice with a targetted *Pkd1* mutation. *Nat Genet*, 1997. **17**(2): 179–81.
50. Piontek, K.B. et al., A functional floxed allele of *Pkd1* that can be conditionally inactivated in vivo. *J Am Soc Nephrol*, 2004. **15**(12): 3035–43.
51. Wu, G. et al., Cardiac defects and renal failure in mice with targeted mutations in *Pkd2*. *Nat Genet*, 2000. **24**(1): 75–8.
52. Yu, S. et al., Essential role of cleavage of polycystin-1 at G protein-coupled receptor proteolytic site for kidney tubular structure. *Proc Natl Acad Sci USA*, 2007. **104**(47): 18688–18693.
53. Adam, M.G. et al., Synaptojanin-2 binding protein stabilizes the notch ligands DLL1 and DLL4 and inhibits sprouting angiogenesis. *Circ Res*, 2013. **113**(11): 1206–1218.
54. Kim, K. et al., Polycystin 1 is required for the structural integrity of blood vessels. *Proc Natl Acad Sci USA*, 2000. **97**(4): 1731–1736.
55. Wu, G. et al., Somatic inactivation of *Pkd2* results in polycystic kidney disease. *Cell*, 1998. **93**(2): 177–88.
56. Muzumdar, M.D. et al., A global double-fluorescent Cre reporter mouse. *Genesis*, 2007. **45**(9): 593–605.
57. Livet, J. et al., Transgenic strategies for combinatorial expression of fluorescent proteins in the nervous system. *Nature*, 2007. **450**(7166): 56–62.
58. Payne, S., De Val, S., and Neal, A., Endothelial-specific Cre mouse models. *Arterioscler Thromb Vasc Biol*, 2018. **38**(11): 2550–2561.
59. Chakraborty, R. et al., Promoters to study vascular smooth muscle. *Arterioscler Thromb Vasc Biol*, 2019. **39**(4): 603–612.
60. Alva, J.A. et al., VE-Cadherin-Cre-recombinase transgenic mouse: A tool for lineage analysis and gene deletion in endothelial cells. *Dev Dyn*, 2006. **235**(3): 759–67.
61. Bazigou, E. et al., Genes regulating lymphangiogenesis control venous valve formation and maintenance in mice. *J Clin Invest*, 2011. **121**(8): 2984–2992.

62. Claxton, S. et al., Efficient, inducible Cre-recombinase activation in vascular endothelium. *Genesis*, 2008. **46**(2): 74–80.
63. Gustafsson, E. et al., Tie-1-directed expression of Cre recombinase in endothelial cells of embryoid bodies and transgenic mice. *J Cell Sci*, 2001. **114**(Pt 4): 671–6.
64. Holtwick, R. et al., Smooth muscle-selective deletion of guanylyl cyclase-A prevents the acute but not chronic effects of ANP on blood pressure. *Proc Natl Acad Sci USA*, 2002. **99**(10): 7142–7147.
65. Kisanuki, Y.Y. et al., Tie2-Cre transgenic mice: A new model for endothelial cell-lineage analysis in vivo. *Dev Biol*, 2001. **230**(2): 230–42.
66. Lepore, J.J. et al., High-efficiency somatic mutagenesis in smooth muscle cells and cardiac myocytes in SM22α-Cre transgenic mice. *Genesis*, 2005. **41**(4): 179–184.
67. Licht, A.H. et al., Endothelium-specific Cre recombinase activity in flk-1-Cre transgenic mice. *Dev Dyn*, 2004. **229**(2): 312–8.
68. Pham, T.H. et al., Lymphatic endothelial cell sphingosine kinase activity is required for lymphocyte egress and lymphatic patterning. *J Exp Med*, 2010. **207**(1): 17–27.
69. Pitulescu, M.E. et al., Inducible gene targeting in the neonatal vasculature and analysis of retinal angiogenesis in mice. *Nat Protoc*, 2010. **5**(9): 1518–34.
70. Srinivasan, R.S. et al., Lineage tracing demonstrates the venous origin of the mammalian lymphatic vasculature. *Genes Dev*, 2007. **21**(19): 2422–32.
71. Jakobsson, L. et al., Endothelial cells dynamically compete for the tip cell position during angiogenic sprouting. *Nature Cell Biology*, 2010. **12**: 943.
72. Coxam, B. et al., *Pkd1* regulates lymphatic vascular morphogenesis during development. *Cell Rep*, 2014. **7**(3): 623–633.
73. Patel, J. et al., Functional definition of progenitors versus mature endothelial cells reveals key SoxF-dependent differentiation process. *Circulation*, 2017. **135**(8): 786–805.
74. Pichol-Thievend, C. et al., A blood capillary plexus-derived population of progenitor cells contributes to genesis of the dermal lymphatic vasculature during embryonic development. *Development*, 2018. **145**(10): dev160184.
75. McMahon, A.P. et al., GUDMAP: The genitourinary developmental molecular anatomy project. *J Am Soc Nephrol*, 2008. **19**(4): 667–671.
76. Zhang, J. et al., Generation of an adult smooth muscle cell-targeted Cre recombinase mouse model. *Arterioscler Thromb Vasc Biol*, 2006. **26**(3): e23–4.
77. Wirth, A. et al., G12-G13-LARG-mediated signaling in vascular smooth muscle is required for salt-induced hypertension. *Nat Med*, 2008. **14**(1): 64–8.
78. Wendling, O. et al., Efficient temporally-controlled targeted mutagenesis in smooth muscle cells of the adult mouse. *Genesis*, 2009. **47**(1): 14–8.
79. Bar, T., Patterns of vascularization in the developing cerebral cortex. *Ciba Found Symp*, 1983. **100**: 20–36.
80. Fantin, A. et al., The embryonic mouse hindbrain as a qualitative and quantitative model for studying the molecular and cellular mechanisms of angiogenesis. *Nat Protoc*, 2013. **8**(2): 418–429.
81. Breier, G. et al., Expression of vascular endothelial growth factor during embryonic angiogenesis and endothelial cell differentiation. *Development*, 1992. **114**(2): 521–532.
82. Raab, S. et al., Impaired brain angiogenesis and neuronal apoptosis induced by conditional homozygous inactivation of vascular endothelial growth factor. *Thromb Haemost*, 2004. **91**(3): 595–605.
83. Ruhrberg, C. et al., Spatially restricted patterning cues provided by heparin-binding VEGF-A control blood vessel branching morphogenesis. *Genes Dev*, 2002. **16**(20): 2684–2698.

84. Fantin, A. et al., Tissue macrophages act as cellular chaperones for vascular anastomosis downstream of VEGF-mediated endothelial tip cell induction. *Blood*, 2010. **116**(5): 829–840.
85. Gerhardt, H. et al., Neuropilin-1 is required for endothelial tip cell guidance in the developing central nervous system. *Dev Dyn*, 2004. **231**(3): 503–9.
86. Sawamiphak, S., Ritter, M., and Acker-Palmer A., Preparation of retinal explant cultures to study ex vivo tip endothelial cell responses. *Nat Protoc*, 2010. **5**(10): 1659–65.
87. Fantin, A. and Ruhrberg, C., The embryonic mouse hindbrain and postnatal retina as in vivo models to study angiogenesis. *Methods Mol Biol*, 2015. **1332**: 177–88.
88. Stalmans, I. et al., Arteriolar and venular patterning in retinas of mice selectively expressing VEGF isoforms. *J Clin Invest*, 2002. **109**(3): 327–36.
89. Stone, J. et al., Development of retinal vasculature is mediated by hypoxia-induced vascular endothelial growth factor (VEGF) expression by neuroglia. *J Neurosci*, 1995. **15**(7 Pt 1): 4738–47.
90. Benedito, R. et al., The notch ligands Dll4 and Jagged1 have opposing effects on angiogenesis. *Cell*, 2009. **137**(6): 1124–35.
91. Hellstrom, M. et al., Dll4 signalling through notch1 regulates formation of tip cells during angiogenesis. *Nature*, 2007. **445**(7129): 776–80.
92. Dorrell, M.I. and Friedlander, M., Mechanisms of endothelial cell guidance and vascular patterning in the developing mouse retina. *Prog Retin Eye Res*, 2006. **25**(3): 277–95.
93. Connolly, S.E. et al., Characterization of vascular development in the mouse retina. *Microvasc Res*, 1988. **36**(3): 275–90.
94. Gariano, R.F. and Gardner, T.W., Retinal angiogenesis in development and disease. *Nature*, 2005. **438**(7070): 960–6.
95. Fruttiger, M., Development of the retinal vasculature. *Angiogenesis*, 2007. **10**(2): 77–88.
96. Ritter, M.R. et al., Three-dimensional in vivo imaging of the mouse intraocular vasculature during development and disease. *Invest Ophthalmol Vis Sci*, 2005. **46**(9): 3021–6.
97. Passaniti, A. et al., A simple, quantitative method for assessing angiogenesis and antiangiogenic agents using reconstituted basement membrane, heparin, and fibroblast growth factor. *Lab Invest*, 1992. **67**(4): 519–28.
98. Auerbach, R. et al., Angiogenesis assays: A critical overview. *Clin Chem*, 2003. **49**(1): 32–40.
99. Hasan, J. et al., Quantitative angiogenesis assays in vivo—A review. *Angiogenesis*, 2004. **7**(1): 1–16.
100. Nicoli, S. and Presta, M., The zebrafish/tumor xenograft angiogenesis assay. *Nat Protoc*, 2007. **2**(11): 2918–23.
101. Ribatti, D. et al., The gelatin sponge-chorioallantoic membrane assay. *Nat Protoc*, 2006. **1**(1): 85–91.
102. Norrby, K., In vivo models of angiogenesis. *J Cell Mol Med*, 2006. **10**(3): 588–612.
103. Auerbach, R. et al., Angiogenesis assays: Problems and pitfalls. *Cancer Metastasis Rev*, 2000. **19**(1–2): 167–72.
104. Akhtar, N., Dickerson, E.B., and Auerbach, R., The sponge/Matrigel angiogenesis assay. *Angiogenesis*, 2002. **5**(1–2): 75–80.
105. Malinda, K.M., In Vivo Matrigel Migration and Angiogenesis Assay, in *Angiogenesis Protocols: Second Ed.*, C. Murray and S. Martin, Eds. 2009, Humana Press: Totowa, NJ. 287–94.
106. Coltrini, D. et al., Matrigel plug assay: Evaluation of the angiogenic response by reverse transcription-quantitative PCR. *Angiogenesis*, 2013. **16**(2): 469–77.
107. Unseld, M. et al., In vivo tube assay: An optimised protocol of the directed in vivo angiogenesis assay by implementing immunohistochemistry. *J Vasc Res*, 2015. **52**(2): 116–26.

108. Kano, M.R. et al., VEGF-A and FGF-2 synergistically promote neoangiogenesis through enhancement of endogenous PDGF-B–PDGFRβ signaling. *J Cell Sci*, 2005. **118**(16): 3759–3768.
109. Au-Yuan, K. et al., In vivo study of human endothelial-pericyte interaction using the matrix gel plug assay in mouse. *J Vis Exp*, 2016. (118): e54617.
110. Kragh, M. et al., In vivo chamber angiogenesis assay: An optimized Matrigel plug assay for fast assessment of anti-angiogenic activity. *Int J Oncol*, 2003. **22**(2): 305–11.
111. Ley, C.D. et al., Angiogenic synergy of bFGF and VEGF is antagonized by angiopoietin-2 in a modified in vivo Matrigel assay. *Microvasc Res*, 2004. **68**(3): 161–8.
112. Nicosia, R.F. and Ottinetti, A., Growth of microvessels in serum-free matrix culture of rat aorta. A quantitative assay of angiogenesis in vitro. *Lab Invest*, 1990. **63**(1): 115–22.
113. Kruger, E.A. et al., Endostatin inhibits microvessel formation in the ex vivo rat aortic ring angiogenesis assay. *Biochem Biophys Res Commun*, 2000. **268**(1): 183–91.
114. Masson, V.V. et al., Mouse aortic ring assay: A new approach of the molecular genetics of angiogenesis. *Biol Proced Online*, 2002. **4**: 24–31.
115. Stiffey-Wilusz, J. et al., An ex vivo angiogenesis assay utilizing commercial porcine carotid artery: Modification of the rat aortic ring assay. *Angiogenesis*, 2001. **4**(1): 3–9.
116. Baker, M. et al., Use of the mouse aortic ring assay to study angiogenesis. *Nat Protoc*, 2011. **7**(1): 89–104.
117. Seano, G. et al., Modeling human tumor angiogenesis in a three-dimensional culture system. *Blood*, 2013. **131**.
118. Belleri, M. et al., Inhibition of angiogenesis by beta-galactosylceramidase deficiency in globoid cell leukodystrophy. *Brain*, 2013. **136**(Pt 9): 2859–75.
119. Bellacen, K. and Lewis, E.C., Aortic ring assay. *J Vis Exp*, 2009. (33): 1564.
120. Blatt, R.J. et al., Automated quantitative analysis of angiogenesis in the rat aorta model using Image-Pro Plus 4.1. *Comput Methods Programs Biomed*, 2004. **75**(1): 75–79.
121. Blacher, S. et al., Improved quantification of angiogenesis in the rat aortic ring assay. *Angiogenesis*, 2001. **4**(2): 133.
122. Aplin, A.C., Nicosia, R.F., The Rat Aortic Ring Model of Angiogenesis, In: *Vascular Morphogenesis: Methods in Molecular Biology (Methods and Protocols)*, D. Ribatti, Ed., 2015, vol. 1214, Humana Press: New York, NY.
123. Freedman, B.S. et al., Modelling kidney disease with CRISPR-mutant kidney organoids derived from human pluripotent epiblast spheroids. *Nat Commun*, 2015. **6**: 8715.
124. Hofherr, A. et al., Efficient genome editing of differentiated renal epithelial cells. *Pflugers Arch*, 2017. **469**(2): 303–311.
125. Kleene, S.J. and Kleene, N.K., The native TRPP2-dependent channel of murine renal primary cilia. *Am J Physiol Renal Physiol*, 2017. **312**(1): F96–F108.
126. Khan, A.A. et al., Transfection of small RNAs globally perturbs gene regulation by endogenous microRNAs. *Nat Biotechnol*, 2009. **27**(6): 549–555.
127. Guilinger, J.P. et al., Broad specificity profiling of TALENs results in engineered nucleases with improved DNA-cleavage specificity. *Nat Methods*, 2014. **11**(4): 429–435.
128. Hockemeyer, D. et al., Genetic engineering of human pluripotent cells using TALE nucleases. *Nat Biotechnol*, 2011. **29**(8): 731–734.
129. Cho, S.W. et al., Analysis of off-target effects of CRISPR/Cas-derived RNA-guided endonucleases and nickases. *Genome Res*, 2014. **24**(1): 132–141.
130. Sigoillot, F.D. and King, R.W., Vigilance and validation: Keys to success in RNAi screening. *ACS Chem Biol*, 2011. **6**(1): 47–60.
131. Kim, H. et al., Ciliary membrane proteins traffic through the Golgi via a Rabep1/GGA1/Arl3-dependent mechanism. *Nat Commun*, 2014. **5**: 5482.

132. Qiu, N., Zhou, H., and Xiao, Z., Downregulation of PKD1 by shRNA results in defective osteogenic differentiation via cAMP/PKA pathway in human MG-63 cells. *J Cell Biochem*, 2012. **113**(3): 967–976.
133. Kazenwadel, J. et al., In vitro assays using primary embryonic mouse lymphatic endothelial cells uncover key roles for FGFR1 signalling in lymphangiogenesis. *PLOS ONE*, 2012. **7**(7): e40497.
134. Denizot, F. and Lang, R., Rapid colorimetric assay for cell growth and survival. Modifications to the tetrazolium dye procedure giving improved sensitivity and reliability. *J Immunol Methods*, 1986. **89**(2): 271–7.
135. Ahmad, S. et al., Cholesterol interferes with the MTT assay in human epithelial-like (A549) and endothelial (HLMVE and HCAE) cells. *Int J Toxicol*, 2006. **25**(1): 17–23.
136. Gratzner, H.G. et al., The use of antibody specific for bromodeoxyuridine for the immunofluorescent determination of DNA replication in single cells and chromosomes. *Exp Cell Res*, 1975. **95**(1): 88–94.
137. Gratzner, H., Monoclonal antibody to 5-bromo- and 5-iododeoxyuridine: A new reagent for detection of DNA replication. *Science*, 1982. **218**(4571): 474–475.
138. Rodriguez, L.G., Wu, X., and Guan, J.-L., *Wound-Healing Assay*, in *Cell Migration: Developmental Methods and Protocols*, J.-L. Guan, Editor. 2005, Humana Press: Totowa, NJ. 23–29.
139. Yue, P.Y.K. et al., A simplified method for quantifying cell migration/wound healing in 96-well plates. *J Biomol Screen*, 2010. **15**(4): 427–433.
140. Lampugnani, M.G., Cell Migration into a Wounded Area In Vitro, in *Adhesion Protein Protocols*, E. Dejana and M. Corada, Eds. 1999, Humana Press: Totowa, NJ. 177–182.
141. Jonkman, J.E.N. et al., An introduction to the wound healing assay using live-cell microscopy. *Cell Adh Migr*, 2014. **8**(5): 440–451.
142. Kang, Y. et al., Knockdown of CD146 reduces the migration and proliferation of human endothelial cells. *Cell Res*, 2006. **16**: 313.
143. Kim, B. et al., Endothelial pyruvate kinase M2 maintains vascular integrity. *J Clin Invest*, 2018. **128**(10): 4543–4556.
144. Liang, C.-C., Park, A.Y., and Guan, J.-L., In vitro scratch assay: A convenient and inexpensive method for analysis of cell migration in vitro. *Nat Protoc*, 2007. **2**: 329.
145. Dormann, D., and Weijer, C.J., Imaging of cell migration. *EMBO J*, 2006. **25**(15): 3480–3493.
146. Beltman, J.B., Marée, A.F.M., and de Boer, R.J., Analysing immune cell migration. *Nat Rev Immunol*, 2009. **9**: 789.
147. Vitorino, P. and Meyer, T., Modular control of endothelial sheet migration. *Genes Dev*, 2008. **22**(23): 3268–3281.
148. Czirok, A., Endothelial cell motility, coordination and pattern formation during vasculogenesis. Wiley interdisciplinary reviews. *Syst Biol Med*, 2013. **5**(5): 587–602.
149. Boyden, S., The chemotactic effect of mixtures of antibody and antigen on polymorphonuclear leucocytes. *J Exp Med*, 1962. **115**: 453–66.
150. Goukassian, D. et al., Overexpression of p27Kip1 by doxycycline-regulated adenoviral vectors inhibits endothelial cell proliferation and migration and impairs angiogenesis. *FASEB J*, 2001. **15**(11): 1877–1885.
151. Guo, H. et al., Comparison of two in vitro angiogenesis assays for evaluating the effects of netrin-1 on tube formation. *Acta Biochimica et Biophysica Sinica*, 2014. **46**(9): 810–816.
152. Liu, J. et al., Caveolin-1 expression enhances endothelial capillary tubule formation. *J Biol Chem*, 2002. **277**(12): 10661–8.
153. DeCicco-Skinner, K.L. et al., Endothelial cell tube formation assay for the in vitro study of angiogenesis. *J Vis Exp*, 2014. (91): e51312–e51312.

154. Guo, S. et al., Assays to examine endothelial cell migration, tube formation, and gene expression profiles. *Methods Mol Biol*, 2014. **1135**: 393–402.
155. O'Connell, K.A. and Edidin, M., A mouse lymphoid endothelial cell line immortalized by simian virus 40 binds lymphocytes and retains functional characteristics of normal endothelial cells. *J Immunol*, 1990. **144**(2): 521–5.
156. Arnaoutova, I. and Kleinman, H.K., In vitro angiogenesis: Endothelial cell tube formation on gelled basement membrane extract. *Nat Protoc*, 2010. **5**: 628.
157. Arnaoutova, I. et al., The endothelial cell tube formation assay on basement membrane turns 20: State of the science and the art. *Angiogenesis*, 2009. **12**(3): 267–74.
158. Montanez, E. et al., Comparative study of tube assembly in three-dimensional collagen matrix and on Matrigel coats. *Angiogenesis*, 2002. **5**(3): 167–72.
159. Connolly, J.O. et al., Rac regulates endothelial morphogenesis and capillary assembly. *Mol Biol Cell*, 2002. **13**(7): 2474–2485.
160. Wolf, K. et al., Collagen-based cell migration models in vitro and in vivo. *Semin Cell Dev Biol*, 2009. **20**(8): 931–941.
161. Morin, K.T. and Tranquillo, R.T., In vitro models of angiogenesis and vasculogenesis in fibrin gel. *Experimental cell research*, 2013. **319**(16): 2409–2417.
162. Korff, T. and Augustin, H.G., Tensional forces in fibrillar extracellular matrices control directional capillary sprouting. *J Cell Sci*, 1999. **112(Pt 19)**: 3249–58.

8 Energy Metabolism, Metabolic Sensors, and Nutritional Interventions in Polycystic Kidney Disease

Sonu Kashyap and Eduardo Nunes Chini

CONTENTS

8.1 INTRODUCTION

Autosomal dominant polycystic kidney disease (ADPKD) is the most common genetic cause of renal insufficiency, affecting more than 600,000 Americans and about 12.5 million people worldwide. ADPKD is caused by mutations to the *PKD1* or *PKD2* gene and is characterized by accumulation of renal cysts that ultimately lead to loss of renal function and renal failure, accounting for 5%–10% of overall end-stage renal disease (ESRD).[1,2] Although the FDA recently approved one drug for ADPKD,[3] the treatment options for management of this disease remain largely supportive, including dialysis and renal transplantation. Thus, understanding the pathophysiology of ADPKD is imperative for the development of newer effective therapies for this genetic disorder.

Cystogenesis in ADPKD is believed to include two phases, cyst initiation and cyst progression. Cyst initiation is thought to be mediated by loss of function of the *PKD1* or *PKD2* gene, which encode for the polycystin-1 (PC1) and polycystin-2 (PC2) proteins, respectively.[2,4] The polycystins complex plays a key role in cystogenesis when mutations to *PKD1* or *PKD2* reduce the levels of PC1 or PC2 below a certain critical threshold that is required to prevent cystogenesis. The subsequent expansion of cysts occurs through a process that involves fluid secretion and cell proliferation. Cyst expansion results in high cystic burden in the kidney, compressing and damaging the surrounding tissue, leading to inflammation, fibrosis, and eventual replacement of the normal tissue, culminating in kidney failure.[2,4] Although the precise mechanisms of cystogenesis still remain to be elucidated, a number of signaling pathways implicated in these phenotype changes have been reported. Interestingly, recent studies indicate a central role of metabolic sensors and energy metabolism in the pathogenesis of this cystic disease.[5–7] In this chapter, we review the role of energy metabolism and metabolic sensors, as well as nutritional manipulations in ADPKD pathogenesis, and discuss future lines of investigation.

8.2 METABOLIC SENSORS IN ADPKD

In eukaryotes, metabolism is tightly controlled by nutrient and energy sensing pathways involving several metabolic sensors such as the mammalian target of rapamycin (mTOR) pathways, AMP-activated protein kinase (AMPK), and sirtuins (Figure 8.1). These metabolic sensors monitor nutrient availability and control

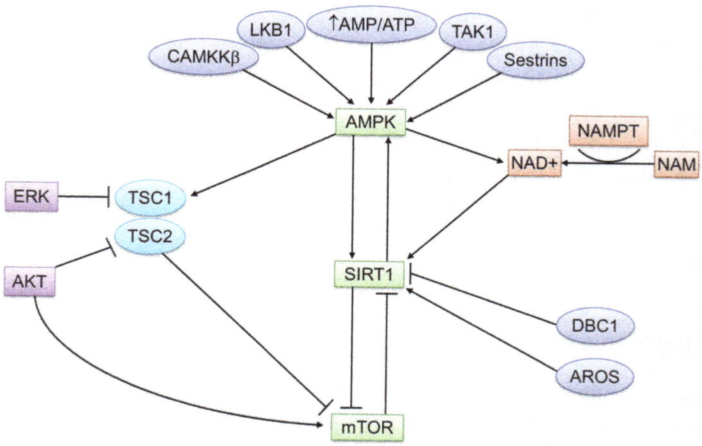

FIGURE 8.1 A simple schematic of metabolic sensors mTOR, AMPK, and SIRT1 and their regulating pathways. A straight arrow indicates stimulatory interactions, whereas a blunt line indicates inhibitory interactions.

metabolic adaptations in cells in response to environmental changes.[8] Interestingly, several studies indicate that these metabolic sensors also play a key role in the pathogenesis of ADPKD.[9–11] mTOR is the central regulator of the nutrient-sensing signaling pathway that controls cellular metabolism, protein synthesis, and cell growth. In fact, the first evidence recognizing the role of metabolism in ADPKD was the finding that mTOR plays a key role in the pathogenesis of this disease.[12] Elevated mTOR activity has been reported in ADPKD and is considered a major driver of cell proliferation.[11] Moreover, the inhibition of this pathway with rapamycin and other rapalogs is protective in experimental PKD.[11,13,14] The mTOR pathway is negatively regulated by the metabolic sensor AMPK, and activation of AMPK also reduces the cystogenesis in ADPKD.[9] Activation of SIRT1, a nicotinamide adenine dinucleotide-dependent (NAD-dependent) deacetylase also plays a role in the pathogenesis of ADPKD, and its inhibition shows a delay in cyst formation and growth.[10,15] These findings indicated that metabolic alterations play a key role in the pathogenesis of ADPKD.

Targeting mTOR pathways, AMPK and sirtuins (SIRT1) have been proposed as potential therapeutic targets in PKD and clinical trials to test some of modulators of these pathways are ongoing.[16] Unfortunately, to date, clinical trials exploring the potential therapeutic effects of mTOR inhibitors have failed to provide clear-cut beneficial effects.[11,17] Therefore, understanding the precise mechanisms that control metabolic sensors and signaling pathways in ADPKD is crucial to developing novel and effective therapies for this disease.

8.2.1 The AMPK Pathway

AMPK, a serine/threonine kinase is a metabolic sensor of cellular energy supply consisting of three subunits: a catalytic α-subunit (α1 and α2), a scaffolding β-subunit (β1and β2), and a regulatory γ-subunit (γ1, 2, and 3).[18,19] The α-subunit has an N-terminus kinase domain that is activated via several upstream kinases and phosphatases. Liver kinase (LKB1), calmodulin-dependent protein kinase β (CaMKKβ), and transforming growth factor (TGF-β)–activated kinase 1 are the upstream kinases that phosphorylate the Thr172 residue of the AMPKα subunit.[19] Activation of AMPK is mediated in part by a small group of stress-induced proteins known as sestrins (Sesn), which lead to AMPK activation and inhibition of the mTOR pathway.[20] AMPK is also regulated by cellular energy status sensed by the ATP/AMP ratio. The γ-subunit binds AMP, ADP, and ATP and regulates the enzyme activity. The binding of ADP or AMP to the γ-subunit increases APMK activity via conformational change, leading to activation of the Ser/Thr kinase domain in the subunit and inhibiting the dephosphorylation of Thr172.[19,21] Upon activation, AMPK exerts a direct effect on many downstream substrates that are involved in transcriptional reprogramming, metabolism, cell cycle regulation, and proliferation, as well as inflammation. For example, under metabolic stress, AMPK inhibits the mTOR pathway and diminishes cell growth and proliferation.[22] Also AMPK is a negative regulator of glycolysis and tumor suppressor which regulates cell growth in unstressed proliferating cells by blocking a shift in metabolism toward aerobic glycolysis (Warburg effect). Therefore, AMPK activators are therapeutically beneficial in metabolic diseases and cancer. Additionally, AMPK phosphorylates and directly inhibits cystic fibrosis transmembrane regulator (CFTR),[23–25] which plays a significant role in cystic growth.[26] AMPK also plays a key role in development of ADPKD by modulating the mTOR pathways.[9] Therefore activating AMPK in ADPKD could provide beneficial effect in cystogenesis.

8.2.1.1 Pitfalls in Understanding a Direct Role of AMPK in ADPKD

Although it has been shown that AMPK activity is decreased in ADPKD, a direct role and regulation of AMPK in this disease is poorly understood (Table 8.1). To date, the role of AMPK in ADPKD has been mostly supported by the use of supra-pharmacological dosage studies with nonspecific indirect agonists. For example, metformin, a well-known indirect activator of AMPK, has been shown, in much higher concentrations than achievable pharmacological dosage in humans, to be effective in decreasing cysts.[9] The mechanism of action of metformin is not completely known, however it is very clear that its effect on AMPK activation is most likely indirect when used at high concentrations. The effects of supra-pharmacological concentrations of metformin are mediated by inhibition of mitochondrial complex-I[27] and increased AMP levels, which may lead to AMPK activation by an allosteric effect and inhibition of cAMP generation by adenylate cyclase.[28] In addition, AMPK-independent effects of metformin have also been documented.[29] Thus several questions remain open in understanding the role of AMPK in ADPKD pathogenesis.

8.2.1.2 Genetic and Pharmacological Manipulations of AMPK in ADPKD

To date, there is no direct evidence implicating AMPK in the pathogenesis of ADPKD. As mentioned above, metformin is not an ideal tool to study the functional role of AMPK; therefore genetic manipulations and more specific and direct activators of AMPK need to be tested in animal models of PKD (Table 8.1). These tools that are available include isoforms and tissues specific knockouts and transgenic animals,[30–32] dominant negative AMPK constructs,[33] and specific pharmacological activators of this enzyme such as A76, PF-064095577, and PF-249[34,35] (Table 8.1). In addition, animal models in which the electron transport activity of complex I is inhibited (e.g., the incorporation of metformin-resistant *Saccharomyces cerevisiae* NADH dehydrogenase NDI1 subunit to the mitochondria[36,37]) will be important to determine the relative role of complex I inhibition versus AMPK activation in ADPKD (Table 8.1). It would also be important to determine the cell-specific isoforms of AMPK that are involved in ADPKD.

8.2.1.3 Little Is Known about the AMPK Regulation in ADPKD

Sestrins are stress-induced proteins and their expression is induced by environmental and metabolic stresses including low nutrient availability.[38] Three isoforms of Sesn exist in mammalian cells, including Sesn1, 2, and 3.[38] One of the most important biological functions of Sesn is regulation of the AMPK-mTOR pathway in which Sesn accumulation activates AMPK and inhibits mTORC1.[38] Therefore, it is important to determine if Sesn also plays a role in AMPK regulation in ADPKD since it has not been explored so far whether or not Sesn is suppressed in the ADPKD. In particular, animal models of ADPKD deficient in sestrins[39,40] would be instrumental

TABLE 8.1
Limitations of the Studies Performed on Metabolism, Metabolic Sensors, and Nutritional Manipulations in ADPKD and Potential Tools or Approaches That Could Be Utilized in Future Studies to Overcome These Limitations

Targets	Limitations	Potential Tools/Approaches
AMPK	• Studies showing a direct role of AMPK in ADPKD are lacking • Use of nonspecific or indirect activator • Role of sestrins in AMPK regulation needs to be explored	• Tissue-specific knockout and transgenic animals[30–32] • Dominant negative AMPK constructs[33] • Specific and direct activators, e.g., A76, PF-064095577, and PF-249[34,35] • Animal models with the incorporation of metformin resistant *Saccharomyces cerevisiae* NADH dehydrogenase NDI1 subunit to the mitochondria[37] • Use of sestrin-deficient animals[39,40]
SIRT1	• Studies limited to rapidly progressive PKD animal models only • Use of nonspecific inhibitors and activators • Regulation of SIRT1 in ADPKD is unknown	• Study SIRT1 deficiency in slow progressive ADPKD animal models[61,62] • SIRT1 specific inhibitor and activators • Use of DBC1 knockout animals[68]
Defective metabolism	• It is not clear that the changes in HK2 expression in ADPKD cause the inhibition of the development of cystic disease, or are the marker of kidney injury improvement • Not all studies confirm defective metabolism in PKD and some report conflicting results • Mitochondrial dysfunction is secondary to kidney injury or part of cystogenesis is still unknown	• Determination of HK2-expressing cells • Improved cell culture conditions[91,92]
Nutritional manipulations	• Long-term dietary restriction might not be for everyone • Studies on role of IGF-1/IGF-R1 are still limited	• Alternative modes mimicking FR • Explore IGF1/IGF-1R pathway in ADPKD • Determination of role of specific calories and macronutrients in nutritional manipulations

to understand the role of AMPK regulators in ADPKD. One would predict that Sesn knockout and activation would suppress and augment PKD cystic growth and kidney dysfunction, respectively.

8.2.2 The SIRT1 Pathway

The NAD^+-dependent deacetylase SIRT1 is the most extensively studied member of the sirtuins family which serves as a metabolic master switch that is part of several physiological pathways.[41,42] SIRT1 uses NAD^+ as a substrate to promote deacetylation of target proteins affecting several cellular processes.[43–45] Sirtuins play an important role in several diseases, specifically those with alterations in metabolism and stress response. Activation of SIRT1 has shown protection against liver steatosis, type II diabetes, and cancer and also has been shown to delay some features of aging.[42–44,46] Although debated, the beneficial effects of dietary interventions might be mediated by SIRT1.[42–44]

To date, only one study, published by Zhou et al., has documented the role of SIRT1 in the pathogenesis of ADPKD.[10] The increased expression and activity of SIRT1 was observed in PKD mutant renal epithelial cells as well as tissues. The authors further proposed that increased SIRT1 expression regulated cystic epithelial cell proliferation and cell death through deacetylation and phosphorylation of Rb and deacetylation of p53, respectively, that promote changes in the equilibrium between cellular proliferation and apoptosis in PKD leading to cystic expansion.[10] In agreement with this, we have also observed that SIRT1 expression increases in ADPKD.[5]

8.2.2.1 Knowledge Gaps in the Study of SIRT1 in PKD

SIRT1 activation has shown multiple beneficial effects in many kidney diseases.[47] For example, SIRT1 activation decreases podocyte and tubular injury in models of acute kidney injury,[48–51] attenuates diabetic albuminuria,[52–55] reduces blood pressure,[56,57] and delays kidney fibrogenesis.[58,59] In contrast, the activation of SIRT1 has been proposed to promote cyst formation in ADPKD.[10] Similarly, the contrasting role of SIRT1 has also been proposed in cancer biology, where SIRT1 acts as both a tumor suppressor and an oncogene.[60] Thus, the role of SIRT1 in ADPKD is most likely context dependent. One explanation for this duality could be SIRT1-induced renal epithelial cell proliferation and decreasing apoptosis in acute injury. Therefore, SIRT1 promotes growth of the epithelium for renal recovery in ADPKD where kidney has abnormal cyst formation and damaging tubules. There are still several questions related to the role of SIRT1 in this cystic disease which remain outstanding (Table 8.1). In particular, the role of SIRT1 deficiency has been limited to models of rapidly progressive PKD. In this regard, the outcomes of SIRT1 deficiency in models of PKD that replicate the time course of ADPKD will be interesting to study (Table 8.1).[61,62] It is important to highlight that knockout of SIRT1 in non-PKD animal models can lead to a decrease in kidney size and weight of about 40% compare with SIRT1 wildtype.[63] Thus, it cannot be ruled out that the effects of SIRT1 knockout in animal models of ADPKD could be mediated by growth retardation independent of the PKD phenotype.

The use of nonspecific inhibitors, such as nicotinamide (NAM),[45] and indirect nonspecific activators of SIRT1, such as resveratrol, in ADPKD studies is another concern.[64,65] NAM is a form of vitamin B3 that in high concentrations can inhibit several NAD-dependent enzymes including sirtuins, CD38, and PARPs.[66] In addition, NAM can promote an increase in the tissue NAD levels, including in the kidney, which is protective against ischemic injury in the kidney.[67] SIRT1 activity is regulated not only by its expression, but also by the availability of its substrate NAD.[68] Interestingly, it has been shown in at least one study that NAD+ levels are much lower in animal models of ADPKD and that expression of the rate-limiting enzyme Nampt is significantly decreased in ADPKD patients and animal models.[69] Of interest is whether or not the increased expression of SIRT1 is a response to the decreased availability of NAD or are the effects of vitamin B3 (nicotinamide) mediated by SIRT1 inhibition or by tissues "NAD" boosting. Thus, the use of nicotinamide as a specific SIRT1 inhibitor is questionable and several aspects of the role of NAD metabolism and sirtuins require exploration in ADPKD.

Similarly, resveratrol does not seem to be a specific SIRT1 activator,[70,71] and it appears that resveratrol promotes SIRT1 activation by indirect mechanisms that involve inhibition of mitochondrial complex I[72] and phosphodiesterases.[73]

8.2.2.2 SIRT1 Regulation in ADPKD Is Unknown

SIRT1 has a complex regulatory network that includes protein interactions that regulate its activation and cellular localization.[68] In addition to regulation of SIRT1 by NAD intracellular NAD+ concentrations, protein deleted in breast cancer 1 (DBC1) is a key regulator of SIRT1 activity. The activation of SIRT1 is mediated by the dissociation of the SIRT1–DBC1 complex, which is dependent on the energy state of the cell.[68] SIRT1 is positively regulated by PKA and AMPK and is independent of NAD+ concentration changes. DBC1 deficiency in mice also leads to SIRT1 activation and the protection against features of metabolic syndrome such as liver steatosis indicating that SIRT1–DBC1 interaction serves as a metabolic sensor. Therefore, to better understand the role of SIRT1 in ADPKD, it is important to investigate regulation of SIRT1 by DBC1 and other endogenous modulators in ADPKD. Another reported cellular regulator of SIRT1 is the active regulator of SIRT1 (AROS), which directly regulates SIRT1 function.[74] However, no data are available on the potential role of AROS in PKD. Other modifications of SIRT1, such as phosphorylation, have been reported and may have roles in its regulation in PKD.

Further validation of the active role of metabolic sensor SIRT1 in cystogenesis and understanding of its regulation in ADPKD is important. Thus, the study of SIRT1 in the pathogenesis of PKD presents a significant opportunity for discoveries on the pathophysiology and therapeutics in PKD.

8.3 ENERGY METABOLISM IN ADPKD

Recently, it has been proposed that ADPKD cells develop an altered metabolic state with extreme dependency on aerobic glycolysis and a low rate of mitochondrial oxidative phosphorylation similar to the Warburg effect observed in cancer cells.[75–77] In the Warburg effect, cells exhibit a drastically higher glucose uptake and predominately produce energy by higher glycolytic rates. Following glycolysis, pyruvate is primarily converted into lactate in cytosol, shunting part of its glucose metabolism to the pentose pathway and the *de novo* synthesis of lipids, leading to the biosynthesis of nucleotides, and biological membranes that are crucial for cell growth and proliferation.[78] The cystic cells also exhibit similar abnormal glucose metabolism with higher glucose dependency, and these metabolic adaptations in ADPKD appear to be mediated by changes in the expression and activity of the key glycolytic enzymes hexokinase 1 and 2 (HK1 and HK2).[79,80] We also observed an increased expression of HK2 in ADPKD.[5] The inhibition of the glycolytic pathway using 2-deoxyglucose (2-DG) rescues experimental ADPKD.[75,80,81] Mitochondrial dysfunction has also been purposed to alter the metabolic pathways that are involved in the pathogenesis of ADPKD.[82,83] Fatty acid beta oxidation is also reported to be defective in cystic cells and potentiated by mitochondrial defects leading to the dysregulation of the peroxisome proliferator-activated receptor (PPAR) family, which plays an important role in fatty acid metabolism.[84,85] Reduced levels of PPAR-α and fatty acid oxidation have been reported in experimental ADPKD.[84,86] PPAR-α activator fenofibrate restores fatty acid oxidation and reduces cystogenesis,[84] and PPAR-Υ agonists rosiglitazone and pioglitazone attenuate fibrogenesis[87,88] in ADPKD.

8.3.1 Gaps in the Energy Metabolic Studies in ADPKD

Although the role of metabolic alterations have been reported in PKD, there are still gaps that exist in these studies (Table 8.1). As mentioned above, the inhibition of glycolysis can prevent the development and progression of PKD, but it is not clear if the changes in HK2 expression cause the inhibition of the development of cystic disease in ADPKD, or if they are the marker of kidney injury improvement. Therefore, it will be necessary to determine which cells in the kidney express HK2 to further determine its role in ADPKD. Similarly, whether mitochondrial dysfunction is secondary to kidney injury or part of the pathogenesis of cystic growth and expansion is not known. To date, not all studies confirmed defective metabolism in PKD and some even reported conflicting results.[5,89,90] This could be attributed to difference across animal and cellular models, cell culture conditions, or the type of mutation carried by the cells used in the specific study. In particular, the use of cell culture in metabolic studies is widespread and may not adequately translate into *in vivo* models. Understanding the limitations of *in vitro* studies can improve the reliability of results. For example, the levels of several metabolic substrates including glucose, pyruvate, amino acids, and oxygen are several-fold higher than concentrations observed *in vivo*.[91] Higher concentrations of nutrients typically found in commercial media can skew the metabolism of cancer cells generating undesired phenotypes. In fact, a recent study in cancer cells clearly demonstrates a role for cell media formulation in cell metabolism studies *in vitro*.[91,92] It has been demonstrated that the culture medium formulation significantly affects the results obtained in commonly used assays, such as those testing colony formation and gene expression in cancer cells. The development of culture media Plasmax, a medium composed of nutrients and metabolites at the concentration normally found in human plasma, has advanced metabolic studies *in vitro*.[92] To date, *in vitro* experiments involving metabolism studies in PKD cells have relied on standard cell culture conditions. Experimental conditions should be carefully considered when designing studies that aim to identify the dependency of cells on specific nutrients or metabolic pathways or when investigating the effects of antimetabolites.

8.4 NUTRITIONAL MANIPULATIONS IN ADPKD

Food/calorie restriction (FR), first described by McCay and Crowell in 1934, showed prolonged life spans of rats when fed a severely reduced calorie diet while maintaining micronutrient levels.[93] FR has been well-tolerated in laboratory animals, including mice, rats, monkeys, and humans, showing advantageous metabolic effects.[8,42,94] In addition, FR has also been reported to decrease oxidative stress and age-related renal dysfunction and prevent the development of ESRD in the rat remnant kidney model.[95,96] Health-span studies published on FR in nonhuman primates (rhesus monkeys) also confirms the health benefits,[97] showing reduction in age-related and all-cause mortality in these animals.[98] Furthermore, data from the "Comprehensive Assessment of Long-Term Effects of Reducing Intake of Energy (CALERIE)" trials also shown that FR leads to favorable results and a better metabolic profile in humans.[99,100] Severe FR also showed protection against breast cancer.[101]

Therefore, FR without malnutrition is a highly effective and reproducible approach that does not involve any genetic manipulations and provides the beneficial effects in aging as well as increased longevity.[15,94] Interestingly, beneficial effects of FR are mediated by several metabolic sensors such as the AMPK, mTOR-S6 K, and SIRT1 pathways.[15,94,102] FR leads to the inhibition of the mTOR-S6 K pathway and activation of AMPK.[8] Furthermore, metabolic sensors regulated by FR also appear to be involved in cystogenesis in PKD. Therefore, pharmacological intervention involving these pathways could mimic some features of FR which might be beneficial in this cystic disease. Our group postulated that FR may provide protection against the development of ADPKD.[5] In a study published in the *Journal of the American Society of Nephrology* (*JASN*), we clearly demonstrate for the first time that FR can ameliorate cystic disease burden in animal models of ADPKD.[5] Our data were independently confirmed and expanded by the group of Dr. Thomas Weimbs.[7] These studies opened new avenues for the understanding of ADPKD pathogenesis and treatment. In particular, we postulated that signaling pathways differentially regulated in the kidneys of *ad libitum* and FR ADPKD animals may provide a clue into the pathophysiology of ADPKD.

Therefore, our study demonstrating a protective role of FR in an ADPKD animal model has significant translational potential and points to a plausible role for long-term dietary interventions in patients who are at risk for ADPKD.

8.4.1 Limitations of Nutritional Manipulations in PKD

FR so far provides the best protective effects in ADPKD and has high translational potential and would provide a safe and inexpensive interventional approach for ADPKD patients, but there are still some limitations (Table 8.1). For example, long-term FR might have deleterious effects on some patients.[95,102] We observed that even mild FR was able to slow the progression of cysts in PKD mice, and therefore it will be crucial to evaluate the exact timing and degree of FR that may provide beneficial effects in ADPKD patients.

Developing alternative therapeutic approaches that will mimic dietary restriction will be of great clinical interest. Beneficial effects of FR appear to be mediated by suppression of IGF-1.[103,104] Our study also demonstrates the possibility of IGF-1/IGF-1R pathway mediated effects of FR.[5] Although IGF-1 has been reported in ADPKD patients and cystic cells,[105,106] the role of IGF-1 in cystogenesis in PKD is still limited. Therefore, the studies that can explore the precise mechanistic role of IGF-1 in ADPKD and its implications in FR will be imperative.[5]

Furthermore, the restriction of a single amino acid, methionine restriction (MR), can decrease IGF-1 levels and also mimic several beneficial effects of FR.[107] However, protein restriction in patients with renal failure and advanced ADPKD failed to show clear benefits.[108] Therefore, it is important to determine the specific role of calories and macronutrients in the nutritional manipulations.

8.5 CONCLUDING REMARKS

Energy metabolism, metabolic sensors, and nutritional manipulations play a key role in the pathogenesis of ADPKD. A large number of animal and cell culture models have been utilized to study the role of energy sensing metabolic sensors in ADPKD, and several molecules have been tested to explore their mechanisms of action. Despite advances, there are still gaps in our understanding of metabolic sensors and their role in the pathogenesis of ADPKD. Therefore, the mechanisms that are involved in these metabolic alterations and through which the metabolic sensors and nutritional manipulations interact with the metabolic and signaling pathways are imperative to study. In this regard, experimental tools and approaches discussed here can provide benefit and further directions to future studies.

REFERENCES

1. Chebib, F.T. and V.E. Torres, Autosomal dominant polycystic kidney disease: Core curriculum 2016. *American Journal of Kidney Diseases*, 2016. **67**(5): pp. 792–810.
2. Torres, V.E., P.C. Harris, and Y. Pirson, Autosomal dominant polycystic kidney disease. *The Lancet*, 2030. **369**(9569): pp. 1287–1301.
3. https://www.accessdata.fda.gov/drugsatfda_docs/nda/2018/204441Orig1s000TOC.cfm
4. Grantham, J.J., J.L. Geiser, and A.P. Evan, Cyst formation and growth in autosomal dominant polycystic kidney disease. *Kidney International*, 1987. **31**(5): pp. 1145–1152.
5. Warner, G. et al., Food restriction ameliorates the development of polycystic kidney disease. *Journal of the American Society of Nephrology: JASN*, 2016. **27**(5): pp. 1437–1447.
6. Rowe, I. and A. Boletta, Defective metabolism in polycystic kidney disease: Potential for therapy and open questions. *Nephrology Dialysis Transplantation*, 2014. **29**(8): pp. 1480–1486.
7. Kipp, K.R. et al., A mild reduction of food intake slows disease progression in an orthologous mouse model of polycystic kidney disease. *American Journal of Physiology-Renal Physiology*, 2016. **310**(8): pp. F726–F731.

8. Cantó, C. and J. Auwerx, Calorie restriction: Is AMPK a key sensor and effector? *Physiology*, 2011. **26**(4): pp. 214–224.
9. Takiar, V. et al., Activating AMP-activated protein kinase (AMPK) slows renal cystogenesis. *Proceedings of the National Academy of Sciences*, 2011. **108**(6): pp. 2462–2467.
10. Zhou, X. et al., Sirtuin 1 inhibition delays cyst formation in autosomal-dominant polycystic kidney disease. *Journal of Clinical Investigation*, 2013. **123**(7): pp. 3084–3098.
11. Ibraghimov-Beskrovnaya, O. and T.A. Natoli, mTOR signaling in polycystic kidney disease. *Trends in Molecular Medicine*, 2011. **17**(11): pp. 625–633.
12. Brook-Carter, P.T. et al., Deletion of the TSC2 and PKD1 genes associated with severe infantile polycystic kidney disease—a contiguous gene syndrome. *Nature Genetics*, 1994. **8**(4): pp. 328–332.
13. Wahl, P.R. et al., Inhibition of mTOR with sirolimus slows disease progression in Han:SPRD rats with autosomal dominant polycystic kidney disease (ADPKD). *Nephrology Dialysis Transplantation*, 2005. **21**(3): pp. 598–604.
14. Shillingford, J.M. et al., The mTOR pathway is regulated by polycystin-1, and its inhibition reverses renal cystogenesis in polycystic kidney disease. *Proceedings of the National Academy of Sciences*, 2006. **103**(14): pp. 5466–5471.
15. Bordone, L. and L. Guarente, Calorie restriction, SIRT1 and metabolism: Understanding longevity. *Nature Reviews Molecular Cell Biology*, 2005. **6**: p. 298.
16. Seliger, S.L. et al., A randomized clinical trial of metformin to treat autosomal dominant polycystic kidney disease. *American Journal of Nephrology*, 2018. **47**(5): pp. 352–360.
17. Ponticelli, C. and F. Locatelli, Autosomal dominant polycystic kidney disease and mTOR inhibitors: The narrow road between hope and disappointment. *Nephrology Dialysis Transplantation*, 2010. **25**(12): pp. 3809–3812.
18. Salt, I. et al., AMP-activated protein kinase: Greater AMP dependence, and preferential nuclear localization, of complexes containing the alpha2 isoform. *Biochemical Journal*, 1998. **334**(Pt 1): pp. 177–187.
19. Kim, J. et al., AMPK activators: Mechanisms of action and physiological activities. *Experimental & Molecular Medicine*, 2016. **48**(4): pp. e224–e224.
20. Budanov, A.V., J.H. Lee, and M. Karin, Stressin' Sestrins take an aging fight. *EMBO Molecular Medicine*, 2010. **2**(10): pp. 388–400.
21. Sanli, T. et al., AMP-activated protein kinase (AMPK) beyond metabolism: A novel genomic stress sensor participating in the DNA damage response pathway. *Cancer Biology & Therapy*, 2014. **15**(2): pp. 156–169.
22. Shaw, R.J. LKB1 and AMP-activated protein kinase control of mTOR signalling and growth. *Acta Physiologica*, 2009. **196**(1): pp. 65–80.
23. Kongsuphol, P. et al., Mechanistic insight into control of CFTR by AMPK. *Journal of Biological Chemistry*, 2009. **284**(9): pp. 5645–5653.
24. Davidow, C.J. et al., The cystic fibrosis transmembrane conductance regulator mediates transepithelial fluid secretion by human autosomal dominant polycystic kidney disease epithelium in vitro. *Kidney International*, 1996. **50**(1): pp. 208–218.
25. Hanaoka, K. et al., A role for CFTR in human autosomal dominant polycystic kidney disease. *American Journal of Physiology-Cell Physiology*, 1996. **270**(1): pp. C389–C399.
26. Li, H., I.A. Findlay, and D.N. Sheppard, The relationship between cell proliferation, Cl− secretion, and renal cyst growth: A study using CFTR inhibitors. *Kidney International*, 2004. **66**(5): pp. 1926–1938.
27. Fontaine, E., Metformin-Induced Mitochondrial Complex I Inhibition: Facts, Uncertainties, and Consequences. *Frontiers in Endocrinology*, 2018. **9**(753).
28. Miller, R.A. et al., Biguanides suppress hepatic glucagon signalling by decreasing production of cyclic AMP. *Nature*, 2013. **494**: p. 256.

29. Foretz, M. et al., Metformin: From mechanisms of action to therapies. *Cell Metabolism*, 2014. **20**(6): pp. 953–966.
30. Mount, P.F. et al., The outcome of renal ischemia-reperfusion injury is unchanged in AMPK-β1 deficient mice. *PLOS ONE*, 2012. **7**(1): p. e29887.
31. Woods, A. et al., Liver-specific activation of AMPK prevents steatosis on a high-fructose diet. *Cell Reports*, 2017. **18**(13): pp. 3043–3051.
32. Lazo-Fernández, Y. et al., Kidney-specific genetic deletion of both AMPK α-subunits causes salt and water wasting. *American Journal of Physiology-Renal Physiology*, 2017. **312**(2): pp. F352–F365.
33. Woods, A. et al., Characterization of the role of AMP-activated protein kinase in the regulation of glucose-activated gene expression using constitutively active and dominant negative forms of the kinase. *Molecular and Cellular Biology*, 2000. **20**(18): pp. 6704–6711.
34. Salatto, C.T. et al., Selective activation of AMPK β1-containing isoforms improves kidney function in a rat model of diabetic nephropathy. *Journal of Pharmacology and Experimental Therapeutics*, 2017. **361**(2): pp. 303–311.
35. Cool, B. et al., Identification and characterization of a small molecule AMPK activator that treats key components of type 2 diabetes and the metabolic syndrome. *Cell Metabolism*, 2006. **3**(6): pp. 403–416.
36. Irwin, M.H., K. Parameshwaran, K., and C.A. Pinkert, Mouse models of mitochondrial complex I dysfunction. *The International Journal of Biochemistry & Cell Biology*, 2013. **45**(1): pp. 34–40.
37. Wheaton, W.W. et al., Metformin inhibits mitochondrial complex I of cancer cells to reduce tumorigenesis. *eLife*, 2014. **3**: pp. e02242–e02242.
38. Lee, J.H., U.-S. Cho, and Karin, M., Sestrin regulation of TORC1: Is sestrin a leucine sensor? *Science Signaling*, 2016. **9**(431): pp. re5–re5.
39. Tao, R. et al., Sestrin 3 protein enhances hepatic insulin sensitivity by direct activation of the mTORC2-Akt signaling. *Diabetes*, 2015. **64**(4): pp. 1211–1223.
40. Peng, M., N. Yin, and Ming, O. Li, Sestrins function as guanine nucleotide dissociation inhibitors for rag GTPases to control mTORC1 signaling. *Cell*, 2014. **159**(1): pp. 122–133.
41. Guarente, L., Sirtuins as potential targets for metabolic syndrome. *Nature*, 2006. **444**: p. 868.
42. Baur, J.A. et al., Dietary restriction: Standing up for sirtuins. *Science*, 2010. **329**(5995): pp. 1012–1013.
43. Guarente, L. and F. Picard, Calorie restriction—the SIR2 connection. *Cell*, 2005. **120**(4): pp. 473–482.
44. Longo, V.D. and B.K. Kennedy, Sirtuins in aging and age-related disease. *Cell*, 2006. **126**(2): pp. 257–268.
45. Cantó, C. and J. Auwerx, Targeting sirtuin 1 to improve metabolism: All you need is NAD(+)? *Pharmacological Reviews*, 2012. **64**(1): pp. 166–187.
46. Li, X. et al., SIRT1 Deacetylates and positively regulates the nuclear receptor LXR. *Molecular Cell*, 2007. **28**(1): pp. 91–106.
47. Guan, Y. and C.M. Hao, SIRT1 and kidney function. *Kidney Diseases*, 2015. **1**(4): p. 258–265.
48. Hasegawa, K. et al., Kidney-specific overexpression of Sirt1 protects against acute kidney injury by retaining peroxisome function. *Journal of Biological Chemistry*, 2010. **285**(17): pp. 13045–13056.
49. Funk, J.A. and R.G. Schnellmann, Accelerated recovery of renal mitochondrial and tubule homeostasis with SIRT1/PGC-1α activation following ischemia-reperfusion injury. *Toxicology and Applied Pharmacology*, 2013. **273**(2): pp. 345–354.
50. Hao, C.-M. and V.H. Haase, Sirtuins and their relevance to the kidney. *Journal of the American Society of Nephrology*, 2010. **21**(10): pp. 1620–1627.

51. Fan, H. et al., The histone deacetylase, SIRT1, contributes to the resistance of young mice to ischemia/reperfusion-induced acute kidney injury. *Kidney International*, 2013. **83**(3): pp. 404–413.
52. Hasegawa, K. et al., Renal tubular Sirt1 attenuates diabetic albuminuria by epigenetically suppressing Claudin-1 overexpression in podocytes. *Nature Medicine*, 2013. **19**: p. 1496.
53. Wen, D. et al., Resveratrol attenuates diabetic nephropathy via modulating angiogenesis. *PLOS ONE*, 2013. **8**(12): p. e82336.
54. Baur, J.A. et al., Resveratrol improves health and survival of mice on a high-calorie diet. *Nature*, 2006. **444**: p. 337.
55. Milne, J.C. et al., Small molecule activators of SIRT1 as therapeutics for the treatment of type 2 diabetes. *Nature*, 2007. **450**: p. 712.
56. Gao, P. et al., Overexpression of SIRT1 in vascular smooth muscle cells attenuates angiotensin II-induced vascular remodeling and hypertension in mice. *Journal of Molecular Medicine*, 2014. **92**(4): pp. 347–357.
57. Miyazaki, R. et al., SIRT1, a longevity gene, downregulates angiotensin II Type 1 receptor expression in vascular smooth muscle cells. *Arteriosclerosis, Thrombosis, and Vascular Biology*, 2008. **28**(7): pp. 1263–1269.
58. He, W. et al., Sirt1 activation protects the mouse renal medulla from oxidative injury. *Journal of Clinical Investigation*, 2010. **120**(4): pp. 1056–1068.
59. Huang, X.-Z. et al., Sirt1 activation ameliorates renal fibrosis by inhibiting the TGF-β/Smad3 pathway. *Journal of Cellular Biochemistry*, 2014. **115**(5): pp. 996–1005.
60. Wilking, M.J. and N. Ahmad, The role of SIRT1 in cancer: the saga continues. *The American Journal of Pathology*, 2015. **185**(1): pp. 26–28.
61. Hopp, K. et al., Functional polycystin-1 dosage governs autosomal dominant polycystic kidney disease severity. *Journal of Clinical Investigation*, 2012. **122**(11): pp. 4257–4273.
62. Kurbegovic, A. and M. Trudel, Progressive development of polycystic kidney disease in the mouse model expressing Pkd1 extracellular domain. *Human Molecular Genetics*, 2013. **22**(12): pp. 2361–2375.
63. Boutant, M. and C. Cantó, SIRT1 metabolic actions: Integrating recent advances from mouse models. *Molecular Metabolism*, 2014. **3**(1): pp. 5–18.
64. Mohar, D.S. and S. Malik, The sirtuin system: The holy grail of resveratrol? *Journal of Clinical & Experimental Cardiology*, 2012. **3**(11): p. 216.
65. Villalba, J.M. and F.J. Alcaín, Sirtuin activators and inhibitors. *BioFactors (Oxford, England)*, 2012. **38**(5): pp. 349–359.
66. Chini, C.C.S., M.G. Tarragó, and E.N. Chini, NAD and the aging process: Role in life, death and everything in between. *Molecular and Cellular Endocrinology*, 2017. **455**: pp. 62–74.
67. Bulluck, H. and D.J. Hausenloy, Modulating NAD+ metabolism to prevent acute kidney injury. *Nature Medicine*, 2018. **24**(9): pp. 1306–1307.
68. Nin, V. et al., Role of Deleted in Breast Cancer 1 (DBC1) Protein in SIRT1 deacetylase activation induced by protein kinase A and AMP-activated protein kinase. *Journal of Biological Chemistry*, 2012. **287**(28): pp. 23489–23501.
69. Hwang, V.J. et al., Anticystogenic activity of a small molecule PAK4 inhibitor may be a novel treatment for autosomal dominant polycystic kidney disease. *Kidney International*, 2017. **92**(4): pp. 922–933.
70. Beher, D. et al., Resveratrol is not a direct activator of SIRT1 enzyme activity. *Chemical Biology & Drug Design*, 2009. **74**(6): pp. 619–624.
71. Pacholec, M. et al., SRT1720, SRT2183, SRT1460, and resveratrol are not direct activators of SIRT1. *Journal of Biological Chemistry*, 2010. **285**(11): pp. 8340–8351.

72. Desquiret-Dumas, V. et al., Resveratrol induces a mitochondrial complex I-dependent increase in NADH oxidation responsible for sirtuin activation in liver cells. *Journal of Biological Chemistry*, 2013. **288**(51): pp. 36662–36675.
73. Park, S.-J. et al., Resveratrol ameliorates aging-related metabolic phenotypes by inhibiting cAMP phosphodiesterases. *Cell*, 2012. **148**(3): pp. 421–433.
74. Kim, E.-J. et al., Active Regulator of SIRT1 cooperates with SIRT1 and facilitates suppression of p53 activity. *Molecular Cell*, 2007. **28**(2): pp. 277–290.
75. Rowe, I. et al., Defective glucose metabolism in polycystic kidney disease identifies a new therapeutic strategy. *Nature Medicine*, 2013. **19**: p. 488.
76. Ishimoto, Y. et al., Mitochondrial abnormality facilitates cyst formation in autosomal dominant polycystic kidney disease. *Molecular and Cellular Biology*, 2017. **37**(24): e00337–17.
77. Lian, X. et al., The changes in glucose metabolism and cell proliferation in the kidneys of polycystic kidney disease mini-pig models. *Biochemical and Biophysical Research Communications*, 2017. **488**(2): pp. 374–381.
78. Lu, J., M. Tan, and Q. Cai, The Warburg effect in tumor progression: mitochondrial oxidative metabolism as an anti-metastasis mechanism. *Cancer Letters*, 2015. **356**(2 Pt A): p. 156–164.
79. Chen, L. et al., Macrophage migration inhibitory factor promotes cyst growth in polycystic kidney disease. *Journal of Clinical Investigation*, 2015. **125**(6): p. 2399–2412.
80. Riwanto, M. et al., Inhibition of aerobic glycolysis attenuates disease progression in polycystic kidney disease. *PLOS ONE*, 2016. **11**(1): p. e0146654.
81. Chiaravalli, M. et al., 2-Deoxy-d-Glucose ameliorates PKD progression. *Journal of the American Society of Nephrology*, 2016. **27**(7): pp. 1958–1969.
82. Hwang, V.J. et al., The cpk model of recessive PKD shows glutamine dependence associated with the production of the oncometabolite 2-hydroxyglutarate. *American Journal of Physiology. Renal Physiology*, 2015. **309**(6): p. F492–F498.
83. Sun, Y. et al., Glutamine metabolism via glutaminase 1 in autosomal-dominant polycystic kidney disease. *Nephrology Dialysis Transplantation*, 2018. **33**(8): p. 1343–1353.
84. Lakhia, R. et al., PPARα agonist fenofibrate enhances fatty acid β-oxidation and attenuates polycystic kidney and liver disease in mice. *American Journal of Physiology-Renal Physiology*, 2018. **314**(1): p. F122–F131.
85. Liu, Y. et al., Rosiglitazone inhibits cell proliferation by inducing G1 cell cycle arrest and apoptosis in ADPKD cyst-lining epithelia cells. *Basic & Clinical Pharmacology & Toxicology*, 2010. **106**(6): pp. 523–530.
86. Hajarnis, S. et al., microRNA-17 family promotes polycystic kidney disease progression through modulation of mitochondrial metabolism. *Nature Communications*, 2017. **8**: p. 14395.
87. Liu, Y. et al., Rosiglitazone inhibits transforming growth factor-β1 mediated fibrogenesis in ADPKD cyst-lining epithelial cells. *PLOS ONE*, 2011. **6**(12): p. e28915.
88. Yoshihara, D. et al., PPAR-γ agonist ameliorates kidney and liver disease in an orthologous rat model of human autosomal recessive polycystic kidney disease. *American Journal of Physiology-Renal Physiology*, 2011. **300**(2): p. F465–F474.
89. Menezes, L.F. et al., Fatty acid oxidation is impaired in an orthologous mouse model of autosomal dominant polycystic kidney disease. *EBioMedicine*, 2016. **5**: pp. 183–192.
90. Padovano, V. et al., Metabolism and mitochondria in polycystic kidney disease research and therapy. *Nature Reviews Nephrology*, 2018. **14**(11): pp. 678–687.
91. Cantor, J.R. et al., Physiologic medium rewires cellular metabolism and reveals uric acid as an endogenous inhibitor of UMP synthase. *Cell*, 2017. **169**(2): pp. 258–272.e17.
92. Vande Voorde, J. et al., Improving the metabolic fidelity of cancer models with a physiological cell culture medium. *Science Advances*, 2019. **5**(1): p. eaau7314.
93. McCay CM, Crowell MF.: Prolonging the life span. *Sci Mon*. 1934. **39**: pp. 405–414.

94. Mair, W. and A. Dillin, Aging and survival: The genetics of life span extension by dietary restriction. *Annual Review of Biochemistry*, 2008. **77**(1): pp. 727–754.
95. Sohal, R.S. and R. Weindruch, Oxidative stress, caloric restriction, and aging. *Science*, 1996. **273**(5271): pp. 59–63.
96. Tapp DC, Wortham WG, Addison JF, Hammonds DN, Barnes JL, Venkatachalam MA. Food restriction retards body growth and prevents end-stage renal pathology in remnant kidneys of rats regardless of protein intake. *Lab Invest*, 1989. **60**: 184–195.
97. Mattison, J.A. et al., Impact of caloric restriction on health and survival in rhesus monkeys from the NIA study. *Nature*, 2012. **489**: p. 318.
98. Colman, R.J. et al., Caloric restriction reduces age-related and all-cause mortality in rhesus monkeys. *Nature Communications*, 2014. **5**: p. 3557.
99. Group, f.t.C.-S. et al., Body-composition changes in the comprehensive assessment of long-term effects of reducing intake of energy (CALERIE)-2 study: A 2-y randomized controlled trial of calorie restriction in nonobese humans. *The American Journal of Clinical Nutrition*, 2017. **105**(4): pp. 913–927.
100. Rickman, A.D. et al., The CALERIE study: Design and methods of an innovative 25% caloric restriction intervention. *Contemporary Clinical Trials*, 2011. **32**(6): pp. 874–881.
101. Michels, K.B. and A. Ekbom, Caloric restriction and incidence of breast cancer. *JAMA*, 2004. **291**(10): pp. 1226–1230.
102. Guarente, L., Calorie restriction and sirtuins revisited. *Genes & Development*, 2013. **27**(19): pp. 2072–2085.
103. Dunn, S.E. et al., Dietary restriction reduces insulin-like growth factor I levels, which modulates apoptosis, cell proliferation, and tumor progression in p53-deficient mice. *Cancer Research*, 1997. **57**(21): pp. 4667–4672.
104. Fontana, L. et al., Long-term effects of calorie or protein restriction on serum IGF-1 and IGFBP-3 concentration in humans. *Aging Cell*, 2008. **7**(5): pp. 681–687.
105. Alvaro, D. et al., Morphological and functional features of hepatic cyst epithelium in autosomal dominant polycystic kidney disease. *The American Journal of Pathology*, 2008. **172**(2): pp. 321–332.
106. Parker, E. et al., Hyperproliferation of PKD1 cystic cells is induced by insulin-like growth factor-1 activation of the Ras/Raf signalling system. *Kidney International*. **72**(2): pp. 157–165.
107. Malloy, V.L. et al., Methionine restriction decreases visceral fat mass and preserves insulin action in aging male Fischer 344 rats independent of energy restriction. *Aging Cell*, 2006. **5**(4): pp. 305–314.
108. Klahr, S. et al., Dietary protein restriction, blood pressure control, and the progression of polycystic kidney disease. Modification of Diet in Renal Disease Study Group. *Journal of the American Society of Nephrology*, 1995. **5**(12): pp. 2037–2047.

9 "Kidney in a Dish" Organoids for PKD

Nelly M. Cruz and Benjamin S. Freedman

CONTENTS

9.1 INTRODUCTION

Two decades after *PKD1* and *PKD2* were first discovered, how these genes normally function to prevent cystogenesis is still not fully understood.[1,2] A major barrier to deciphering PKD mechanistically has been a lack of human cellular models that faithfully recapitulate PKD-specific cystogenesis from tubules. Rodent and other vertebrate models replicate certain features of PKD, but do not fully genocopy or phenocopy human PKD or its treatment.[3,4] Animal models are furthermore highly complex, placing constraints on experimental approaches and throughput. *In vitro*, polarized epithelial cells in three-dimensional cultures can form hollow spheroids, sometimes called "cysts," but such structures arise even in nonmutant cells and are therefore not PKD-specific.[5,6]

To bridge this gap, human kidney organoid cultures have recently emerged as a new system for studying PKD.[7–11] This "kidney in a dish" organoid system has several strengths:

1. Human cysts form from kidney tubules in a PKD-specific way.
2. It is a flexible, defined-component system readily accessible to microscopy and experimental perturbation for extended periods of time.

3. *PKD1* and *PKD2* are endogenously expressed, and can be monitored with specific antibodies, or mutated using gene-editing systems.
4. It is amenable to automation and high-throughput screening using liquid-handling robots.

Here, we will briefly review the human kidney organoid system and its components, focusing on the aspects most relevant to PKD. We will then proceed to outline step-by-step protocols for generating kidney organoids and using them to reconstitute PKD cystogenesis.

9.2 KIDNEY ORGANOID CULTURES

The first techniques to differentiate human pluripotent stem cells (hPSC) stepwise into kidney organoids have recently been established.[8,12–14] hPSC include both human embryonic stem (ES) cells, which are cultured from blastocyst-stage embryos, and human induced pluripotent stem (iPS) cells, which are reprogrammed from somatic cells.[15,16] CHIR99021, a small molecule inhibitor of GSK3β and related kinases, is typically applied to hPSC to induce differentiation into the renal lineage. Differentiation from hPSC to kidney organoid typically takes approximately 3 weeks. A variety of different culture geometries are possible, ranging from adherent cultures similar to a typical monolayer, to spheroid aggregates of dissociated cells. Side-by-side comparisons of organoid differentiation protocols demonstrate that the cell types that emerge in these systems are similar regardless of the particular reagents or geometry used.[10,11,17,18]

In general, translucent, tubule-like structures begin to appear by phase contrast microscopy ~10 days after induction of nephron lineage differentiation with CHIR99021. These continue to grow for an additional week or so until they are fully mature. The hallmark feature of mature kidney organoids is the presence of nephron-like arrays with contiguous segments of distal tubules, proximal tubules, and podocytes along a distal-to-proximal axis.[8,12–14] A simple quality-control step for kidney organoid cultures is to stain the terminally differentiated structures for these three nephron segment cell types and confirm that they are present in the proper arrangements. Organoid tubules exhibit tissue-specific transport functions and responses to injury.[8,12,13] Podocytes, on the other hand, form tightly clustered aggregates with specialized basal junctions.[8,19] In addition to hPSC, kidney organoids can also be differentiated from cultured nephron progenitor cells, and from mouse pluripotent stem cells, although these have not yet been used to study PKD.[20–22]

While the hPSC system offers many advantages over conventional cell and organismal models of PKD, it also has limitations.[8,10] For instance, kidney organoids are relatively simple compared with the intact kidney, lack perfusion, and arise without bona fide vasculature or respiration. In addition, mature collecting ducts, which are an important source of PKD cysts *in vivo*, are absent in human kidney organoids and have yet to be successfully differentiated.[8,11,17,18,23] It should also be noted that hPSC differentiation and disease modeling work is both expensive and labor-intensive. For the initiate, it may be most helpful to confer or collaborate with an experienced laboratory. Such limiting factors must therefore be considered and balanced against the unique advantages of this system for studying PKD when determining whether or not to perform an experiment in the organoid background.

9.3 HUMAN PLURIPOTENT STEM CELLS WITH PKD MUTATIONS

Undifferentiated hPSC are immortal in culture, and can be established as clonal cell lines relatively easily.[15,16] Thus, hPSC lines provide a practically unlimited source of diverse cell types with naturally occurring or gene-edited mutations associated with disease states such as PKD.[7,8,24] Such mutant hPSC can be differentiated *de novo* into many lineages, which is advantageous because PKD affects many different organs.

Organoids derived from different individuals form with different efficiencies and may therefore be challenging to compare directly for outcomes related to disease.[8,10] This concern may be partially ameliorated by utilizing gene-editing systems, such as CRISPR/Cas9, to introduce mutations on a uniform genetic background.[8,25] Although gene editing is helpful in reducing variability between cell lines, gene editing may also have off-target effects, and even isogenic cell lines can exhibit significant differences unrelated to the disease mutation.[11] Phenotypes must therefore be assessed carefully, using multiple cell lines and repeated, independent trials.

Notably, hPSC in the undifferentiated state are polarized, ciliated epithelial cells that endogenously express *PKD1* and *PKD2*, providing quick and easy readouts for certain genotypic or phenotypic outcomes. To model PKD in organoids, we have established a cohort of hPSC with defined mutations in PKD-related genes including *PKD1*, *PKD2*, and *PKHD1*.[7,8,10] Other laboratories have also established PKD hPSC from patient cells.[26,27] In our experience, gene-edited cells are superior to patient-derived cells, and WA09 (H9) ES cells (WiCell) and WTC-11 iPS cells (Coriell Biorepository, GM25256) are optimal genetic backgrounds in which to perform most "PKD in a dish" experiments.[10] In addition to PKD lines, it may also be useful to cultivate hPSC with mutations in non-PKD genes, as a negative control. For instance, *PODXL*$^{-/-}$ hPSC are gene edited in a similar way to the PKD lines but do not exhibit cystogenesis phenotypes, rather they express a defect in podocyte junctional migration.[8,19]

9.4 ASSAYS FOR PKD IN ORGANOID MODELS

Experiments in the human PKD organoid system have revealed several disease-relevant phenotypes.[7–10] These include defective expression or localization of PKD gene products, increased formation of cysts from kidney tubules, and defects in the ability of the organoids to compact extracellular matrix. Cystogenesis phenotypes have thus far been limited to lines with biallelic, loss-of-function mutations in *PKD1* or *PKD2*. It is not yet known whether hPSC with heterozygous mutations in these genes, such as those found naturally in autosomal dominant PKD patients, have an increased tendency to form cysts, nor is it clear whether *PKHD1* mutations have a similar cystic phenotype. Such experiments should be carried out compared to isogenic negative controls (no mutations) and positive controls (e.g., *PKD1*$^{-/-}$).

For the purposes of modeling PKD, the differentiation system that we have used most successfully involves differentiating the organoids initially in adherent cultures for 3 weeks, after which the tubular structures are microdissected off the plate and transferred into suspension culture for an additional 2 weeks to form cysts

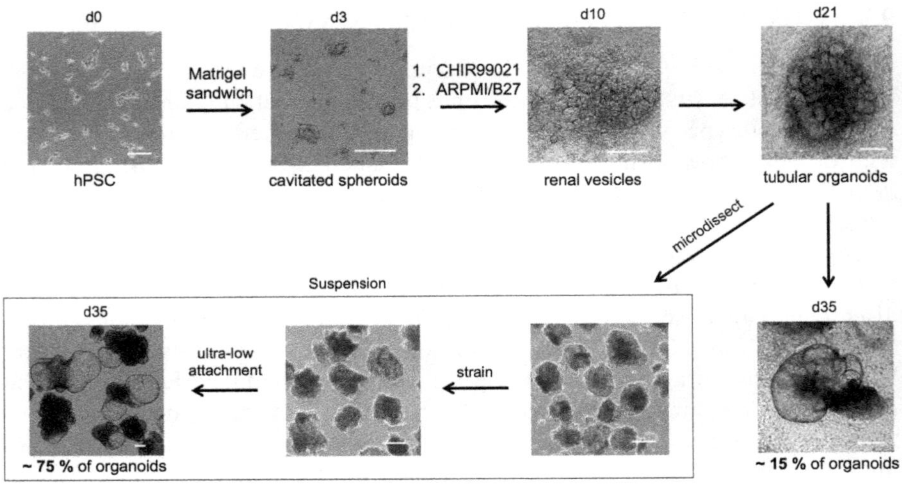

FIGURE 9.1 Summary of protocol for kidney organoid differentiation and PKD cystogenesis assay. hPSCs with $PKD1^{-/-}$ or $PKD2^{-/-}$ mutations are seeded at low density in Matrigel-coated plates. The next day, mTeSR1 containing Matrigel is used to feed the cells, resulting in the coating of the cells with a thin layer of extracellular matrix that promotes cavitated spheroid formation. A CHIR99021 pulse is used to induce the spontaneous formation of kidney organoids. Epithelial structures start to appear at day 10, and by day 21 successful cultures will contain three-dimensional tubular organoids. These can be cultured for 14 additional days, in which case a minority of organoids will form cysts, or microdissected at day 21 and placed in suspension cultures to promote robust cystogenesis. Scale bars, 200 μm.

(Figure 9.1).[10] The second step of transferring into suspension culture greatly increases the number of cysts that form. However, it is not absolutely necessary, as cysts will also form from PKD organoids left for the same amount of time in adherent cultures, although to a lesser degree.[8]

In the following step-by-step protocols, we will focus on the initial differentiation of hPSC into organoids, the generation of PKD cysts in suspension cultures, and analysis of these structures for phenotypes relevant to PKD.

9.5 STEP-BY-STEP METHODS

9.5.1 Differentiation of hPSC into Kidney Organoids

The starting material for the production of kidney organoids is hPSC (either ES or iPS) seeded at low density on tissue-culture plates precoated with a suitable matrix. We will describe an adherent culture differentiation protocol developed in our laboratory, which has been shown to produce PKD phenotypes in human kidney organoids.[8] Twenty-one days after plating hPSC to initiate the organoid differentiation protocol, successful cultures will have three-dimensional tubular structures, including podocytes, proximal tubules, and distal tubules in proper proximal-to-distal order and surrounded by other types of cells such as endothelial cells, stromal cells, and neurons.[8]

Setting up duplicate plates for each experiment is recommended, so that one plate can be fixed and stained for detection of nephron markers and confirmation that the three-dimensional structures are kidney organoids, while the other plate can be maintained further in culture to produce PKD cysts. With experience, kidney organoids can be identified by live phase microscopy because of their characteristic morphology consisting of distinct, translucent tubular structures. Each organoid is ~250 μm in diameter, and ~100 of these structures form per well of a 24-well plate.

Kidney organoids with PKD mutations will start to form cysts in adherent culture conditions ~28 days after plating hPSC (Figure 9.1). Typically, only one cyst forms per organoid, and cystogenesis rate is ~10% to 30% of all organoids in the well, depending on the genetic background of the cell line used and the particular clone. The cysts will expand in size over time, often surpassing the size of the organoids from where they originate. Organoids with cysts can be kept up to 4 weeks longer in adherent culture, but over time, the cultures become overgrown by other types of cells and the organoids start to deteriorate. Unhealthy organoids have a darker appearance and fuzzy, less-defined tubules.

Materials

- Feeder-free hPSC (50%–80% confluent)
- Matrigel (Corning #356231) or Geltrex (Thermo Fisher #A1413302)
- Accutase (StemCell Technologies #07920)
- mTeSR1 (StemCell Technologies #85850) supplemented with penicillin-streptomycin (Thermo Fisher #15140122)
- PBS pH 7.4 (Thermo Fisher #10010-049)
- Advanced RPMI (Thermo Fisher #12633020) supplemented with 1× GlutaMAX (#35050061) and Penicillin-Streptomycin (Thermo Fisher #15140122)
- "RB" media: Advanced RPMI supplemented with 1× GlutaMAX, penicillin-streptomycin (Thermo Fisher #15140122) and 1× B27 (#17504-044)
- ROCK inhibitor Y27632 (Tocris #1254), 10 mM stock solution in DMSO
- CHIR99021 (StemGent #04-004), 10 mM stock solution in DMSO
- Tissue culture plates (we prefer Midwest Scientific #TP92024 TPP Tissue Culture Plates)

Protocol

a. Coat a 24-well tissue culture plate with 1% Geltrex or 1.7% Matrigel by incubating the plate at 37°C for 1 h. The plate can be coated in advance and kept at 4°C for up to a week.
b. Dissociate cells with Accutase for 5–10 min at 37°C.
c. Collect dissociated cells and centrifuge at 300 rcf for 4 min at room temperature.
d. Carefully aspirate the Accutase and resuspend cell pellet in 1 mL of mTeSR supplemented with 10 μM ROCK inhibitor Y-27632. Pipette up and down 5–10 times with a P1000 pipetteman to generate a single-cell suspension.
e. Use preferred method to count cells and determine the concentration of the single-cell suspension.

f. Use mTeSR supplemented with 10 μM ROCK inhibitor to dilute cells to the desired concentration for plating. The optimal cell density will vary depending on the cell line and a range should be tested. As a starting point, plate 4 different cell densities in a 24-well plate (6 wells per cell density), at a range of 10,000 cells/well to 40,000 cells/well.
g. Tilt a coated 24-well plate 45° and aspirate the coating media. Add diluted cells to the plate, 500 μL/well. The suggested cell densities are a starting point, but certain hPSC lines may require lower or higher densities and they should be adjusted as necessary.
h. Inspect the plate under a light microscope. The cells should appear evenly spread across the well. If the cells appear aggregated, rock the plate back and forth several times to disperse them evenly across the well. Incubate at 37°C overnight.
i. During the morning of day 2, replace the medium of each well with 500 μL of ice-cold mTeSR supplemented with 1.5% Geltrex or 2.25% Matrigel. Incubate at 37°C overnight.
j. The next day (day 3), replace the medium of each well with 500 μL of mTeSR (prewarmed to room temperature).
k. During the evening of day 4, aspirate the mTeSR1 in each well and replace with 1 mL of Advanced RPMI supplemented with 1× GlutaMAX, penicillin-streptomycin, and with 12 μM CHIR99021 to induce differentiation. Incubate for 36 h at 37°C.
l. After 36 h in CHIR99021 (morning of day 6), replace the media with RB (1 mL/24-well). Aspirate and drip on the media gently from this point forward in the protocol, as the structures are delicate and can be easily disrupted.
m. Replace the RB media after 2 days and subsequently every third day after that. Discrete nests of translucent, convoluted tubular structures should emerge by day 18 at the latest in at least one of the conditions. Each nest of tubules counts as a separate "organoid." Different cell lines have different tendencies to make organoids. If organoids consistently do not emerge, troubleshoot cell number and concentration or supplier of CHIR99021 and other factors. When troubleshooting, utilize the WA09 ES cells and WTC-11 iPS cells as positive controls for organoid differentiation. Some hPSC lines may benefit from addition of 10 ng/mL Noggin (Peprotech #120-10C) or 2 ng/mL BMP4 (R&D 314-BP) in addition to the CHIR99021 during the induction step. We have not found it necessary to add fibroblast growth factor 9 or other factors described in alternative kidney organoid differentiation protocols.

9.5.2 Picking Organoids for Suspension Culture

Organoids can be "picked" (purified by microdissection) from their adherent plates and cultured in suspension. Suspension culture greatly enhances cystogenesis in PKD organoids and allows for longer culture periods. Organoids can be kept for months in this manner, since the organoids are initially relatively small (~250 μm in diameter), and they do not have necrotic cores that other organoid systems show during long-term cultures. PKD organoids will begin to form translucent cysts within 2 weeks of suspension culture

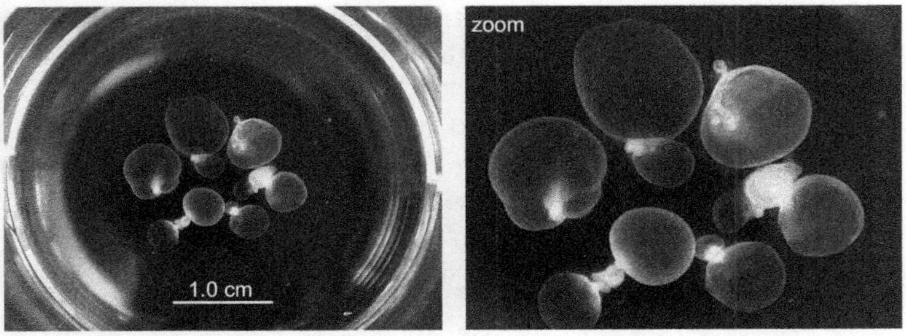

FIGURE 9.2 PKD organoids in 6-well (3.5 cm) dishes after several months of suspension culture. Fluid-filled translucent cysts continue to expand over time and can be visualized without a microscope. Scale bar, 1.0 cm.

(Figure 9.1). The cysts continue to expand in suspension cultures and reach diameters of millimeters and even centimeters in the subsequent months (Figure 9.2).

Reagents

- Adherent culture of mature human kidney organoids (day 21 after plating)
- 23 G needle and 1 mL syringe
- 40 μm cell strainer (Corning #352340)
- 50 mL conical tube
- 6-well ultralow attachment plate (Corning #3471)
- RB medium: Advanced RPMI 1640 Medium (Thermo Fisher #12633012) supplemented with 1× B-27 (Thermo Fisher #17504044), 1× GlutaMAX (Thermo Fisher #35050061), and 1× penicillin-streptomycin (Thermo Fisher #15140122)

Equipment

- Phase microscope
- P1000 pipette

Step-by-Step Protocol

a. Identify kidney organoids on an inverted phase contrast light microscope. Make adjustments necessary for sitting ergonomically and comfortably at the microscope. Ideally, use a microscope with ample working distance between the sample stage and the condenser unit.
b. Use a 23 G needle attached to a 1 mL syringe to detach organoids from the adherent culture. Hold the syringe like a pencil and use the cap to bend the needle in a 45°C angle, if necessary, to facilitate its manipulation. Tightly circumscribe the organoid, tracing its edges with the needle tip. Gently scrape away the cells surrounding the organoid. Touch lightly the base of the organoid with the needle and detach from the plate.

c. Repeat for all the desired organoids in the well. Swirl the plate in a tight circular motion to accumulate all the detached organoids in the center of the well.
d. Place a 40 μm cell strainer on a 50 mL conical tube, inverted so that the outer mesh is accessible (it will resemble a hat). Using a P1000 tip or a sterile plastic transfer pipette, collect all the organoids and place on the outer mesh of the 40 μm strainer.
e. Rinse the organoids by pipetting 1 mL of RB media onto the strainer and letting it drip through. Repeat twice. This step removes cell debris and non-organoid cells that were detached and transferred during the procedure.
f. Carefully invert the strainer containing the organoids into a 6-well ultralow attachment plate containing 1 mL of RB media. Pipette 500 μL of media throughout the surface of the strainer to flush all the organoids into the plate. Repeat one additional time to make sure all the organoids have been transferred from the strainer to the plate.
g. Gently rock the plate back and forth to disperse the organoids throughout the well and place in a 37°C incubator. Typically, 30 organoids are cultured in a 6-well plate containing 2 mL of media. Culturing too many organoids in a 6-well plate is not recommended, as they will collide with each other and merge into multi-organoid structures.

9.5.3 Feeding Organoids in Suspension Culture

Organoids in suspension culture should be fed once per week. If the media shows signs of becoming acidified (as indicated by the media turning from reddish-pink to yellow), feed more frequently.

Step-by-Step Protocol

a. Recline the plate at a 45° angle in the cell culture hood or biosafety cabinet. The plate can be resting on top of a 50 mL serological pipet, for example, to maintain the desired angle for 5 min. After 5 min, all the organoids will collect at the bottom edge of the well due to gravity.
b. Use a P1000 to manually remove media from the plate and discard, avoiding the organoids. Leave ~500 μL media in the plate.
c. Slowly add 2 mL of fresh RB media, drop by drop to not disturb the organoids. Rock the plate back and forth gently to disperse the organoids and put back in the 37°C incubator.

9.5.4 Fixation and Immunofluorescence Analysis of Organoids

Reagents

- Paraformaldehyde (16% aqueous solution, Electron Mycroscopy Sciences #15710)
- Phosphate buffered saline, pH 7.4 (Thermo Fisher #10010023)
- Blocking Buffer: 1× PBS, 5% normal goat serum, 0.3% Triton X-100

- Antibody Dilution Buffer: 1× PBS, 1% BSA, 0.3% Triton X-100
- DAPI (Thermo Fisher #D1306)

Step-by-Step Protocol A: Adherent Organoids

Organoids can be fixed and stained in their original adhesion culture plates. Alternatively, if confocal imaging is desired, the organoids can be picked and re-plated in glass bottom plates (Mat Tek #P96GC-1.5-5-F) coated with Geltrex or Matrigel. For most applications, No. 1.5 is the preferred coverslip thickness. The organoids should be fixed 1 or 2 days after re-plating. Longer culture times will lead to the spreading and flattening of the tubules and the three-dimensional architecture of the organoid will be lost. The volumes below are for a 24-well plate, and should be scaled up or down depending on the type of plate used.

a. Dilute fresh 16% paraformaldehyde (PFA) 1:1 with PBS to obtain an 8% solution. Remove half of the media in the well to be fixed (500 μL). Slowly add an equal amount of 8% PFA (500 μL) directly to the remaining media in the well, to reach a final concentration of 4% PFA. Rock the plate back and forth gently to mix. This method is preferred over removing the entire media and adding 4% directly to the cells because the organoid architecture is better preserved and cysts present in the culture are less likely to deflate.
b. Incubate the plate containing the PFA for 15 min at room temperature.
c. Gently aspirate the PFA and discard according to the appropriate chemical waste disposal guidelines.
d. Tilt the plate at a 45° angle and add 1 mL of PBS slowly, drop by drop, to the edge of the plate. Avoid pipetting directly on top of the cells.
e. Aspirate the PBS and repeat twice for a total of 3 PBS washes. The fixed plate can be left in PBS, sealed with parafilm, and stored at 4°C for later use.
f. Remove PBS and add 300 μL of blocking buffer. Incubate for 1 h at room temperature.
g. Remove blocking buffer and add 300 μL of primary antibody solution (antibodies of choice in antibody dilution buffer). Incubate overnight at 4°C.
h. Remove antibody dilution buffer and add 1 mL PBS. Aspirate the PBS and repeat 2 more times for a total of 3 PBS washes.
i. Remove PBS and add 300 μL of secondary antibody solution (antibodies of choice and 2 μg/mL DAPI in antibody dilution buffer). Incubate overnight at 4°C.
j. Remove antibody dilution buffer and add 1 mL PBS. Aspirate the PBS and repeat 2 more times for a total of 3 PBS washes. Leave PBS from the last wash in the plate. The plate is ready for imaging.

Step-by-Step Protocol B: Suspension Organoids

Organoids can also be fixed and imaged in suspension. The protocol for staining organoids in suspension is similar to the one described for adhesion cultures, but extra care needs to be taken when removing and adding solutions. Organoids are allowed to pellet to the bottom of the tube by gravity. Solutions are removed by manual pipetting and added slowly each time.

a. Transfer organoids to a 1.5 mL tube with a P1000 pipette or a transfer pipette. For transferring cystic organoids, cut the transfer pipette with a clean blade or scissors to make the opening wide enough.
b. Let the organoids settle to the bottom of the tube for 5 min. Then aspirate media, leaving ~100 μL of media.
c. Add 100 μL of 8% PFA (final concentration will be 4% PFA). Gently swirl the tube to mix. Incubate for 15 min at room temperature, slightly flicking the tube to suspend the organoids 2 times during the incubation period.
d. Remove most of the PFA, avoiding disturbing the organoids in the tube. Leave ~50 μL of liquid.
e. Slowly add 500 μL of PBS. Incubate for 5 min to allow the organoids to collect at the bottom of the tube.
f. Remove most of the PBS, leaving around 50 μL. Repeat steps e-f twice for a total of 3 PBS washes.
g. Remove PBS and add 200 μL of blocking buffer. Incubate for 1 h at room temperature.
h. Remove blocking buffer and add 200 μL of primary antibody solution (antibodies of choice in antibody dilution buffer). Incubate overnight at 4°C, in a rocking platform set up at a low speed setting.
i. The next day, place the tubes in a rack and let the organoids collect at the bottom of the tube. Remove antibody dilution buffer and add 500 μL PBS. Aspirate the PBS and repeat 2 more times for a total of 3 PBS washes, letting the organoids settle at the bottom of the tube each time.
j. Remove PBS and add 200 μL of secondary antibody solution (antibodies of choice and DAPI in antibody dilution buffer). Incubate overnight at 4°C, rocking gently.
k. Remove antibody dilution buffer and add 500 μL PBS. Aspirate the PBS and repeat 2 more times for a total of 3 PBS washes. Leave the organoids in the PBS. The organoids are ready for imaging.
l. Organoids can be stored in PBS at 4°C. For imaging, transfer the organoids using a P1000 or a transfer pipet into a glass bottom dish. Make the pipette or pipette tips wider, if necessary, by using a clean blade or sharp scissors to cut them.

9.5.5 Quantification of Cystogenesis

For most experiments, the cystogenesis rate can be assessed by quantifying the number of cystic organoids (i.e., organoids with one or more cysts) over time. Although most organoids typically produce only a single cyst, they may occasionally be multicystic, and in such cases, it may be useful to also quantify the number of cysts per organoid. In addition, the diameters or the area of cysts can be measured. We recommend taking images of the organoids, as these constitute the raw data for quantification. Microscope images can be analyzed morphometrically using ImageJ or another image-processing program.

Suspension cultures can be particularly challenging to image. By swirling the plate in a circular motion while horizontal, organoids can be collected to the center of the well for imaging. Usually, several pictures stitched together will be needed to capture the entire group of organoids into a large image. If using a microscope with a motorized stage, adjust the stage movement speed to slow as needed to prevent the organoids from moving too much between snapshots. Cystic organoids tend to float adrift and do not collect as easily to the center of the well. After swirling the plate and collecting most organoids to the center of the well, place the plate in the microscope stage. Then, use a sterile pipette tip to gently push any adrift organoids to the center with the rest of the organoids.

To assess a phenotype, it is important to compare control and mutant organoids differentiated side by side and on multiple separate occasions. It is also critical to include more than one mutant, more than one gRNA, and more than one isogenic control line in the cohort. Technical replicates from a single experiment (e.g., different wells containing organoids from a single differentiation) should be summed, not averaged. Such replicates are useful for assessing internal variability within a single experiment, but are irrelevant for assessing reproducibility from one experiment to the next. Rather, biological replicates should be used for all averaging and statistical analyses.

Biological replicates include either different cell lines with similar mutations that are differentiated side-by-side in a single experiment, or the same cell line differentiated on two different occasions. The total pooled set of biological replicates should include multiple cell lines for each genotype, as mentioned earlier, to avoid idiosyncrasies of any particular line. General strategies for successfully differentiating and phenotyping iPS-derived kidney cells should be followed, such as starting with multiple cell lines in each experiment and staggering experiments so that a steady pipeline of organoid production is maintained in the laboratory.[9]

9.5.6 Preparation of Protein Lysates from Kidney Organoids

Reagents

- RIPA Lysis and Extraction Buffer (Thermo Fisher #89900)
- Complete, mini, EDTA-free protease inhibitor cocktail (MilliporeSigma #11836170001)
- PhosSTOP phosphatase inhibitor (MilliporeSigma #4906845001)
- Benzonase nuclease (MilliporeSigma #E1014)

Equipment

- Microcentrifuge for centrifugation at 4°C

Step-by-Step Protocol

a. Preparing protease inhibitor and phosphatase inhibitor stocks
    - Complete, mini, EDTA-free protease inhibitor cocktail: Resuspend one tablet in 1.5 mL RIPA buffer for 7× protease inhibitor stock
    - PhosSTOP phosphatase inhibitor: Resuspend one tablet in 500 μL RIPA buffer for 20× phosphatase inhibitor stock

b. Prepare the lysis buffer. Aliquot the amount needed of RIPA buffer into a tube. Before using the buffer, add the following: 1× protease inhibitor, 1× phosphatase inhibitors, and Benzonase (250 U/mL). Mix by pipetting or vortexing the tube and keep in ice. Once the protease and phosphatase Inhibitors have been added to the lysis buffer, the buffer should be used within the next 4 h. Do not reuse leftover lysis buffer—use fresh protease and phosphatase inhibitors each time. Benzonase will lose activity rapidly after adding to the buffer and should be added right before use.

c. Pick the organoids and place in a 1.5 mL centrifuge tube. Let them settle in the bottom of the tube for 5 min. We typically start with a least 30–50 organoids to obtain sufficient material for downstream applications such as Western blot.

d. Remove media by manual pipetting and add 500 μL of ice-cold 1× PBS. Let organoids settle and remove the PBS. Repeat PBS wash once.

e. Centrifuge the organoids at 1000 rpm for 4 min at 4°C. Remove the PBS and resuspend the pellet with ice-cold lysis buffer. Use 2 μL of lysis buffer for each organoid picked and pipette up and down to thoroughly mix. Incubate the lysates on ice for 30 min, vortexing briefly every 5 min. This incubation step ensures that cells are fully lysed and that the proteins are completely solubilized.

f. Centrifuge the tubes at 14,000 rpm for 10 min at 4°C to pellet unwanted cellular debris.

g. Transfer the supernatant from each tube into a new, chilled 1.5 mL microcentrifuge tube. This is your final protein lysate.

h. At this point, the protein lysates can be stored at −20°C or −80°C for future use, or proceed directly to measure the protein content with standard protein quantification methods such as the bicinchoninic acid (BCA) method (e.g., Pierce BCA Protein Assay Kit, Catalog #23225).

9.5.7 Immunoblot Analysis of PKD Proteins

The *PKD1* gene product, polycystin-1, is a difficult protein to analyze by immunoblot. Nevertheless, immunoblots can be used very successfully to quantitatively assess

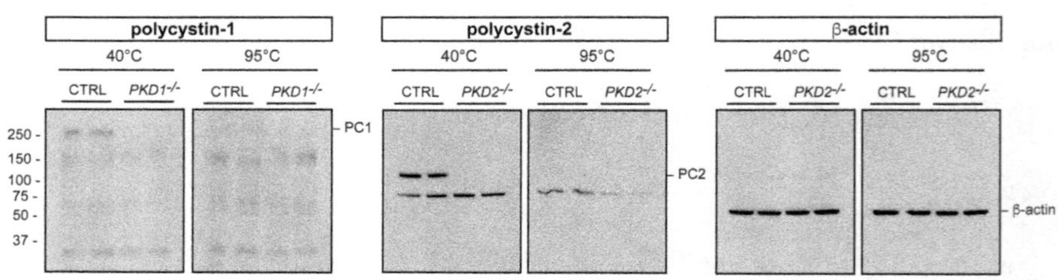

FIGURE 9.3 Immunoblots of PC1 and PC2 in genome-edited PKD hPSC or isogenic controls. Lysates were heated at 40°C or 95°C for 5 min, ran in a 4%–20% SDS-PAGE gel, transferred to a PVDF membrane, and blotted for PC1 (sc-130554), PC2 (sc-10376), or β-actin (Cell Signaling #4970). A nonspecific band is also observed in the PC2 blot (~80 kD).

levels of both polycystin-1 and the *PKD2* gene product, polycystin-2 (Figure 9.3). This analysis can be performed in undifferentiated hPSC or in differentiated organoids.

Reagents

- 4× Laemmli sample buffer (Bio-Rad #1610747 or equivalent)
- 10× Tris-glycine running buffer (25 mM Tris, 192 mM glycine, 0.1% SDS, pH 8.3)
- 1× Transfer buffer (25 mM Tris, 192 mM Glycine, 0.01% SDS, pH 8.3), prechilled at 4°C
- 1× TBST (20 mM Tris, 500 mM sodium chloride, 0.1% Tween 20, pH 7.5)
- SDS-PAGE 4%–12% gradient gel
- Blocking buffer: 5% dry nonfat milk powder diluted in TBST. Make fresh with each use
- Prestained protein standards (Precision Plus Protein Kaledioscope Prestained Protein Standards, Bio-Rad #1610375, or equivalent)
- PVDF membrane (Bio-Rad #1620177)
- Blot filter paper (Bio-Rad #1703932)
- Pierce ECL Western blotting substrate (Thermo Fisher #32106)
- Methanol
- Ponceau S stain (Sigma #P7170)

Equipment

- Power supply
- Electrophoresis chamber for protein gels
- Protein transfer equipment
- Chemiluminescence imaging system or X-ray film developer
- Rocker or shaker

Step-by-Step Protocol

a. Pipette the appropriate volume of protein lysate into a clean microcentrifuge tube. Load a significant mass of protein, at least 30 μg per lane of a 10-well gel. It is recommended to load at least one positive control (e.g., undifferentiated hPSC, which express the *PKD1* and *PKD2* genes) and one negative control (cells with biallelic knockout mutations in the gene of interest).
b. Add 4× sample buffer to each lysate for a final concentration of 1× sample buffer. Load all lanes of the gel, with the same volume, to ensure that the electrophoresis runs evenly across the gel. Adjust the volumes of each sample as necessary using 1× sample buffer (sample buffer diluted with PBS) to make them the same final volume.
c. Lysate samples are not boiled. Instead, heat them at 37°C–40°C for 5 min prior to loading. This is crucial for successful PC1 and PC2 blots (Figure 9.3).
d. Spin down the samples and protein standard tubes briefly (~30 s) and proceed to load them on the gel. Gradient gels are recommended (4%–20%)

for quantitative analysis. Load 1× sample buffer in any empty lane to ensure the samples run evenly.

e. Run the protein gel at 100–125 V until the ~20 kDa band in the protein standards ladder approaches the bottom of the gel and the top bands are well separated. Place the gel in a container with 1× transfer buffer.
f. Cut a piece of PVDF membrane to the same size of the running gel. Notch one corner of the membrane to help you remember which side of the membrane was next to the gel in the sandwich you will assemble in later steps.
g. Wet the PVDF membrane for 1 min in methanol. Use a clean pair of forceps when handling the membrane.
h. Transfer the membrane to a container with deionized water to rinse. Make sure the membrane is completely submerged in the water (it is hydrophobic, so it will tend to float).
i. Transfer the membrane to 1× transfer buffer (you can place it in the same container as the gel) and leave it until ready for transfer.
j. Assemble the transfer sandwich according to the transfer method of choice. Make sure there are no air bubbles between any layers of the sandwich; use a roller or a pipet to gently roll out any bubbles. Use the notched corner of the membrane and the location of the prestained ladder on the gel to track which side of the membrane is touching the gel (the "protein side" of the membrane). If using the wet/tank blotting method, transfer at 80 V for 2 h or 35 V overnight.
k. When the transfer is finished, disassemble the sandwich. If the transfer was successful, the prestained protein standards ladder should have transferred to the membrane. To verify that all proteins transferred properly, you can briefly stain the gel or membrane in Ponceau S (stains total protein). This will reveal any regions of uneven transfer, blemishes, and bubbles. Ponceau S stain can be removed from the membrane by rinsing with deionized water.
l. Incubate the membrane protein side up in blocking buffer on a rocker or shaker for 1 h at room temperature. Apply a sufficient volume of blocking buffer to completely cover the membrane.
m. Remove the blocking buffer and add primary antibody solution (antibody of choice diluted to the proper working concentration in blocking buffer). Incubate in a rocker or shaker overnight at 4°C. After incubation is complete, diluted antibody solution can be saved at –20°C and reused for a few additional blots. We use the following antibodies:
   - Polycystin-1 antibody: Santa Cruz Biotechnology sc-130554, 1:2000
   - Polycystin-2 antibody: Santa Cruz Biotechnology sc-25749, 1:1000 or (G-20) Santa Cruz Biotechnology sc-10376, 1:1000
n. Wash the membrane with 1× TBST 4 times, 5 min for each wash in a rocker or shaker at room temperature.
o. Incubate the membrane in secondary antibody solution (HRP-conjugated antibody against the species in which the primary antibody was raised, diluted appropriately in blocking buffer) on a rocker or shaker at room temperature for 1 h.
p. Wash the membrane with 1× TBST 4 times, 5 min for each wash in a rocker or shaker at room temperature.

q. To prepare the western detection solution, mix in a tube 1 mL of each of the two substrate solutions from the Pierce ECL Western blotting substrate kit. Using a clean pair of forceps, place the membrane protein-side face up in a clean container or on a piece of plastic wrap.
r. Pour the mixed detection solution onto the membrane (protein-side face up) and let sit for 1 min. Make sure that the entire surface of the membrane is evenly coated.
s. After 1 min, remove the detection solution. The membrane is now ready for developing, using a chemiluminescence imaging system or film. Try a short and a longer exposure. The PC1 blot may need a long exposure; if no bands are detected, use an ECL substrate with higher sensitivity. Compare the positive and negative controls to evaluate whether or not the blot worked. Typically, three bands are observed >250 kDa for polycystin-1, and polycystin-2 bands at ~110 (monomer) and >250 kDa (multimer) (Figure 9.3).
t. After developing, the membrane can be blotted for a loading control protein such as β-actin or GAPDH without the need of stripping the blot, as long as their molecular weight is not similar. Rinse briefly with TBST and proceed to incubating with the other antibody.
u. Blots can and should be quantified by taking images of exposures in the linear range of band intensities (not overexposed or underexposed) and estimating the image intensity of each lane using computational software such as ImageJ's magic wand tool. The same rigor rules that apply to quantification of images of cysts also apply to quantification of immunoblots, that is to say, experiments should be repeated in biological replicates, on multiple occasions and with multiple cell lines, to obtain a pool of data on which to perform statistical analysis.

9.5.8 Preparation of RNA Lysates from Kidney Organoids

The miRNeasy Micro Kit from Qiagen is recommended, although other commercially available kits can be used as well. The miRNeasy Micro Kit is design for purification of total RNA from small amount of cells and tissues.

Reagents

- miRNeasy Micro Kit (Qiagen #217084)
- 30 G needles and 1 mL syringes
- Chloroform
- Sterile, RNase-free pipette tips
- Sterile, RNase-free 1.5 mL centrifuge tubes

Equipment

- Microcentrifuge for centrifugation at 4°C and room temperature

Step-by-Step Protocol

a. Pick the organoids and place in a 1.5 mL centrifuge tube. Let them settle in the bottom of the tube for 5 min.

b. Remove media by manual pipetting and add 500 μL PBS. Let organoids settle and remove the PBS. Repeat PBS wash once.
c. Resuspend the kidney organoids into 500 μL Qiazol. Pass through 30 G syringe needle 5–10 times to completely disrupt and homogenize the organoids.
d. Follow the miRNeasy Micro Kit Handbook protocol.

9.6 TECHNICAL CONSIDERATIONS

- Include appropriate controls for experiments, such as isogenic cell lines with no PKD mutations differentiated alongside mutant cell lines. Failed differentiations can produce cyst-like structures in lieu of organoids. This phenomenon is not specific to PKD and should not be confused with PKD cyst formation. In PKD cyst formation, organoids form normally at first and only later do cysts begin to appear. Having controls differentiate side by side with mutants is the best recipe for assessing general problems with the protocol.
- Every cell line is different, even among clones of the same genetic background. For accurate disease modeling, utilize an isogenic cohort of cell lines including multiple control clones and multiple mutants. This will help you distinguish *bona fide* genetic phenotypes from stochastic idiosyncrasies of a particular cell line.
- Similar to the above, it can be helpful to produce clones on two different genetic backgrounds. To avoid off-target effects, the mutants should be generated using tools that target at least two different genetic loci, independently (e.g., two different guide RNA sequences for each gene).
- Organoids must be "picked" (microdissected) carefully and tightly to avoid dragging along nonkidney cells. Organoids from some hPSC lines are likely to be more difficult to pick than others, due to differences in the composition of the cultures. We recommend starting with H9 ES cells as these form organoids that can be separated from the surrounding mesenchyme with relative ease.
- Use ultralow attachment plates for organoid suspension cultures. Organoids can attach to regular non-tissue culture treated plates.
- Individual organoids can be placed in 96-well ultralow attachment plates, one organoid per well. This facilitates following individual organoids over time, if desired. Using round-bottom ultralow attachment plates may facilitate the removal of media during feedings.

ACKNOWLEDGMENTS

PKD research in the Freedman Laboratory is supported by NIH Awards K01DK102826, R01DK117914, and UG3TR002158, a gift from the Northwest Kidney Centers to the Kidney Research Institute, the Lara Nowak Macklin Research Fund, and start-up funds from the University of Washington. The authors are inventors on patent applications related to human kidney organoid differentiation and modeling of PKD in this system.

REFERENCES

1. Ward, C. J. 1994. The polycystic kidney disease 1 gene encodes a 14 kb transcript and lies within a duplicated region on chromosome 16. The European Polycystic Kidney Disease Consortium. *Cell* 78(4):725.
2. Mochizuki, T., G. Wu, T. Hayashi et al. 1996. PKD2, a gene for polycystic kidney disease that encodes an integral membrane protein. *Science* 272(5266):1339–42.
3. Lu, W., X. Fan, N. Basora et al. 1999. Late onset of renal and hepatic cysts in Pkd1-targeted heterozygotes. *Nat Genet* 21(2):160–1.
4. Walz, G., K. Budde, M. Mannaa et al. 2010. Everolimus in patients with autosomal dominant polycystic kidney disease. *N Engl J Med* 363(9):830–40.
5. Neufeld, T. K., D. Douglass, M. Grant et al. 1992. In vitro formation and expansion of cysts derived from human renal cortex epithelial cells. *Kidney Int* 41(5):1222–36.
6. Carone, F. A., S. Nakamura, R. Bacallao et al. 1995. Impaired tubulogenesis of cyst-derived cells from autosomal dominant polycystic kidneys. *Kidney Int* 47(3):861–8.
7. Freedman, B. S., A. Q. Lam, J. L. Sundsbak et al. 2013. Reduced ciliary polycystin-2 in induced pluripotent stem cells from polycystic kidney disease patients with PKD1 mutations. *J Am Soc Nephrol* 24(10):1571–86.
8. Freedman, B. S., C. R. Brooks, A. Q. Lam et al. 2015. Modelling kidney disease with CRISPR-mutant kidney organoids derived from human pluripotent epiblast spheroids. *Nat Commun* 6:8715.
9. Freedman, B. S. 2015. Modeling kidney disease with iPS cells. *Biomark Insights* 10(Suppl 1):153–69.
10. Cruz, N. M., X. Song, S. M. Czerniecki et al. 2017. Organoid cystogenesis reveals a critical role of microenvironment in human polycystic kidney disease. *Nat Mater* 16:1112–1119.
11. Czerniecki, S. M., N. M. Cruz, J. L. Harder et al. 2018. High-throughput screening enhances kidney organoid differentiation from human pluripotent stem cells and enables automated multidimensional phenotyping. *Cell Stem Cell* 22(6):929–40.
12. Morizane, R., A. Q. Lam, B. S. Freedman et al. 2015. Nephron organoids derived from human pluripotent stem cells model kidney development and injury. *Nat Biotechnol* 33(11):1193–200.
13. Takasato, M., P. X. Er, H. S. Chiu et al. 2015. Kidney organoids from human iPS cells contain multiple lineages and model human nephrogenesis. *Nature* 526(7574):564–8.
14. Taguchi, A., Y. Kaku, T. Ohmori et al. 2014. Redefining the in vivo origin of metanephric nephron progenitors enables generation of complex kidney structures from pluripotent stem cells. *Cell Stem Cell* 14(1):53–67.
15. Thomson, J. A., J. Itskovitz-Eldor, S. S. Shapiro et al. 1998. Embryonic stem cell lines derived from human blastocysts. *Science* 282(5391):1145–7.
16. Takahashi, K., K. Tanabe, M. Ohnuki et al. 2007. Induction of pluripotent stem cells from adult human fibroblasts by defined factors. *Cell* 131(5):861–72.
17. Wu, H., K. Uchimura, E. L. Donnelly et al. 2018. Comparative analysis and refinement of human PSC-derived kidney organoid differentiation with single-cell transcriptomics. *Cell Stem Cell* 23(6):869–81 e8.
18. Freedman, B. S. 2019. Better being single? Omics improves kidney organoids. *Nephron* 141(2):128–32.
19. Kim, Y. K., I. Refaeli, C. R. Brooks et al. 2017. Gene-edited human kidney organoids reveal mechanisms of disease in podocyte development. *Stem Cells* 35(12):2366–78.
20. Brown, A. C., S. D. Muthukrishnan, and L. Oxburgh. 2015. A synthetic niche for nephron progenitor cells. *Dev Cell* 34(2):229–41.
21. Tanigawa, S., A. Taguchi, N. Sharma et al. 2016. Selective in vitro propagation of nephron progenitors derived from embryos and pluripotent stem cells. *Cell Rep* 15(4):801–13.

22. Li, Z., T. Araoka, J. Wu et al. 2016. 3D culture supports long-term expansion of mouse and human nephrogenic progenitors. *Cell Stem Cell* 19(4):516–529.
23. Taguchi, A., and R. Nishinakamura. 2017. Higher-order kidney organogenesis from pluripotent stem cells. *Cell Stem Cell* 21(6):730–46 e6.
24. Cruz, N. M., and B. S. Freedman. 2018. CRISPR gene editing in the kidney. *Am J Kidney Dis* 17(6), 874–83.
25. Jinek, M., K. Chylinski, I. Fonfara et al. 2012. A programmable dual-RNA-guided DNA endonuclease in adaptive bacterial immunity. *Science* 337(6096):816–21.
26. Ameku, T., D. Taura, M. Sone et al. 2016. Identification of MMP1 as a novel risk factor for intracranial aneurysms in ADPKD using iPSC models. *Sci Rep* 6:30013.
27. Thatava, T., A. S. Armstrong, J. G. De Lamo et al. 2011. Successful disease-specific induced pluripotent stem cell generation from patients with kidney transplantation. *Stem Cell Res Ther* 2(6):48.

10 Rodent Autosomal Dominant Polycystic Kidney Disease Models

Sara J. Holditch, Raphael A. Nemenoff, and Katharina Hopp

CONTENTS

10.1 INTRODUCTION

Since the identification of the autosomal dominant polycystic kidney disease (ADPKD) genes nearly 20 years ago, much progress has been made in understanding the natural disease history and underlying molecular pathomechanisms. In large part, this can be attributed to the development and characterization of rodent PKD models. Rodent PKD models have allowed for systematic dissection of different disease causes and the identification of abnormal signaling pathways. For example, rodent models have laid the foundation for our current understanding of the genetic mechanisms and the genetic interactions driving cystogenesis. Embryonic lethality accompanied by renal cyst development of polycystin-1 (PC1)– or polycystin-2 (PC2)–null mice, as discussed in Section 10.4, indicate that complete loss of PC1 or PC2 is incompatible with life and that renal cystogenesis is driven by alterations in ADPKD protein function.[1,2] Beyond these fundamental observations, rodent models have provided the foundation for two distinct genetic mechanisms driving focal renal cyst initiation. The *Pkd2*$^{WS25/-}$ mouse, which contains a hypermutable allele resulting in somatic loss of PC2 expression, suggests that focal cystic lesions are triggered by a two-hit mechanism: a germline loss-of-function allele and a somatic mutation to *Pkd1* or *Pkd2* in the renal epithelium.[3] In contrast, the *Pkd1*$^{RC/RC}$ mouse, which harbors a hypomorphic *Pkd1* mutation, suggests that focal cystic disease can be triggered by reduced but functional levels of PC1 in the renal epithelium.[4] The generation of, and further insights generated through the use of both of these models are discussed in Section 10.5. Additionally, the use of rodent models has revealed a complex network of genetic interactions driving cystogenesis. This includes interactions of known "cystogenes" (e.g., *Pkd1/Pkd2*,[5] *Pkd1/Pkd2/Pkhd1*,[6] *Hnf1b/Pkd1/Pkd2*,[7] and *Pkd2/Bicc1*[8]), as well as unknown modifier genes which augment disease severity in various mouse backgrounds (Section 10.8).[9]

Apart from revealing key genetic interactions, studies of rodent PKD models have also highlighted a critical window for PC1 and PC2 expression, in which the rate of disease progression upon protein loss depends on the age of the mouse. PC1 and PC2 loss prior the end of neonatal renal development results in rapidly progressive disease, whereas ADPKD protein loss thereafter triggers slowly progressive adult-onset PKD (Section 10.4).[10–12] Furthermore, rodent models have illuminated the relationship between kidney injury and PKD progression—specifically, that ischemia-reperfusion plays a key role in cyst progression, attributed to altered epithelial cell integrity during the injury-repair process.[13] These observations have revised our understanding of PKD processes, and renal injury is now considered a "third hit" contributing to cyst progression, but not necessarily cyst initiation.[14] Relevant to this point, recent studies have also shown that activation of the adaptive and innate immune system plays a key role in renal cyst progression and that signaling feedback loops between interstitial cells and the extracellular matrix components (i.e., fibrosis) are essential modulators of disease.[15–19]

Furthermore, advances in our understanding of PKD have been made through the increasing awareness of a group of recessive, primarily childhood diseases known as ciliopathies[20] and the characterization of different ciliopathy models (Section 10.7). To this point, the roles of primary cilia and planar cell polarity (PCP) in cystogenesis were discovered using rodent PKD models. The importance of these concepts to PKD progression is discussed in detail in Chapter 14. In short, studies using conditional rodent models of genes necessary for cilia biogenesis or function have reliably resulted in PKD,[21] and multiple *Pkd1* and *Pkd2* models have demonstrated ciliary defects upon ADPKD protein loss.[4,22,23] Likewise, defects in PCP have resulted in an increase in renal tubular diameter instead of proper tubular elongation, supporting cyst expansion. Similar observations have been made in *Pkd1*- and *Pkd2*-mutant mice.[24–26]

Beyond the insights of genetic interactions, renal injury, and cilia proteins, these models provide the foundation to understand disrupted cellular signaling networks associated with, or supportive of, cyst progression, which are excellently reviewed in the literature.[27–29] As a result, rodent models have become the workhorse for preclinical

PKD studies, resulting in their use in many preclinical trials, some of which are summarized in this chapter or in other reviews.[30–32] As most of this book focuses on outlining detailed insights and protocols regarding ADPKD pathomechanisms, this chapter instead highlights the key rodent models that have enabled our fundamental understanding of PKD, organized by the methodology of how they were generated. Furthermore, key PKD-related characteristics are outlined for each rodent model. It is important to note, that beyond this chapter, there are many exceptional reviews on rodent PKD models,[9,27,33–37] and unfortunately, it is impossible to list in detail all models established to date. Therefore, this chapter outlines the models most frequently used among PKD researchers and models that were instrumental for pathomechanistic studies.

10.2 GENERATION OF GENOME ENGINEERED RODENT MODELS

Genetically engineered rodent models (GERMs) are widely recognized as a vital tool to study human disease. This holds true for ADPKD, where GERMs have proven critical to delineate the natural disease history and elucidate molecular/cellular mechanisms of pathogenesis. Moreover, GERMs have built the foundation for preclinical testing of therapeutic compounds with the promise of alleviating clinically relevant disease pathologies. In recent years, the use of engineered nucleases (ZNFs, TALENs, CRISPER/Cas9)[38] have replaced classical methodologies to produce GERMs. However, none of the current PKD models were generated through these novel approaches, and therefore, the technical aspects of using engineered nucleases will not be reviewed here. It is important to note that not all PKD models are founded on the basis of genetic engineering. Instead, some models arose through spontaneous mutations, chemical induction, or insertional mutagenesis. This is particularly the case for PKD rat models and various models mimicking syndromic forms of human PKD, classified as ciliopathies.[9,21] These will be briefly discussed in Section 10.7 of this chapter.

There are two classical techniques to produce GERMs. The first is pronuclear injection, typically used to generate transgenic animals, and the second is modification of embryonic stem (ES) cells through homologous recombination, generating knockout or knockin animals. Both techniques have been predominantly used in mice because of the inherent difficulty in manipulating the rat genome and the instability of rat ES cells following injection.[39] Pronuclear injection is typically used to generate models addressing the phenotypic consequence of overexpressing an endogenous gene. This approach has been used for both *Pkd1* and *Pkd2* models, as discussed in Section 10.6 of this chapter. Here, exogenous/linearized DNA is injected into the pronuclei, containing genetic material from the sperm head, of a single-cell–stage fertilized egg. The injected egg is then transferred into the oviduct of a pseudopregnant foster dam. Typically, 10%–20% of pups born to the dam will have integrated the exogenous DNA construct into their genome and hence become transgenic founder animals. It is important to note that each founder pup is unique, as transgenes nearly always insert as concatemers of varying numbers, randomly throughout the genome, both significantly contributing to transgene expression. Hence, multiple founders must show identical phenotypes to rule out any insertion site–mediated phenotype. Typically, successful transgene integration is assessed by polymerase chain reaction (PCR) and/or Southern blot, and founder animals are maintained by interbreeding. To date, there are many commercial and academic resources available for generating transgenic animals, costing ~\$3000 to \$6000 with a turnaround of ~8–16 weeks. Detailed protocols outlining the generation process are available online.[40]

The majority of GERMs used in PKD research are generated by genetically altering the endogenous PKD locus in ES cells versus overexpression of PKD genes as described above. This is primarily because of the fact that disease is thought to be caused by loss-of-gene function. Indeed, there are no known dominant-negative or gain-of-function mutations identified in ADPKD, autosomal recessive (AR)PKD, or ciliopathy patients to date.

ES cells, even after genetic manipulation, can be introduced into a preimplantation embryo in which they contribute to all subsequent cell linages including the germ line, creating a chimeric pup. Breeding of chimeras results in animals heterozygous for the ES-cell gene alteration, which can be bred to homozygosity and maintained as such. Gene targeting of ES cells is possible because mammalian cells possess housekeeping enzymes necessary for DNA repair. These host enzymes facilitate recombination of exogenous DNA and the endogenous host-chromosome if the targeting construct (incoming DNA) is homologous to the endogenous locus. Since the frequency of homologous recombination is relatively low, positive and negative selection cassettes are often included in the targeting construct. The positive selection cassette is typically located close to the DNA sequence aimed to be introduced into the endogenous locus and flanked by homologous DNA sequences. The negative selection cassette typically lies outside of the homologous sequence and is lost if successful recombination occurs (Figure 10.1a). If the targeting construct integrates randomly, then the complete sequence, including positive and negative selection cassettes, will be integrated. As a result, treating ES cells with chemical compounds active to both negative and positive selection will kill any cell with inappropriate recombination. However, treatment with both compounds will allow survival of ES cells that have undergone correct homologous recombination because the negative selection cassette will be lost (Figure 10.1a).

Gene targeting constructs come in many flavors, dependent upon the researchers' question, but for PKD research, three general types are of primary focus: knockout, knockin, and conditional/inducible mutagenesis constructs. Knockout gene targeting constructs typically contain exon–intron sequences homologous to the endogenous gene, where one or multiple exons are replaced by a positive selection cassette ending in a stop codon (Figure 10.1a). Successful homologous recombination will result in premature transcriptional termination of their targeted allele and mimic a loss-of-function allele. Classical knockout alleles of *Pkd1* or *Pkd2* are briefly discussed in Section 10.4. However, classical knockout models of *Pkd1* or *Pkd2* have not aided tremendously in our understanding of ADPKD pathomechanisms because of their embryonic lethality in homozygosity and their very mild phenotype in heterozygosity. However, in some instances, the endogenous exon is not only replaced with a positive selection cassette but also with a reporter gene transcribed in place of the endogenous gene (Figure 10.1b). Models such as these are insightful in that they allow for characterization of endogenous gene expression patterns by assessing reporter gene expression through various staining techniques. In the case of ADPKD/ARPKD,

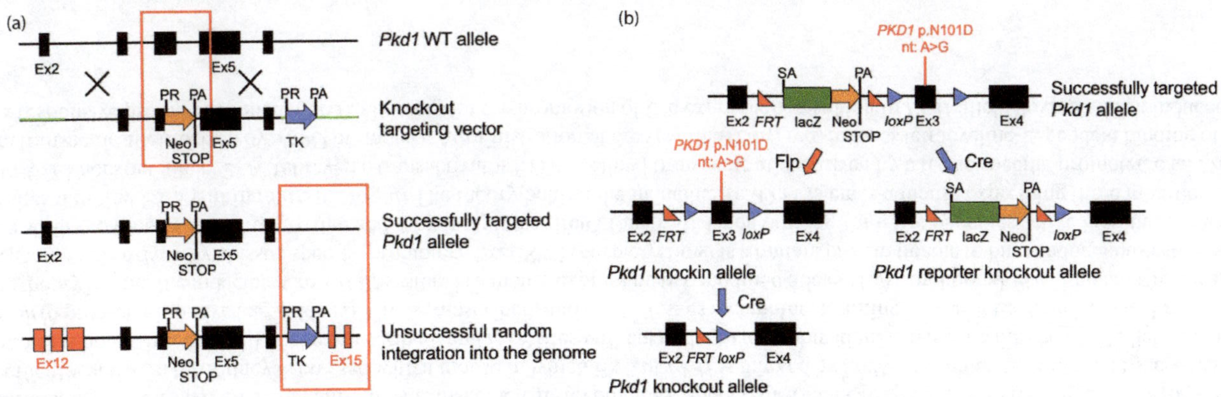

FIGURE 10.1 Targeting strategies in generating genetically engineered rodent models. (a) Targeted disruption of the endogenous *Pkd1* locus using a classical knockout strategy. The knockout targeting vector contains part of exon (Ex) 3 through Ex 6 homologous (black) to *Pkd1*, flanked by the insertion vector backbone (green). Within Ex 3 through Ex 6, the targeting vector contains a positive selection cassette (Neomycin, [Neo, Orange]) replacing part of Ex 4/Intron 4. The positive selection cassette contains its own promoter (PR), polyadenylation signal (PA), and stop sequence (STOP). The vector also contains a negative selection cassette (e.g., thymidine kinase, [TK, blue]), outside of the homologous sequences. The vector is shown undergoing homologous recombination with the endogenous copy of *Pkd1* as would occur in embryonic stem (ES) cells. Upon successful targeted recombination, the negative selection cassette is lost. However, if random integration into a nontargeted locus (red exons) occurs, both selection cassettes remain, and the negative selection cassette is expressed, killing ES cells under positive and negative selection. (b) Targeted *Pkd1* knockout allele for the generation of (I) a conditional knockin or (II) a knockout mouse and (III) a conditional *Pkd1* reporter mouse, utilizing both *loxP* (blue) and *FRT* (red) recombination sites. The targeted allele initially functions as a knockout allele, with *FRT* sites flanking a reporter gene (e.g., β-galactosidase, [lacZ, green]) fused to a positive selection cassette (Neo, orange), a splice acceptor site (SA), PA, and STOP signal. The genomic sequence of Ex 3 has been mutated to mimic the human *PKD1* mutation p.N101D. (I) Upon exposure to FLP recombinase the two *FRT* sites recombine removing the lacZ-Neo-STOP cassette and leave a mutated Ex 3 *Pkd1* knockin allele. (II) Next the knockin mouse can be crossed with a mouse expressing Cre. This mating will result in the recombination of the remaining *loxP* sites, removing exon 3, creating a conditional knockout animal. (III) Exposing the original animal with intact *lox*P sites flanking Ex 3 of *Pkd1* to Cre will result in recombination of the *loxP* sites and result in a *Pkd1* knockout reporter mouse. This is an important distinction from the initial targeting allele, as it remains a possibility that the original targeted allele did not generate a complete knockout if the SA of the lacZ-Neo-STOP cassette is not successfully recognized by the spliceosome.

(Continued)

(c)

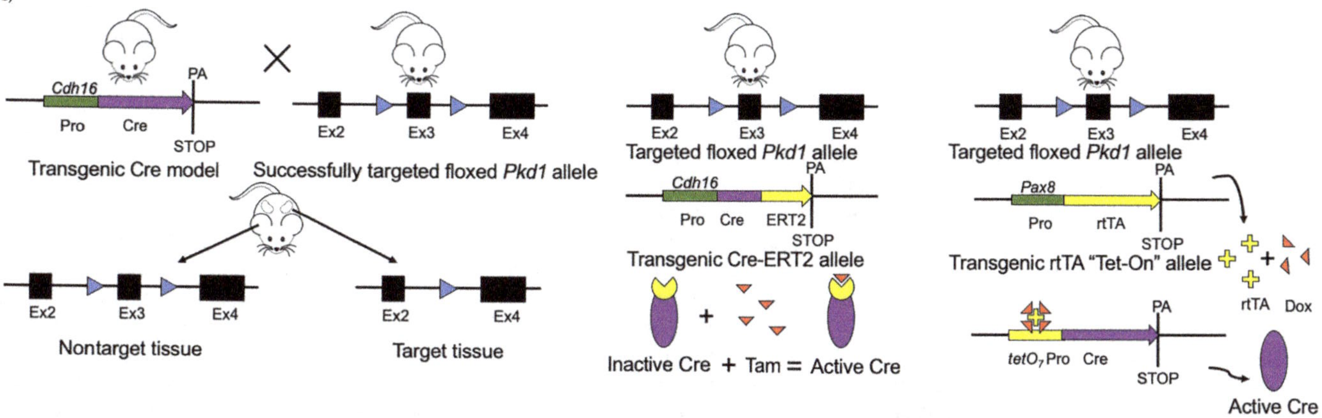

FIGURE 10.1 (Continued) Targeting strategies in generating genetically engineered rodent models. (c, *Left*) An example of the strategy in creating a kidney specific *Pkd1* conditional knockout animal. A transgenic Cre-recombinase model, in which Cre (purple) is driven by the *Cdh16* promoter (Pro, green), specifically active in the kidney is crossed with a mouse in which Ex 3 of *Pkd1* is flanked by *loxP* sites (blue). In nontarget tissues (e.g., skin), the *Cdh16* promoter is not active, Cre will not be transcribed, and *loxP* sites will not undergo recombination, leaving a wildtype *Pkd1* allele. In target tissues where the *Cdh16* promoter is active (e.g., kidney), Cre is transcribed, and the *loxP* sites recombine, resulting in loss of Ex 3, and a *Pkd1* knockout allele. (c, *Middle*) The theory behind the inducible Cre-ERT2 system is a mouse expressing two modified alleles (1. A conditional *Pkd1* knockout allele and 2. a Cre-ERT2 transgenic allele driven by a tissue specific promoter). CreERT2 (purple/yellow) is a mutant protein unable to bind endogenous estrogen; however, it retains the ability to bind synthetic tamoxifen and 4-hydroxytamoxifen (Tam, red). Upon binding Tam, Cre becomes active facilitating recombination of the *loxP* sites flanking Ex 3 within *Pkd1*. (C, *Right*) The theory behind the inducible Tet-On system is a mouse expressing three modified alleles (1. A conditional *Pkd1* knockout allele; 2. A Tet reverse transactivator [rtTA, yellow] transgenic allele driven by a tissue specific promoter, e.g., *Pax8* [green]; and 3. A Cre transgenic allele driven by a *tetO* promoter). Administration of doxycycline (Dox, red) induces tetracycline-dependent binding of rtTA to the tetracycline-responsive promoter element (tetO), allowing for the promotion of Cre expression, resulting in *Pkd1* allele floxing and an induced knockout.

reporter mice are available for *Pkd1*,[41,42] *Pkd2*,[43] and *Pkhd1*,[44] which have provided valuable insight about where in the kidney or body these genes are expressed and where their function may play an important role.

Knockin targeting constructs are very similar to knockout constructs, with the exception that instead of removing a segment of endogenous DNA, the endogenous sequence is altered, or a novel sequence is introduced. Here, the positive selection cassette is typically placed in the intronic region of the endogenous gene in close proximity to where the sequence is altered, and it is typically spliced out when the gene is transcribed (Figure 10.1b). *Pkd1/Pkd2* models generated through this method will be discussed in detail in Section 10.5.

The most frequently used GERMs in ADPKD research are conditional or inducible models. As the name implies, in these models, the introduced genetic alteration can be conditionally activated or induced. Often, and in the majority of cases for ADPKD, the genetic alteration is a loss-of-gene function (e.g., loss of exonic region). Conditional or inducible loss-of-endogenous sequence can be achieved by the use of site-specific recombinase systems. The two recombinase systems predominantly used either in combination or alone are Cre-*loxP* and Flp-*FRT*. The systems function by introducing a set of two 34-base-pair DNA sequences, known as either *loxP* or *FRT* sites. These sites flank the genomic region to be conditionally modified and are recognized by either Cre or Flp recombinase, respectively. In the presence of these recombinases, the sequence flanked by *loxP* or *FRT* sites will be recombined, resulting in excision or inversion of the flanked genomic sequence, dependent upon the directionality of the 34-base pair recognition site. A typical targeting construct to generate conditional/inducible GERMs for ADPKD research consists of intron(s) or exon(s) sequence(s) homologous to *Pkd1* or *Pkd2* where *loxP* sites are introduced into the intronic region(s) flanking one or multiple exons (Figure 10.1b). If introduced into ES cells, these constructs generate a mouse model that has two *loxP* sites in the intronic region of endogenous *Pkd1* or *Pkd2*, which generally is considered a minimal DNA alteration and ignored by the spliceosome, thereby generating wildtype *Pkd1* or *Pkd2* mRNA and protein. However, if the mice are bred to homozygosity or crossed with a knockout model generating $Pkd1^{loxP/loxP}$ or $Pkd1^{loxP/null}$ mice and Cre recombinase is expressed ubiquitously, then the *loxP* sites at the endogenous locus will recombine, resulting in a global loss-of-function allele and a knockout animal (Figure 10.1b).

The advantage of a conditional/inducible system is that the expression of Cre recombinase does not have to be ubiquitous, but instead can be induced either at specific times, in specific cell types, or when a specific drug is administered. This is achieved by crossing GERMs containing *loxP*-flanked exons with mouse lines that express Cre recombinase driven by a promotor that is active under the above-mentioned conditions. The system is considered a conditional knockout if Cre recombinase is expressed by a tissue-specific promoter or a promoter active at a specific time point in development (Figure 10.1b, c). In contrast, if the expression of Cre recombinase is dependent on administration of a substance, it is considered an inducible knockout. For ADPKD, three inducible systems are available based on the administration of tamoxifen, doxycycline, or interferon. The key difference among them is that for interferon inducible systems, Cre recombinase expression is global, whereas the tamoxifen/doxycycline system can be combined with a tissue/developmental-specific promoter. Furthermore, tamoxifen/interferon inducible Cre expression is achieved by crossing *loxP*-containing GERMs with a single transgenic line that produces an active Cre recombinase in response to drug administration. Conversely, doxycycline-inducible systems require two transgenes. The exact mechanism of how each of these systems function is outlined in Figure 10.1c and available conditional/inducible ADPKD models are discussed in Section 10.4. To date, a vast repertoire of transgenic mice expressing Cre recombinase dependent upon tissue or temporal promoters is commercially available. Additionally, similarly to transgenic GERMs, many companies and academic institutions are able to generate knockout/knockin GERMs, including *loxP/FRT* knockins, by manipulating and transplanting ES cells (>$5,000+, >12 weeks). Exact protocols on how to generate GERMs are available online.[45]

10.3 PHENOTYPES ASSOCIATED WITH POLYCYSTIC KIDNEY DISEASE PROGRESSION IN MURINE MODELS

In an attempt to increase the relevance and clinical translation of academic findings and therapeutics tested in animal models, characterization of PKD presentation in rodents has tried to closely follow clinical standards. Further, as different genotypes and genetic backgrounds generate significant diversity in murine PKD phenotypes, characterization and standardization of clinical markers and fundamental sequela across models are warranted. The following is an overview of commonly reported phenotypes in murine PKD models that are used as surrogates for established clinical parameters.

The most frequently used gross phenotype measurement of PKD severity in rodent models is percent kidney weight (KW) corrected for body weight (BW, %KW/BW). Percent KW/BW serves as a surrogate for the clinical parameter height adjusted total kidney volume (htTKV), which is formally defined both by the Food and Drug Administration and the European Medicines Agency as a prognostic enrichment biomarker for progressive decline in renal function.[46] Percent KW/BW calculations are quantified at time of animal sacrifice by measuring the gross weight of right and left kidneys divided by the whole-body weight of the animal. However, the validity/robustness of the normalizing factor, body weight, has recently come into question in light of the impact that diet and body mass index (BMI) have on disease progression.[47,48] For example, food restriction has been found to suppress cystogenesis in murine models,[47] and the annual percent change in htTKV has been found to increase with increasing BMI in early-stage ADPKD patients.[48] Hence, researchers using rodent models have recently started to report %KW to heart weight (HW) or %KW to femur or body length as a surrogate of clinical htTKV if weight loss is associated with the interventional group (e.g., due to modified diets or toxicity of the administered compound) (Figure 10.2a). Similar to PKD severity, polycystic liver disease (PLD) severity in rodent models is traditionally measured using %liver weight (LW)/BW. To date, no studies have been published showing that changes to the normalization factor, BW, effect PLD severity.

Cystic index, cyst area, or cyst volume and cyst number are also widely reported histological measurements of PKD/PLD severity in murine models. The former is

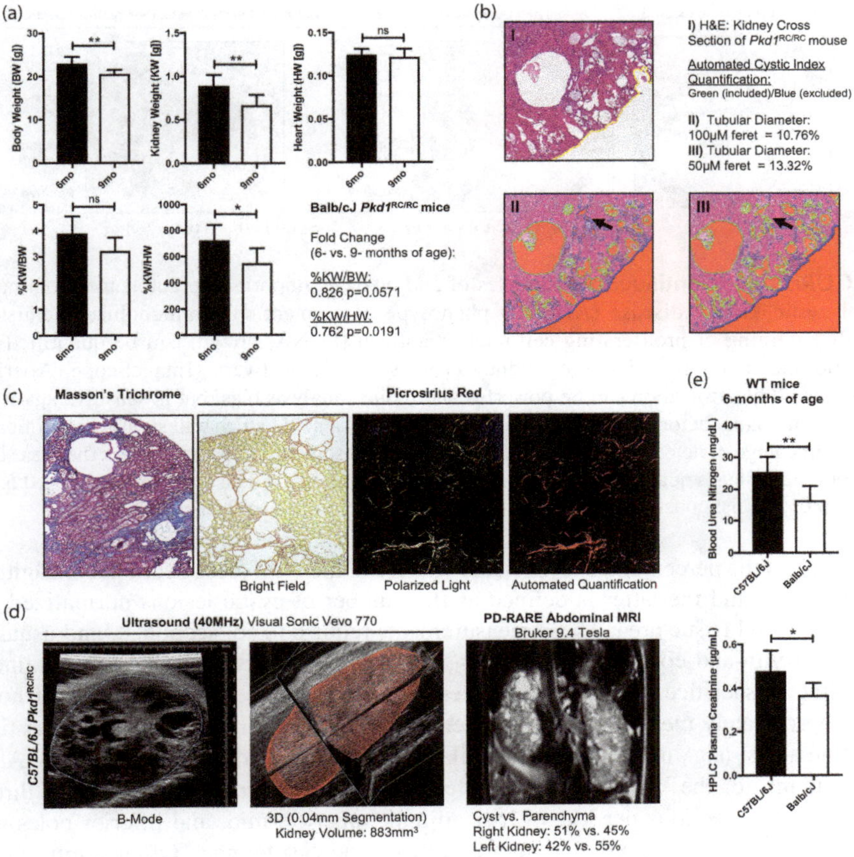

FIGURE 10.2 Examples of and nuances important to autosomal dominant polycystic kidney disease (ADPKD) phenotype assessment. (a) Balb/cJ *Pkd1*$^{RC/RC}$ mice aged to 6 and 9 months (mo) were measured at study termination for body weight (BW), kidney weight (KW), and heart weight (HW) to assess changes in kidney mass as a correlate of disease progression. In this case, BW was significantly impacted by health of the animals, but diet or age could be other confounding factors. HW remained unchanged. As a consequence, HW normalization (%KW/HW) increases the sensitivity of detecting changes in PKD progression, compared with BW normalization (%KW/BW). (b) Representative images showing automated quantification of cyst index using a macro developed for the Nikon *NisElements* software. The images highlight the sensitivity of quantification depending on tubular diameter settings, (I) Hematoxylin & Eosin (H&E) stained kidney cross section of a *Pkd1*$^{RC/RC}$ mouse, (II) Tubular diameter set to 100 μM feret as a threshold to categorize a tubule as cystic, and (III) augmented cystic index with tubular diameter set to 50 μM feret. (c) Masson's Trichrome and Picrosirius Red staining (bright field and polarized light) of cystic kidney cross sections. (*Far right*) An example of automated quantification of fibrosis highlighted by Picrosirius Red staining visualized under polarized light. (d) Representative *in vivo* small animal imaging techniques, Ultra Sound (40 Mhz) and PD-RARE Abdominal MRI (Bruker 9.4 Tesla). (e) Significantly different blood urea nitrogen (BUN) and plasma creatinine measurements in age-matched wildtype (WT) strains highlighting strain effects in routine renal function measurements. (*Continued*)

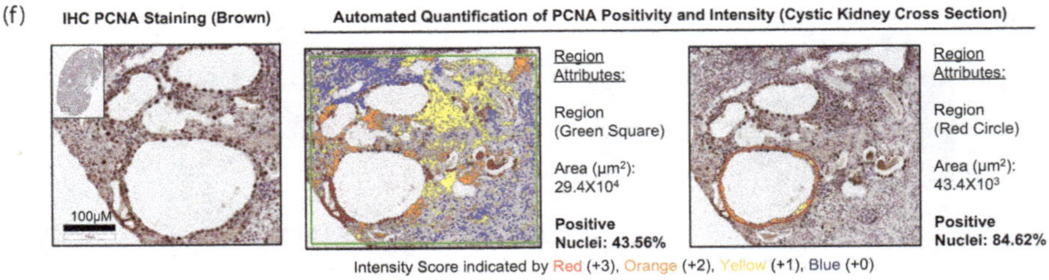

FIGURE 10.2 (Continued) Examples of and nuances important to autosomal dominant polycystic kidney disease (ADPKD) phenotype assessment. (f) Immunohistochemistry (IHC) staining of proliferating cell nuclear antigen (PCNA, brown) can be quantified in cystic and noncystic tubules of kidney cross sections. Software (ImageScope, Aperio) automated quantification can be powerful and reduce analysis bias, but is sensitive again to the normalizing factor of area or the number of nuclei in a field, which will significantly impact the percentage of detected positive nuclei. Detailed and standardized methods for the area size or number of cysts quantified should be in place when reporting data such as these. $P < 0.5$: *, $P < 0.01$: **, ns: nonsignificant, Statistics: Nonparametric Man-Whitney test.

defined as the percentage of cystic area relative to the total cross-sectional or sagittal tissue area and the latter is defined as the number of cystic lesions normalized to the analyzed tissue area. These measurements require fixed, sectioned, and usually hematoxylin-and-eosin stained sections of renal or hepatic tissues, which are obtained at time of sacrifice. Unfortunately, there is no consensus in the literature of how to appropriately measure these parameters. More importantly, what characteristics define a cyst (e.g., minimum size or dilation width is without clear definition). At a minimum, for the kidney, it is suggested to measure and report the mean of three independent sections per kidney, per animal (superior, mid, and inferior poles) to obtain an accurate representation of cyst index and cyst number. Often, a minimum feret tubular diameter of 50–100 μm, reflecting a 5–10-fold dilation compared with a normal mouse tubule, is used to define a cyst. For the liver, sections of each lobe should be averaged, with cystic lesions defined by the pathological finding of a von Meyenburg complex (marked bile duct proliferation and biliary hamartomas). It is important to note that these measurements have an inherent bias, requiring analysts (e.g., pathologists or researchers) to be blinded. For reduced technician bias and increased reproducibility, there are computer-based morphometry software programs available, such as ImageJ, that can be used to calculate cyst index and cyst number automatically (Figure 10.2b). Again, proper sectioning and staining of histological samples, careful definition of cystic lesions, and blind analyses are important for accuracy and reproducibility in data interpretation.

Similar to cystic index, PKD researchers often report renal fibrotic index as a surrogate of disease severity. In this case, the area occupied by collagen fibers is quantified and averaged over the total kidney area minus the cystic area. Again, it is suggested to evaluate three independent sections per kidney, per animal (superior, mid, and inferior poles). The two most common stains used for this analysis are Masson's trichrome or Picrosirius red (Figure 10.2c). Masson's trichrome is a three-color stain in which nuclei appear brown, cytoplasm red, and collagen blue. Because of its three-colored nature, it can be difficult to employ computerized morphometric assessment methods to automate quantification of fibrotic areas. Hence, researchers as well as pathologists often use this method only for visual assessment of fibrotic severity, which are then classified categorically (mild, moderate, severe). Furthermore, multiple studies have shown that trichrome stains may not be sensitive at milder levels of fibrosis but are sensitive to the duration of formalin fixation,[49] which introduces an important variable in pathological studies. Alternatively, the Picrosirius red stain under bright field microscopy identifies collagen fibers as red and nuclei/cytoplasm as yellow. Picrosirius red staining can also be examined under polarized light where it detects specifically collagen type I and III filaments and creates a high signal-to-noise ratio image ideal for computational analysis (Figure 10.2c). This staining method is more sensitive to fine levels of fibrosis and believed to be highly reproducible;[49] hence, it should be the preferred method for accurate quantification of renal fibrosis.

The major pitfall of the above-described gross histopathological parameters for measuring PKD/PLD severity in murine models is that they are one-time measurements determined at time of sacrifice. They are unable to reflect disease progression and ignore possible differences in disease onset, severity at study start, and variability in cyst growth among study animals. To circumvent these limitations and facilitate repeat *in vivo* measurements, ultrasound (US) and magnetic resonance imaging (MRI) have recently become more frequently implemented in murine PKD studies. Murine US TKV measurements are commonly obtained from a three-dimensional reconstruction of the kidney utilizing B-mode images at a step size of 0.05–0.5 mm, measuring total kidney area every 0.5–1 mm (Figure 10.2d). The most commonly used instrument for small-animal US imaging is the Vevo instrument series from Visual Sonics. Published MRI data of murine TKV have been obtained using 7–16 Tesla horizontal or vertical small animal MRI or nuclear magnetic resonance machines, respectively[50–52] (Figure 10.2d). Typically, T-1 or T-2 weighted images are acquired at a slice size of 0.5–1 mm thickness, and kidney volume is calculated by summing the products of kidney area measurements obtained through stereology.[53] The advent of implementing these pre and post intervention *in vivo* imaging techniques within rodent PKD research is that it improves the sensitivity of the data obtained in preclinical trials and hence translatability of the results to the clinic. Furthermore, quantitative MRI techniques such as diffusion weighted imaging (DWI), magnetization transfer imaging (MTI), blood oxygen level dependent (BOLD) imaging, magnetic resonance elastography (MRE), and texture analyses are now being tested in ADPKD patients for their suitability in characterizing healthy renal parenchyma from cystic and fibrotic lesions. These techniques, with inherently increased sensitivity in evaluating renal composition,[54,55] will hopefully provide more accurate correlations with estimated glomerular filtration rate (eGFR) decline than those currently founded on age, baseline eGFR, and htTKV. Some of these quantitative MRI measurements, in particular DWI[56] and MTI,[57] have been tested in murine models, highlighting their ease of translatability to small animals. The key benefit of implementing *in vivo* TKV measurements and quantitative MRI techniques in rodent PKD research is the power to monitor TKV, cyst burden, and gross renal pathological changes (e.g., fibrosis) throughout the course of a genetic or pharmaceutical intervention. Progressive measurements of disease within the same animal significantly increases the sensitivity of the calculated interventional effect and allows researchers to determine the impact of the intervention on early versus late disease.

The principal measurement of disease progression or treatment effect in ADPKD patients is glomerular filtration rate (GFR) or serum creatinine based estimated GFR (eGFR) as a marker of renal function. Similarly, noninvasive measurements of kidney function such as blood urea nitrogen (BUN) and creatinine measurements (serum, plasma, urine) are ubiquitously used in rodent models of ADPKD. BUN is typically evaluated using a colorimetric assay designed to detect urea nitrogen levels in plasma obtained either sequentially from cheek or retro-orbital bleeds, or terminally by cardiac puncture. Creatinine levels are also determined by either colorimetric assay or HPLC using plasma or sequentially collected urine (e.g., spot urine or 24 h urine collected in metabolic cages). As in patients, BUN and creatinine levels are influenced by age and gender. Further, small animal model BUN levels are also influenced by mouse strain and thus require age-, sex-, and strain-matched controls (Figure 10.2e). Jackson Laboratory provides a physiological data summary sheet for each of their wildtype strains, which includes normal BUN and creatinine values for male and female mice at different ages as a reference and guideline for assay accuracy. However, as for ADPKD patients, BUN and creatinine levels in rodents lack the sensitivity to detect changes in early disease progression as their values only change once significant renal damage has occurred.[17,58] An alternative and likely more sensitive way of determining renal function is through transdermal measurements of the clearance of fluorescein isothiocyanate (FITC)-labeled sinistrin. This method has been evaluated once in an ADPKD rodent model using an optical imaging protocol.[59] Here, FITC-labeled sinistrin was injected via tail-vein and fluorescence clearance was monitored by acquiring images of the anesthetized rats' ear every two minutes. To date, MediBeacon has developed a wearable monitor utilizing the same principle of monitoring fluorescence clearance but allows longitudinal measurements without restraining the animal. This continuous renal function monitor has shown high sensitivity in many rodent AKI and CKD studies.[60,61] However, its ability to detect changes in renal function in early versus late stage PKD has not been evaluated in murine ADPKD models.

In addition to kidney function, blood pressure (BP) is another physiological parameter that is regularly monitored in ADPKD patients, as hypertension is considered a significant independent risk factor for progression to end-stage renal disease (ESRD). While BP is not commonly measured in ADPKD rodent models, noninvasive tail-cuff[62–66] or invasive mean arterial BPs[67,68] are reported in some studies. In the case of tail-cuff BP, conscious animals are restrained in artificially heated, size-dependent holders and a pressure sensor plus occlusion cuff is threaded on to the base of the tail. Usually, rodents are acclimatized in a series of trials ($n > 3$) with the machine, restraint, tail-cuff, and volume pressure recording sensors prior to recording BP measurements. However, most small animals maintain some degree of anxiety and stress while restrained, artificially elevating their heart rate. An additional disadvantage of the method is that tail-cuff measurements may not reflect central arterial pressure appropriately, as tail pressure is sensitive to temperature, blood volume status, vascular tone, sympathetic tone, hypotension, or vasoactive substances.[69] More accurate measurements of BP can be obtained by invasive measurements of mean arterial BP (iMAP), recorded through a strain gauge manometer or a wireless electronic signal (radio telemetry). In the first case, a fluid-filled catheter is implanted into the femoral or carotid artery, connected to a pressure transducer, and subsequently linked to a data acquisition system for measuring iMAP. Aside from the morbidity and mortality of the surgery, such systems require a heparin lock and daily flushing to prevent clotting of the catheter, single housing of animals to prevent accidental removal of the catheter, and are not well suited for prolonged recording, as the catheter patency diminishes and signal dampening occurs with time.[70] Hence, very few studies have used this method in PKD research.[67,68] In contrast, implanted radiotelemetry units are accepted as the gold standard for monitoring MAP in conscious, freely moving mice, providing sensitive, accurate, and longitudinal data. In this method, a fluid-filled catheter is inserted either into the left carotid artery with subcutaneous placement of a wireless transmitter or into the abdominal aorta with intraperitoneal placement of the transmitter. While considered the gold standard, this wireless probe system can be highly expensive, requires surgical skill for implantation, and has similar rates of mortality and morbidity as the tethered system.[71] Nonetheless, it has been more often used in PKD research.[66,68,72] Detailed step-by-step methods for utilizing the outlined BP measurement systems, as well as advantages and disadvantages beyond what is mentioned above, are reviewed here.[73]

In addition to the clinically relevant macrophenotypes of animal models discussed above, there are a number of microphenotypes shared among PKD models that are commonly assayed. Most importantly, this includes the proliferative and apoptotic potential of tubular and cyst-lining renal epithelial cells, which have been shown to be dysregulated in multiple models.[74,75] While both of these parameters can be measured in renal single-cell suspensions using flow cytometry or in tissue homogenates using Western blotting, they are typically and more accurately analyzed in frozen or fixed kidney sections in order to align the measurements with anatomical features (e.g., tubule origin, cystic vs. nondilated tubules, or cortex vs. medulla). Tubular proliferation is most commonly assayed using antibody-based immunohistochemistry (IHC) or immunofluorescent (IF) protocols against bromodeoxyuridine (BrdU) or 5-ethynyl-2′-deoxyuridine (EdU), proliferating cell nuclear antigen (PCNA), or Ki-67 antigen. BrdU and EdU are thymidine analogues that are incorporated into newly synthesized DNA strands of replicating cells. Mice receive an intraperitoneal injection of BrdU or EdU (100 mg/kg) ~24 h prior sacrifice in order to label actively dividing cells. The incorporated BrdU/EdU can subsequently be detected in sectioned kidneys using specialized IHC/IF protocols and kits. Similarly, PCNA, a DNA clamp protein essential for replication, and Ki-67, a nuclear protein required to maintain chromosome integrity after nuclear envelope disassembly, are also specifically expressed in replicating cells and can be detected by standard IHC/IF protocols. In contrast, apoptotic cells can be detected by multiple different techniques, reviewed in detail by Zhou and Li.[76] However, terminal transferase-mediated dUPT nick end-labeling (TUNEL) staining or detection of cleaved-caspase products are most commonly used for detection of apoptosis among PKD researchers. The TUNEL assay detects DNA breaks formed in the last phase of apoptosis and requires an *in situ* staining protocol. Caspase-3 and -7 are effector caspases that once cleaved, cleave multiple structural and regulatory proteins initiating apoptosis and can be detected by standard IHC/IF protocols. Independent of the assay used for determining proliferation or apoptosis measurements, the key to accurately describe these phenotypes in PKD models lies in the method of quantification (Figure 10.2f). As is the case with cystic or fibrotic indices, there is unfortunately no consensus on how to appropriately measure these indices. Once again, it is suggested to evaluate these parameters on multiple kidney sections ($n > 3$). Furthermore, it is important to count proliferative or apoptotic epithelial cells and not dilute the results by quantifying proliferative or apoptotic cells of the interstitium, unless the research question requires it. Therefore, automated computational protocols are often not suited for analysis of these parameters. Additionally, it is often helpful to provide information about the location of proliferating or apoptotic indices as these can hint toward possible mechanisms. Here, indices are often recorded for noncystic tubules versus cystic tubules and/or different nephron segments by costaining with markers such as peanut agglutinin (PNA)/Aqp1: proximal tubule, THP: loop of Henle, or Dolichos biflorus agglutinin (DBA)/Aqp2: collecting duct.

Additional microphenotypes assayed in murine models largely depend on the research question asked. However, some of the more common phenotypes include assessment of kidney injury, inflammatory state, and cyclic adenosine monophosphate (cAMP) levels. Kidney injury has been shown to be a key driver of PKD progression in multiple rodent PKD models that have been exposed to ischemia-reperfusion surgery[77,78] and is often assayed by IHC, IF, immunoblot, or qPCR detecting Kim-1 (kidney injury molecule-1)[79–82] and/or NGAL (neutrophil gelatinase-associated lipocalin).[81–83] Inflammation has also been linked to PKD progression. However, it remains unclear whether inflammation is a driver or a bystander of the disease. Commonly, inflammation is assessed by qPCR analysis of proinflammatory cytokines (IL-1, IL-6, TNF-α) or MCP-1 (monocyte chemoattractant protein-1) in whole renal tissue homogenates.[84] Last, cAMP-driven mechanisms are believed to be central to the pathogenesis of PKD[85] and have been a major target for therapeutic interventions.[86] As such, cAMP levels are frequently measured in whole-kidney homogenates using ELISA assays as a correlative measure of disease severity.[87–89]

It should be stated that there is a plethora of additional phenotypes that have been characterized in the various PKD models. However, the ones described above most closely mimic the clinical parameters assayed in patients. To reemphasize, no matter the described phenotype, when characterizing novel rodent PKD models or the effect of intervention therapies, the key to increasing mechanistic insight and clinical translatability is the inclusion of, and comparison to, the appropriate strain, age, and gender-matched controls; correctly powered analysis groups; statistically appropriate replicas; and standardized, normalized analyses, to account for variations in phenotypes influenced by tissue harvest, fixation, or staining methods.

10.4 CONDITIONAL/INDUCIBLE AUTOSOMAL DOMINANT POLYCYSTIC KIDNEY DISEASE MODELS

Traditional knockout mice resulting in complete loss of murine *Pkd1* or *Pkd2* are embryonically lethal and typically die around embryonic day (E) 14.5–18.5.[1,2] Hence, they are not suitable for interventional or mechanistic studies relevant to adult onset ADPKD. Nonetheless, these models have provided valuable insight into the function and expression patterns of PC1 and PC2, the *Pkd1* and *Pkd2* gene products, respectively. The first described *Pkd1* knockout model, $Pkd1^{tm1Jzh}$ ($Pkd1^{del34}$, $Pkd1^{-/-}$)[1], provided a detailed sequence of developmental renal abnormalities including the observation that kidneys grow normally until E14.5, after which they presented with multifocal microscopic dilations in the proximal tubules, followed by progressive dilation and cyst formation in the collecting ducts. This sequence highlights that PC1 is not required during nephrogenic induction but critical to the establishment of tubular architecture, which was also confirmed for PC2[2]. Further, $Pkd1^{-/-}$ and $Pkd2^{-/-}$ mice present with pancreatic ductal cysts as early as E13.5, but not hepatic cysts.[1,2] The lack of polycystic liver disease (PLD) in these knockout mice was explained by biliary cyst formation being induced exclusively in postnatal livers, a hypothesis that was confirmed in multiple conditional models as well as the $Pkd2^{tm1Som}$ ($Pkd2^{WS25}$) model.[3] In addition, *Pkd1*- and *Pkd2*-knockout animals have demonstrated that both PC1 and PC2 are required for vascular development. For instance, null animals present with disorganized myocardial trabeculation, thinning of the myocardial wall/blood vessels, and arterial/ventricular septal defects resulting in body edema and focal hemorrhages.[2,90,91] Furthermore, the placenta of $Pkd1^{-/-}$ and $Pkd2^{-/-}$ animals has been shown to have significantly decreased vascular branches compared with wildtype, and these mice present with curved spine, abnormalities in long-bone formation, and delayed cartilage differentiation, highlighting a role of PC1/PC2 in placental and skeletal development.[90,92,93] $Pkd2^{-/-}$ mice also revealed that PC2 is required for heterotaxia, as many null mice displayed *situs inversus* affecting multiple organs.[43] Last, *Pkd1/Pkd2* knockout animals highlight that both genes, genetically as well as mechanistically, interact. For instance, *trans*-heterozygous mice (e.g., $Pkd1^{+/-}$ $Pkd2^{+/-}$) present with cystic lesions as early as 3 months (mo) of age, whereas single heterozygous mice ($Pkd1^{+/-}$ or $Pkd2^{+/-}$) remain without cystic lesions until 12 mo of age.[5] Many more ADPKD knockout models have been developed than mentioned above and a more comprehensive list is available in the book *Kidney Development, Disease, Repair and Regeneration*[94] and from Mouse Genome Informatics (MGI, http://www.informatics.jax.org/).

In order to better understand the pathomechanism underlying PC1/PC2 loss in adult mice, and to have relevant models for translational studies, spatiotemporal conditional and inducible knockout mice of *Pkd1/Pkd2* have been generated. Initially, researchers focused on kidney-specific gene knockouts using the conditional Cre-*loxP* system outlined in Section 10.2. In these cases, $Pkd1^{flox/flox}$ or $Pkd2^{flox/flox}$ mice are crossed with mice transgenic for the Cre-recombinase gene cloned into a promoter system of a second gene typically expressed exclusively within the kidney. Hence, Cre recombinase is only expressed and active, and *Pkd1/Pkd2* disrupted (i.e., knocked-out) wherever the endogenous gene of the given promoter is being actively transcribed (e.g., the kidney). This approach has been utilized implementing many different promoter systems as outlined in Table 10.1. Of note, the resulting PKD phenotype is dependent upon the time during embryonic development the promoter system becomes active, and in which kidney segment it is being transcribed. For example, the promoter of gamma-glutamyl transferase-1, *Ggt1,* starts being transcribed at E7, and expression is limited to the developing cortical tubular epithelium. *Ggt1*-Cre;$Pkd1^{tm3.1Jzh}$ (*Ggt1*-Cre;$Pkd1^{flox/flox\ del2-6}$)[95] mice present with rapid formation of

TABLE 10.1
Conditional and Inducible Autosomal Dominant Polycystic Kidney (ADPKD) Models

Target Gene	PKD Model	Method	Cre-Recombinase Model	MGI ID (Cre Driver)	Timing (Cre Driver)	Target Tissue (Cre Driver)	Key Phenotypes (PKD Model)
				Conditional ADPKD Models			
Pkd1	*Pkd1*^{tm2Ggg}	Cond.	Tg(*Aqp2*-cre)3Dek	MGI:4820838	E13.5	Collecting duct, testes, vas deferens	3-fold increased BUN, 4-fold increased %KW/BW (PN28). Mean survival 8 Wks.[226]
Pkd1	*Pkd1*^{tm2Ggg}	Cond.	Tg(*ATP6V1B1*-cre)45Rnel	MGI:4820423	E12.5	Collecting duct intercalated cells, male reproductive tract, brain, intestine, uterus	No cysts[226,227]
Pkd1	*Pkd1*^{tm2Som}	Cond.	Tg(*Col1a1*-cre)1Bek	MGI:3574761	E12.5	Bone marrow stromal cell, osteoblasts	Renal fibrosis. Renal and pancreatic cysts (PN42). Growth retardation. Abnormal bone development, abnormal adipose tissue development, abnormal parathyroid hormone levels.[228]
Pkd1, *Kif3a*	*Kif3a*^{tm2Gsn}, *Pkd1*$^{tm3.1Jzh}$	Cond.	Tg(*Col2a1*-cre)10Amc	MGI:2176264	E12.5	Cells of chondrogenic lineage during embryogenesis and postnatally	Aggressive renal cysts. Relatively mild skeletal phenotype. PN14 maximum survival.[229,230]
Pkd1	*Pkd1*$^{tm3.1Jzh}$	Cond.	*Twist2*$^{tm1(cre)Dor}$	MGI:3044412	E9.5	Mesoderm derived tissues	Premature closure of synchondroses at the cranial base, impaired postnatal growth.[231]

(Continued)

TABLE 10.1 (*Continued*)
Conditional and Inducible Autosomal Dominant Polycystic Kidney (ADPKD) Models

Target Gene	PKD Model	Method	Cre-Recombinase Model	MGI ID (Cre Driver)	Timing (Cre Driver)	Target Tissue (Cre Driver)	Key Phenotypes (PKD Model)
Pkd1	*Pkd1*^{tm3Jzh}	Cond.	Tg(*Ggt1*-cre)M3Egn	MGI:3811269	PN7	Cortical tubular epithelium of the kidney	Renal cysts (PN10). Increased renal tubule apoptosis, renal failure. Median survival PN28.[95,232]
Pkd1	*Pkd1*^{tm2Ggg}	Cond.	Tg(*Hoxb7*-cre)13Amc	MGI:2675121	E9.5	Collecting duct, ureter, developing gut, spinal cord	Gross renal cysts, 6-fold increased in %KW/BW, 3-fold increased BUN (PN7). Mean survival PN15.[233,234]
Pkd1, *Prkcsh*, *Sec63*, *Pkhd1*	*Pkd1*^{tm2Som}, *Prkcsh*$^{flox/flox}$, *Pkd1*^{tm1Som}, *Sec63*^{tm1Som}, *Pkhd1*^{tm1Som}, *Pkd1*$^{F/H}$-*BAC*	Cond.	Tg(*Cdh16*-cre)91Igr	MGI:2665300	E10.5	Ureteric bud, mesonephric tubules, Wolffian duct, Mullerian duct	Gross renal cysts (PN14). 15-fold increased %KW/BW, 6-fold increased BUN. Mean survival PN28.[6,96]
Pkd1	*Pkd1*$^{tm2.1Ggg}$	Cond.	*Meox2*$^{tm1(cre)Sor}$	MGI:2176174	E5, E7	E5 epiblast, E7 primitive ectoderm	Renal cysts. Midgestational and neonatal lethality.[92]
Pkd1	*Pkd1*^{tm2Ggg}	Cond.	Tg(*MMTV*-cre)4Mam	MGI:2176166	E13.5	Mammary epithelial cells, female germline, salivary gland	Mild infrequent renal and hepatic cysts.[235]

(Continued)

TABLE 10.1 (*Continued*)
Conditional and Inducible Autosomal Dominant Polycystic Kidney (ADPKD) Models

Target Gene	PKD Model	Method	Cre-Recombinase Model	MGI ID (Cre Driver)	Timing (Cre Driver)	Target Tissue (Cre Driver)	Key Phenotypes (PKD Model)
Pkd1	$Pkd1^{tm2Ggg}$	Cond.	Tg(*Nes*-cre)Wme	MGI:4412413	E12.5	Central and peripheral nervous system, kidney, heart	60% renal cystic index (PN49). 3-fold increased %KW/BW, 3-fold increased BUN.[236]
Pkd1	$Pkd1^{tm1a(EUCOMM)Hmgu}$, $Pkd1^{tm2Ggg}$	Cond.	Tg(*Pax2*-cre)1Akg	MGI:3046196	E13.5	Central nervous system, mesonephric tubules, Wolffian ducts, Müllerian ducts	Reproductive tract dilation of efferent duct and coiling defect in the epididymis.[41]
Pkd1	$Pkd1^{tm2Som}$	Cond.	Tg(*Pkhd1*-cre)1Igr	MGI:3794109	E12.5	Kidney, liver	4-fold increased %KW/BW, 3-fold increased BUN, 80% cyst index (PN24).[6,96,237,238]
Pkd1	$Pkd1^{tm2.1Ggg}$	Cond.	Tg(*Tek*-cre)1Ywa	MGI:2450311	E7.5	Cells of endothelial lineage	No gross abnormalities in heart or kidney (E16.5). Increased embryonic demise >E18.5, abnormal placental histology, polyhydramnios.[92,239]
Pkd2	$Pkd2^{tm1.1Gwu}$	Cond.	Tg(*Vil1*-cre)997Gum	MGI:2448639	E12.5	Acinar cells	Renal cysts, renal enlargement, kidney failure. Pancreatic and liver cysts 2–4 mo. Median survival 4–6 mo.[240]
Pkd2	$Pkd2^{tm3Som}$	Cond.	Tg(*Pkhd1*-cre)1Igr	MGI:3794109	E12.5	Renal collecting ducts, bile ducts	Renal cysts (PN14).[6,24,96]

(*Continued*)

TABLE 10.1 (*Continued*)
Conditional and Inducible Autosomal Dominant Polycystic Kidney (ADPKD) Models

Target Gene	PKD Model	Method	Cre-Recombinase Model	MGI ID (Cre Driver)	Timing (Cre Driver)	Target Tissue (Cre Driver)	Key Phenotypes (PKD Model)
Pkd2	*Pkd2*$^{flox/flox}$	Cond.	*E2f1*Tg(Wnt1-cre)2Sor	MGI:5485027	E13.5	Cranial neural crest	No renal/hepatic phenotype. Shortened snout, malocclusion, dome-shaped skull vault, curved spine.[93]
Pkd2	*Pkd2*$^{tm1.1Tjwt}$	Cond.	*Meox2*$^{tm1(cre)Sor}$	MGI:2176174	E5, E7	E5 epiblast, E7 primitive ectoderm	Renal and pancreatic cysts, *situs inversus*, edema (neonatal), dextrocardia. Neonatal lethality.[92,241–243]
				Inducible ADPKD Models			
Pkd1	*Pkd1*$^{tm1.1Djmp}$	Tam.	Tg(*Cdh16*-cre/ERT2*) F427Djmp	MGI:3641108	*Induced	Renal tubules, collecting ducts, Loops-of-Henle, distal tubules	Tam. postnatal, developing kidney, rapid renal cyst formation; Tam. developed kidney, slow renal cyst formation.[12]
Pkd1	*Pkd1*^{tm3Jzh}	IFNα, IFNβ, dsRNA	Tg(*Mx1*-cre*) 1Cgn	MGI:2176073	*Induced	Liver, spleen, heart, kidney	Tam. postnatal, developing kidney, renal cyst formation; Tam. developed kidney, no renal cysts.[244]
Pkd1	*Pkd1*^{tm2Som}	Dox.	Tg(*Pax8*-rtTA2S*M2)1Koes	MGI:3709326	*Induced	Proximal tubule, distal tubule, collecting duct, fetal and adult kidney	Dox. at PN11, 12, 13, aggressively enlarged polycystic kidneys at PN21.[245]

(Continued)

TABLE 10.1 (*Continued*)
Conditional and Inducible Autosomal Dominant Polycystic Kidney (ADPKD) Models

Target Gene	PKD Model	Method	Cre-Recombinase Model	MGI ID (Cre Driver)	Timing (Cre Driver)	Target Tissue (Cre Driver)	Key Phenotypes (PKD Model)
Pkd1, *Kif3a*	$Pkd1^{tm2Som}$, $Kif3a^{tm2Gsn}$	Tam.	Ndor1Tg(UBC-cre/ERT2*)1Ejb	MGI:3707333	*Induced	All tissues	Tam. at PN28-P35, progressive and slow-forming cystic bile ducts and increased % liver weight/BW.[22]
Pkd2	$Pkd2^{f3/-}$	IFNα, IFNβ, dsRNA, polyI:C	Tg(*Mx1*-cre*)1Cgn	MGI:2176073	*Induced	All tissues	Renal, hepatic, pancreatic cysts.[246]
Pkd2	$Pkd2^{tm1.1Tjwt}$	Tam.	Tg(*CAG*-cre/Esr1*)1Lbe	MGI:2677720	*Induced	Bile ducts	Hepatic cysts[247]
Pkd2	$Pkd2^{f3/-}$	Tam.	Tg(*Pdx1*-cre/Esr1*)35.10Dam	MGI:2684321	*Induced	Pancreatic duct, endocrine cell development	Pancreatic cysts[246]

Abbreviations: BUN: Blood urea nitrogen; KW: kidney weight; BW: body weight; Tam: tamoxifen; Dox: doxycycline; Cond: conditional; >: greater than; mo: Months; Wks: weeks; PN: postnatal day; E: embryonic day.

proximal and distal nephron cysts, with a 28-day postnatal (PN) survival. Similarly, kidney-specific cadherin 16 (*Cdh16* or Ksp) promoted Cre recombinase in *Cdh16*-Cre;*Pkd1*$^{tm2.1Som}$ (Ksp-Cre;*Pkd1*$^{flox/flox\ del2-4}$)[96] mice, results in exon excision at E11.5. Ksp-Cre;*Pkd1*$^{flox/flox\ del2-4}$ mice have a 21-day PN survival, and cyst formation in the thick ascending limb through the collecting duct. While conditional models such as these survive embryonic development, the vast majority of kidney-specific conditional ADPKD models present with rapidly progressive disease resulting in early postnatal lethality. Unfortunately, the shortened life span of conditional models does not support preclinical trial development or mechanistic studies of adult disease.

In attempts to overcome the rapidly progressive cyst formation and early lethality, inducible ADPKD mouse models were generated. Most of the seminal work published in ADPKD using this system utilized the Ksp-CreER^{T2} inducible model,[97] which as explained in Section 10.2, relies on a fusion protein of Cre recombinase and the estrogen receptor ligand-binding domain, activated by tamoxifen. Ksp-CreER^{T2};*Pkd1*$^{tm2.1Djmp}$ (Ksp-CreER^{T2};*Pkd1*$^{flox/floxdel\ 2-11}$) mice demonstrated that this system results in loss of PC1 expression in megalin^{+} proximal tubules, aquaporin-2^{+} collecting ducts, and within the Tamm-Horsfall protein^{+} loops-of-Henle and distal tubules.[98] Key mechanistic studies using this system revealed that the timing of *Pkd1* gene disruption during renal development results in a PKD severity gradient. Specifically, gene inactivation at a time of ongoing nephrogenesis (prior to PN14) produces grossly enlarged and severely cystic kidneys within the first month postinactivation. Alternatively, gene inactivation after PN14, results in slowly progressive cystogenesis over several months.[10–12] Additionally, the origin of cystogenesis switches from cortical and medullary cysts to exclusively medullary cysts in early versus late loss of *Pkd1*, respectively.[11] Together, these studies demonstrate distinct functions of PC1 during renal development versus the adult kidney, as well as a previously unrecognized developmental switch in renal maturation, capable of impacting disease severity. It is important to highlight that the efficiency of *Pkd1* loss by tamoxifen administration is both age and dose dependent. For example, one study reports that during nephrogenesis, when tamoxifen is administered at PN1 versus PN8, the percentage of *Pkd1* deletion drops from 18% to 1%.[11] Conversely, another study reports that if tamoxifen is administered at PN4 versus 4 mo of age, the percentage of *Pkd1* deletion changes from 7% to 20%, respectively.[12] Further, multiple studies have shown that the percentage of *Pkd1* deletion depends on the dose of tamoxifen.[11,14] The higher the dose, the higher the percentage of *Pkd1* loss, and the more severe the renal disease. This observation holds true independent of renal development.

Another example of an inducible ADPKD mouse model is the *Pax8*^{rtTA};*TetO*-cre;*Pkd1*$^{flox/flox}$ or *Pax8*^{rtTA};*TetO*-cre;*Pkd2*$^{flox/flox}$ system first described using the *Pkd1*$^{tm2.1Som}$ or *Pkd2*^{tm3Som} (*Pkd2*$^{flox/flox\ del3-4}$) model,[22,96,99] which results in *Pkd1*/*Pkd2* loss after doxycycline administration, as described in Section 10.2. This model has been useful in determining the impact of cilia loss on cyst formation upon adult *Pkd1* or *Pkd2* gene inactivation (*Pkd1*$^{tm2.1Som}$, *Pkd2*^{tm3Som}),[22] the impact of *Pkd1*/*Pkd2* gene inactivation on cilia length (*Pax8*^{rtTA}; *TetO*-cre; *Pkd1*$^{tm2.1Som}$; *Arl13b*-EGFP^{tg} or *Pax8*^{rtTA}; *TetO*-cre; *Pkd2*^{tm3Som}; *Arl13b*-EGFP^{tg}),[100] and the induction of severe PKD by knockout of *Tsc1* (*Pax8*^{rtTA};LC-1; *Tsc1*$^{flox/flox\ del17-18}$ [LC-1: *luc*; P_{tet}bi-1;*cre*],[101] and [*Tsc1*^{tm1Djk}][102]).[103]

Beyond the two inducible models mentioned above, there are additional models described in the PKD literature that are outlined in Table 10.1, together with their respective phenotypes. It is important to note, that conditional/inducible ADPKD models have been the workhorse of PKD researchers in order to elucidate key pathomechanisms of disease progression. Furthermore, temporal control of *Pkd1* disruption has provided the opportunity to study the role of signaling pathways unique to cyst initiation, separate from cyst expansion.[104–107]

10.5 *PKD1/PKD2* KNOCKIN/HYPOMORPHIC AUTOSOMAL DOMINANT POLYCYSTIC KIDNEY DISEASE MODELS

In addition to the frequently used conditional and inducible knockout ADPKD models, a few knockin models exist, characterized by global expression of a mutant protein. Knockin models can be subdivided into two groups: those that contain a single amino acid change in *Pkd1* and those that have reduced *Pkd1* or *Pkd2* gene expression.

10.5.1 Single Amino Acid Knockin Models

The first *Pkd1* knockin model published, the *Pkd1*$^{tm1.1Fqi}$ (*Pkd1*$^{V/V}$)[108] mouse, aimed to investigate the functional and pathophysiological role of PC1 cleavage by introducing the *Pkd1* p.**T**3041V point mutation (*PKD1* p.**T**3049V) at the G-protein-coupled receptor proteolytic (GPS) HL**T** cleavage site. *Pkd1*$^{V/V}$ mice are viable at birth, a distinction from *Pkd1* knockout mice. Compared with *Pkd1*$^{V/+}$ or wildtype (WT), *Pkd1*$^{V/V}$ mice have decreased BW starting at PN9, increased %KW/BW at PN4, and a 50% survival rate by PN21. Renal cysts are observed as early as PN1. which progressively increase in number and size and predominantly arise from the distal tubule and collecting duct. Further, *Pkd1*$^{V/V}$ mice present with bile duct dilations, but normal pancreatic anatomy. Together, this data highlight that PC1 cleavage is not essential for proper embryonic development, nor the integrity of the proximal tubule segment. However, PC1 cleavage is necessary for proper distal tubule, collecting duct, and bile duct function postnatally. Beyond these observations, the model has been invaluable in demonstrating that PC1 cleavage is necessary for proper PC1 trafficking[109] and has since been used in preclinical studies.[110,111]

Another knockin model that is widely used is the *Pkd1*$^{tm1.1Pcha}$ (*Pkd1*$^{RC/RC}$) mouse.[4] The model was established to mimic a mutation found in ADPKD patients, *PKD1* p.R3277C, which was hypothesized to be hypomorphic. For example, patients with typical ADPKD were found to be homozygous for the mutation, while patients with early onset or *in-utero* ADPKD were found to have the *PKD1* p.R3277C mutation *in trans* with a truncating *PKD1* allele.[112,113] Indeed, biochemical analyses of the mutant allele in *Pkd1*$^{RC/RC}$ mice confirm that the PC1 p.R3277C mutation reduces GPS-cleavage efficiency, resulting in improper protein folding and approximately ~40% of mature PC1. In line with this observation, the model mimics human PKD pathophysiology precisely. *Pkd1*$^{RC/null}$ mice have a PN28 median survival and develop renal cysts as early as E16.5. Furthermore, *Pkd1*$^{RC/null}$ mice exhibit significantly elevated %KW/BW at birth and augmented blood urea nitrogen (BUN) levels at PN12 compared to WT. Analogously, *Pkd1*$^{RC/RC}$ mice show no survival difference relative to WT animals and develop slowly progressive PKD from birth characterized by significantly elevated %KW/BW and BUN at 3 mo and 9 mo of age, respectively. In both models, the *Pkd1*$^{RC/null}$ and the *Pkd1*$^{RC/RC}$ mouse, renal cysts arise predominantly in the proximal tubule prior to PN12, and in the collecting duct subsequently. This observation is in line with the switch in the functional role of PC1 between the developing and the adult kidney, as discussed in Section 10.4. Last, *Pkd1*$^{RC/RC}$ mice develop mild ductal plate malformations at 12 mo of age; however, no true polycystic liver disease, as seen in ADPKD patients, is observed. Of note, disease severity and extra-renal manifestations have been shown to be strain dependent, as elaborated on in Section 10.8. The data generated in the *Pkd1*$^{RC/RC}$ model emphasize that focal renal cystic disease can develop despite the presence of functional PC1. This finding is contrary to the historically held two-hit hypothesis in which individual somatic mutations to the WT allele are thought to be required for focal cystic lesions. The *Pkd1*$^{RC/RC}$ model has since been used in mechanistic[17,82,114–117] as well as preclinical studies.[47,52,57,87,88]

The last ADPKD knockin model developed by single amino acid substitution is the recently published *Pkd1*$^{tm1.1Jcal}$ (*Pkd1*$^{deltaL/deltaL}$) mouse.[118] The model mimics the *PKD1* p.L4132del patient mutation located in a putative G-protein binding region. *In vitro*, this mutation has been shown to significantly decrease PC1-stimulated G-protein-dependent signaling.[118] Surprisingly, *Pkd1*$^{deltaL/deltaL}$ mice are embryonically lethal and behave like *Pkd1* knockout animals (see Section 10.4), even though the authors show that the mutation does not affect PC1 expression, cleavage, maturation, and PC1/PC2 complex formation. Interestingly, the authors show that PC1^{deltaL}/PC2 complexes exhibit significantly impaired PC2 channel activity, suggesting that PC1-mediated G-protein signaling impacts PC2 channel activity and is essential for viability.

10.5.2 Knockin Models with Reduced PC1/PC2 Expression

Two knockin models with reduced PC1 expression have been published; the *Pkd1*^{tm1Djmp} (*Pkd1*$^{nl/nl}$)[119] and the *Pkd1*^{tm1Hung} (*Pkd1*$^{L3/L3}$)[120] mouse. In both, reduction of endogenous WT PC1 levels is achieved by intronic insertion of a positive selection cassette that interferes with RNA processing. *Pkd1*$^{nl/nl}$ mice have 13%–20% correctly spliced *Pkd1* transcript and *Pkd1*$^{L3/L3}$ mice are reported to have 20%–25% correctly spliced *Pkd1* transcript. Both homozygous models survive embryogenesis, exhibit growth retardation as neonates, and show little to no liver or pancreatic disease despite globally aberrant *Pkd1* splicing. Interestingly, *Pkd1*$^{nl/nl}$ present with cardiac abnormalities and dissecting aortic aneurysms, while the cardiovascular system in *Pkd1*$^{L3/L3}$ mice develops normally. Further, in these animals, survival and PKD severity are influenced by the amount of remaining functional WT *Pkd1* transcript. For example, by 2 mo of age, 85% of *Pkd1*$^{nl/nl}$ mice have died, whereas, by 6 mo of age, 85% of *Pkd1*$^{L3/L3}$ mice have died, but both models develop grossly enlarged cystic kidneys predominantly arising from the collecting duct. Of note, *Pkd1*$^{nl/nl}$ disease severity has been shown to be influenced by the background strain (see Section 10.8). While both models are genetically and phenotypically similar, the *Pkd1*$^{nl/nl}$ model has been used more widely in key mechanistic[121–123] and preclinical trials studies.[124–126]

Similar to the *Pkd1*$^{nl/nl}$ and *Pkd1*$^{L3/L3}$ models, the *Pkd2*^{tm1Gwu} (*Pkd2*$^{nf3/nf3}$) mouse[127] globally expresses ~33% of the *Pkd2* transcript due to intronic insertion of a neomycin selection cassette impacting RNA splicing. Interestingly, the phenotype of *Pkd2*$^{nf3/nf3}$ is much milder than that of the *Pkd1* models discussed above, and more similar to currently published ARPKD models (see Section 10.7). *Pkd2*$^{nf3/nf3}$ mice are viable beyond 12 mo of age, and proximal tubular cysts form between 6 and 9 mo of age. Additionally, animals present with pancreatic and bile duct cysts.

The most widely used ADPKD model with reduced *Pkd2* protein expression is the *Pkd2*^{tm1Som} (*Pkd2* WS25*)* model[3]. In this model, the mutant allele (WS25) is generated by exonic insertion of a neomycin selection cassette, which endogenously undergoes intragenic homologous recombination at two independent sites. Hence, in *Pkd2*$^{WS25/-}$ mice, each cell is either WT or null for PC2, making this a mosaic model where PC2 levels are reduced globally due to the mixture of WT and null cells. *Pkd2*$^{WS25/-}$ mice present reliably with progressive cystogenesis as early as 1 mo of age without a decrease in survival until >12 mo compared with WT.[2,128] The renal pathology of this model is very variable in severity and presents as focal cystic lesions in the outer medulla and cortex. *Pkd2*$^{WS25/-}$ mice also develop pancreatic cysts but present without cardiac abnormalities.[2] Last, 100% of *Pkd2*$^{WS25/-}$ present with PLD, mimicking clinical ADPKD features, a rarity among orthologous ADPKD rodent models.[129] Focal progressive PKD and reliable PLD are two key reasons why the *Pkd2*$^{WS25/-}$ model has become popular for preclinical[130–136] and mechanistic studies, despite its unique and genetically dissimilar pathomechanism from human disease.[115,137,138]

10.6 TRANSGENIC AUTOSOMAL DOMINANT POLYCYSTIC KIDNEY DISEASE MODELS

Transgenic mice allow analyses of the histopathological effects resulting from overexpression of endogenous gene products. This can be helpful in understanding the pathomechanisms of diseases caused by gain-of-function or dominant-negative mutations. Furthermore, transgenic mice allow detailed evaluation of expression patterns of the endogenous protein if the transgene used to generate the animal contains an expression tag. These models can also be valuable in identifying upstream or downstream pathway substrates or components of the gene of interest, which are undetected by loss-of-function analysis. While no *PKD1/PKD2* gain-of-function mutation has been identified as of yet, a limited number of ADPKD transgenic animals have been generated in order to better understand PKD-associated pathomechanisms.

The first transgenic *Pkd1* model published was the h*PKD1*(TPK1-3) mouse, in which a 108 kbp region containing human *PKD1* and the flanking *TSC2* gene were cloned from a P1-derived artificial chromosome and injected into a mouse pronucleus, resulting in the TPK1, TPK2, and TPK3 independent founder lines.[139] The purpose of the model was to assess the function of human *PKD1* in a mouse and evaluate the phenotypic consequence of altering PC1 levels. In respect to the former, the authors crossed *Pkd1*^{tm1Jzh};h*PKD1* (*Pkd1*$^{del34/+}$;h*PKD1*)[1] mice with *Pkd1*$^{del34/+}$ mice in order to see whether or not *Pkd1*$^{del34/del34}$;h*PKD1* survive. Human *PKD1* was able to rescue the embryonic lethal phenotype of *Pkd1*$^{del34/del34}$ mice, suggesting that human PC1 is functionally active in the murine setting. In respect to the latter, the authors made the surprising observation that two of the founder lines, TPK1 and TPK3, reliably presented with progressive PKD and PLD. These novel observations revealed that not only loss of PC1 but also PC1 overexpression can result in PKD/PLD pathology. Furthermore, *Pkd1*$^{del34/del34}$;h*PKD1* animals, while rescued, still presented with PKD/PLD phenotypes at adult age (5–11 m), possibly due to lower levels of the exogenous rescue hPC1 relative to basal levels of endogenous murine PC1. Both of these phenotype descriptions highlight that a delicate dosage of PC1 expression is critical to maintain renal and liver morphology. At the time, this concept was a shift in paradigm, contradicting accumulating evidence that ADPKD was recessive at the cellular level, and required a second-somatic hit (loss of the WT allele) for cyst initiation.[3,140,141]

A second *Pkd1* transgenic model, the *Pkd1*(TAG) mouse (Tg(Pkd1)6Mtru; Tg(Pkd1)18Mtru, Tg(Pkd1)26Mtru; 6, 18, 26 reflecting different founder lines),[64] was generated from a modified *Pkd1*-bacterial artificial chromosome (BAC) overexpressing PC1 extra-renally and within renal tissues 2–15-fold above endogenous levels. Comparable to the initial transgenic model described above, overexpression of PC1 alone (without TSC2) resulted in renal insufficiency, renal cysts, renal fibrosis, and calcium deposits similar to nephrolithiasis. These mice also developed hepatic fibrosis and ~15% intrahepatic cysts of the bile ducts, impacting females preferentially. Furthermore, *Pkd1*(TAG) mice presented with cardiac pathologies such as severe left ventricular hypertrophy, aortic arch distention, valvular stenosis, and intracranial aneurysms. Considering the phenotype sequela of the *Pkd1*(TAG) mouse, it is interesting to note that even though PC1 is overexpressed which is not correlative to the human disease, the model mimics human pathology better than many other murine *Pkd1* models with reduced expression or loss-of-function alleles. For example, very few ADPKD models present with PLD in females preferentially, and none of the models reliably develop intracranial aneurysms, which have an ~11% prevalence in ADPKD patients.[142]

Next to the two transgenic *Pkd1* animals described above, an additional transgenic *Pkd1* model was developed with the aim to mimic naturally occurring human mutations. The Tg(Pkd1*)39Mtru (*Pkd1*$_{extra}$) model[143] overexpresses the extracellular domain of PC1 (up to amino acid F3043) exclusively, and develops similar phenotypes as other *PKD1* transgenic models. Further, PC1$_{extra}$ protein was found to be expressed in similar patterns as the endogenous protein, and was found to interact with PC2. These findings highlight that the extracellular domain fragment of PC1 has all the necessary regulatory domains for proper expression and that a crosstalk between PC1 and PC2 likely contributes to the ADPKD pathomechanism.

Last, HA- and Flag-tagged WT and GPS-cleavage mutant *Pkd1* transgenic animals were generated from a BAC in order to better evaluate PC1 function and localization, as well as the importance of PC1 cleavage to protein functionality, *Pkd1*$^{F/H}$-BAC (WT) and *Pkd1*^{L3040H}-BAC (GPS-cleavage mutant).[144] Using both of these models, studies have shown that only GPS-cleaved PC1 is functionally active because of proper passage out of the endoplasmic reticulum, and that PC1/PC2 interaction is required for proper PC1 trafficking. Interestingly, transgenic WT *Pkd1*$^{F/H}$-BAC mice did not present with PKD phenotypes by 12 mo of age compared with the above described models, but were able to rescue embryonic lethality of *Pkd1*$^{null/null}$ mice, suggesting the *PKD1*$^{F/H}$ protein is functionally active. This discordant finding among *Pkd1* transgenic models is likely associated with differences in transgene copy number and levels of functional protein, highlighting again that PC1 dosage is critical in the PKD pathomechanism. It also highlights the importance of analyzing multiple transgenic founder lines as expression of the linearized gene construct can differ drastically based on whether random integration occurred at an epigenetically active or silent site.

Similarly dissonant, two independent groups have generated transgenic mice containing human *PKD2* cloned into a pCAGGS expression vector, Tg(CAG-PKD2)#Hwl[145] and Tg(CAG-PKD2)#Wu.[146] Tg(CAG-PKD2)#Wu mice do not present with cystic phenotypes in the kidney, liver, or pancreas even when aged to 12 mo as homozygotes (h*PKD2*$^{Tg/Tg}$). Conversely, hemizygous h*PKD2*$^{Tg/+}$ Tg(CAG-PKD2)#Hwl mice present with renal cystic disease as early as 2 mo of age. A description of the cyst burden in organs other than the kidney was not included in the initial publication. Of note, both models were able to rescue the embryonic lethal phenotype of *Pkd2*$^{null/null}$ mice, highlighting the functionality of human PC2 in mice.

Another transgenic *PKD2* model, expressing human *PKD2* cloned from a BAC, PKD2-Y,[147] also presented with renal abnormalities albeit much milder than reported in the Tg(CAG-PKD2)#Hwl model. PKD2-Y mice aged to greater than 12 mo develop tubulopathy with microscopic cysts and proteinuria, and exhibit a disorganized renal cortex. Interestingly, genomic analysis of PKD2-Y revealed polyploidy, mitotic instability, and centrosome overduplication, suggesting PC2 plays a role in centrosome duplication and/or the M-phase of cellular division.[148]

Last, a transgenic rat model expressing a truncated human *PKD2* gene from amino acid 1-703, mimicking multiple human mutations, has been described.[149] Interestingly, synthesis of this truncated PC2 led to renal cysts, retinal degeneration, and stunted proximal tubule cilia, suggesting that overexpression of mutant PC2 may trigger a similar PKD-associated pathomechanism as overexpression of WT PC2. This is in line with the observations made in the Tg(Pkd1*)39Mtru ($Pkd1_{extra}$) model,[143] but contradictory to the phenotypes observed in the PC1 mutant expressing *Pkd1*^{L3040H}-BAC[144] mice, which did not present with PKD-associated phenotypes.

Together, these models highlight that transgenic animals and overexpression studies can be informative in understanding pathomechanisms, protein localization, and functionality of human mutations. However, it is important to be mindful that the method of generating overexpression models, and the potential for uncontrolled variables such insertion site, copy number, expression level, gender, and species may impact results.

10.7 AUTOSOMAL RECESSIVE POLYCYSTIC KIDNEY DISEASE (ARPKD) AND PKD-RELATED CILIOPATHY MODELS

Autosomal recessive polycystic kidney disease (ARPKD), caused by mutations to *PKHD1*, is the recessive, rarer form of nonsyndromic PKD, with a global prevalence 20-40-fold less than ADPKD.[150,151] ARPKD has *in utero* to juvenile onset and worse disease severity than ADPKD. Unlike ADPKD where cysts are thought to arise from any nephron segment,[152] ARPKD cystic kidneys retain their shape and are primarily fusiform dilations of the collecting duct. Additionally, while hepatic cysts are rare in children with ADPKD but increase in frequency with age,[153] hepatic ductal malformations, congenital hepatic fibrosis (CHF), and hepatic cysts are universally present in ARPKD children.[154]

Despite our clinical knowledge of ARPKD, detailed pathomechanistic studies and preclinical trials are limited. One reason is that while multiple ARPKD knockout mice have been developed (Table 10.2), the majority of them do not mimic the renal phenotype observed in patients, although all of the models develop pancreatic duct dilations and reliable fibrocystic liver disease including ductal plate malformations. The observed renal phenotypes can be subdivided into "no renal abnormalities" ($\mathit{Pkhd1}^{\mathrm{tm1Som}}$ [$\mathit{Pkdh1}^{\mathrm{del4/del4}}$],[155] $\mathit{Pkhd1}^{\mathrm{tm1Rbu}}$ [$\mathit{Pkdh1}^{\mathrm{del40/del40}}$][156]), "dilation of the proximal tubules at advanced age and phenotypes preferentially observed in female animals" ($\mathit{Pkhd1}^{\mathrm{tm1Cjwa}}$ [$\mathit{Pkdh1}^{\mathrm{del2/del2}}$],[157] $\mathit{Pkhd1}^{\mathrm{tm2Cjwa}}$ [$\mathit{Pkdh1}^{\mathrm{LSL/LSL}}$],[158] $\mathit{Pkhd1}^{\mathrm{C642*}}$[159]), and "dilations of the proximal tubules, loop of Henle, and collecting ducts at young age" ($\mathit{Pkhd1}^{\mathrm{tm1.1Ggg}}$ [$\mathit{Pkdh1}^{\mathrm{del3\text{-}4/de3\text{-}14}}$],[160] $\mathit{Pkhd1}^{\mathrm{tm1Sswi}}$ [$\mathit{Pkdh1}^{\mathrm{lacZ/lacZ}}$],[44] $\mathit{Pkhd1}^{\mathrm{tm1Gwu}}$ [$\mathit{Pkdh1}^{\mathrm{del15\text{-}16,GFP/\ del15\text{-}16,GFP}}$][161]). One suggested reason for the discrepancy between mouse and human ARPKD phenotype presentation is that *Pkhd1* undergoes complex splicing in mice, allowing for smaller mRNA/protein products to be generated by the mutant allele each with possible remaining functions. However, the literature on the splicing patterns of *Pkhd1* is conflicting, and the recent study of the $\mathit{Pkdh1}^{\mathrm{LSL/}}$ ^{LSL} mouse, with complete loss of *Pkhd1* transcripts after exon 2, suggests other mechanisms are likely at play.[158,162]

In addition to the above-engineered ARPKD mouse models, a rat model with a spontaneous mutation to *Pkhd1* exon 36, resulting in exon skipping, was discovered in a colony of WT Crj:CD/SD rats at the Charles River Japan production facility (PCK rat).[163] The mutation is inherited recessively and homozygous rats present with progressive, focal renal cystic lesions of the thick ascending loops of Henle, distal tubules, and collecting ducts from ~20 days of age onward. Additionally, mild bile duct dilations are observed neonatally, which develop into large hepatic cysts with advancing age. There are no pancreatic abnormalities noted in the PCK rat. Interestingly, the PCK rat displays sexually dimorphic disease distinct from the above-described *Pkhd1* mouse models, with more severe renal disease in males compared with females[163] and a more prominent liver pathology in females compared with males.[164] This is in line with several other models of syndromic PKD, discussed below, in which males display more severe PKD than females, such as the jck mouse[165] and the Han:SPRD, Cy/+ rat.[166] Furthermore, this sexually dimorphic disease also mimics observations of ADPKD patients, where renal disease is known to be more severe in males and liver disease more prominent in females.[167–169] Similarly, the pathohistological features of the PCK rat parallel ADPKD more than ARPKD. For example, PCK rats present with slowly progressive focal cystic renal disease originating from multiple tubular origins and hepato-fibrotic cystic disease. This is in contrast to the fusiform dilations of the renal collecting ducts and CHF observed in most ARPKD patients.[163] For this reason and others, the PCK rat has been a veritable workhorse for the testing of clinical evaluations of ADPKD-centric therapies (e.g., vasopressin V2 antagonists [Tolvaptan][170,171] and somatostatin agonists[172]). However, even though none of the *Pkhd1* models mimic clinical ARPKD, they have provided

TABLE 10.2
Autosomal Recessive Polycystic Kidney Disease (ARPKD) and Ciliopathy Models

ARPKD Mouse Models	MGI ID	Affected Gene	Mutant Type	Human Disease	Key Phenotypes
Pkhd1^{tm1Som} (*Pkhd1*^{del4})[99]	MGI:3774015	*Pkhd1*	Exon 4/intron 4 replacement by neomycin selection cassette. Two transcripts: (a) lacking part of exon 4 and exon 5, (b) lacking exon 4	ARPKD	No cyst formation (renal, hepatic). Abnormal renal epithelium morphology.
Pkhd1^{tm1Rbu} (*Pkdh1*^{del40})[156]	MGI:3664756	*Pkhd1*	Exon 40 replacement by lacZ-pgk-neomycin selection cassette. Transcript lacking exon 40	ARPKD	No renal cysts or impaired kidney function. Abnormal bile duct morphology and development, liver cysts and fibrosis.
Pkhd1^{C642*}[159]	N/A	*Pkhd1*	CRISPR: Exon20, C642*	ARPKD	Heterozygous and homozygous mice do not have noticeable renal or liver abnormalities. Heterozygotes aged to 1.5 Yrs develop prominent biliary cystic livers, along with liver inflammation and fibrosis.
Pkhd1^{lacZ}[44]	MGI:5430994	*Pkhd1*	Exon 1–3 replacement by lacZ gene with a nuclear localization and a floxed neomycin selection cassette.	ARPKD	Renal interstitial fibrosis and progressive renal cysts (PN45). Abnormal hepatic bile duct morphology, liver fibrosis, abnormal gall bladder morphology, and pancreatic cysts.
Pkhd1^{tm2Cjwa} (*Pkhd1*^{LSL})[158]	MGI:5438324	*Pkhd1*	Intron 2 insertion of *loxP*-STOP-*loxP* (LSL) cassette. No transcript generated.	ARPKD	Dilated proximal tubules (PN90, in females) and kidney cysts (only in females). Liver cysts (PN90), and liver fibrosis (PN28).
Pkhd1^{tm1Cjwa} (*Pkhd1*^{del2})[157]	MGI:3814174	*Pkhd1*	Exon 2 replaced by floxed neomycin selection cassette. Transcript lacking exon 2.	ARPKD	Females develop cysts in the outer medulla and inner cortex (9 mo). Dilated bile ducts (PN28) progress into enlarged cystic and fibrotic livers by 12–15 mo of age.

(Continued)

TABLE 10.2 (*Continued*)
Autosomal Recessive Polycystic Kidney Disease (ARPKD) and Ciliopathy Models

ARPKD Mouse Models	MGI ID	Affected Gene	Mutant Type	Human Disease	Key Phenotypes
Pkhd1$^{tm1.1Ggg}$ (*Pkhd1*$^{del3-4}$)[160]	MGI:3759215	*Pkhd1*	Exon 3–4 flanked by *loxP* sites. Transcript lacking exon 3–4.	ARPKD	Enlarged cystic kidneys (PN90). Abnormal bile duct development, bile duct hyperplasia, biliary cysts, inflammation and fibrosis. Pancreatic cysts observed in 33% of mice aged to 9 mo. Neonatal lethality in a subset of mice due to respiratory failure (PN1), 29% of expected mice survive birth.
Pkhd1^{tm1Gwu} (*Pkdh1*$^{del15-16,GFP}$)[161]	MGI:3826780	*Pkhd1*	Exon 15/16 replacement by GFP/neomycin selection cassette.	ARPKD	Dilated fusiform tubules, renal cysts, renal fibrosis, and renal necrosis. Liver cysts, fibrosis, and increased necrosis. Dilated pancreatic duct. Animals survive >12 mo.
Pkhd1^{Flox67HA} [177]	N/A	*Pkhd1*	Exon 2 flanked by *loxP* sites.	ARPKD	After Cre-mediated deletion, homozygous *Pkhd1*$^{\Delta 67}$ mice are completely normal.
ARPKD Rat Models	**MGI ID**	**Affected Gene**	**Mutant Type**	**Human Disease**	**Key Phenotypes**
Pkhd1^{pck}/Crl (PCK)[163,178]	N/A	*Pkhd1*	N/A	ARPKD	Progressive cystic enlargement of the kidneys and liver. Renal cysts develop as a focal process from the thick ascending loops of Henle, distal tubules, and collecting ducts. Distorted biliary tree and bile duct dilation.

(Continued)

TABLE 10.2 (*Continued*)
Autosomal Recessive Polycystic Kidney Disease (ARPKD) and Ciliopathy Models

PKD-Associated Ciliopathy Mouse Models	MGI ID	Affected Gene	Mutant Type	Human Disease	Key Phenotypes
Arl3$^{Gt(OST263303)Lex}$ (*Arl3*$^{Gt(neo)1Lex}$)[248]	MGI:3623141	*Arl3*	Intron 1 insertion of gene trap vector (splice acceptor + neomycin selection cassette)	JBTS	Abnormal kidney development, renal cysts. Abnormal bile duct morphology and inflammation, distended gallbladder, pancreatic cysts. Maximum survival PN21.
Bcl2^{tm1Sjk} (*Bcl-2*$^{-}$, *BclII*^{tm1Sjk})[249–252]	MGI:1857134	*Bcl2*	Exon 3 replacement with neomycin selection cassette.	N/A	Delayed kidney development, small and pale or grossly enlarged cystic kidneys. Dilated distal and proximal tubules. Severe renal failure at 1–8 Wks of age. Cardiovascular, reproductive, metabolic, immune, and craniofacial abnormalities. Maximum survival of 19 Wks.
Bicc1$^{jcpk\text{-}bpk}$(bpk, jcpk/bpk)[104,253–256]	MGI:1856693	*Bicc1*	Spontaneous; Ttwo base pair insertion in exon 22	Nonsyndromic cystic renal dysplasia (CYSRD)	Proximal and distal tubule cysts, abnormal bile duct morphology.
Bicc1^{jcpk} (jcpk)[9,254,257–262]	MGI:1856920	*Bicc1*	Chemically induced; G to A transition in the splice acceptor site of exon 3, skipping of exon 3, premature stop	Nonsyndromic cystic renal dysplasia (CYSRD)	Proximal and distal tubule cysts, abnormal bile duct morphology.
Cys1^{cpk} (ck, cpk)[192,263]	MGI:1856831	*Cys1*	Spontaneous tandem deletion of 12 bp and 19 bp in exon 1, premature stop	Boichis disease (NPHP and liver fibrosis)	Renal cysts, liver inflammation, and pancreatic cysts.

(Continued)

TABLE 10.2 (*Continued*)
Autosomal Recessive Polycystic Kidney Disease (ARPKD) and Ciliopathy Models

PKD-Associated Ciliopathy Mouse Models	MGI ID	Affected Gene	Mutant Type	Human Disease	Key Phenotypes
Dzip1l^{warpy} (ENU15embryo:006, Marawarpina, wpy) [264]	MGI:5563494	*Dzip1l*	Chemically induced. *Dzip1 l* c.1123C>T, p.Q375*	PKD5, a form of ARPKD	Kidney cysts (E15.5). Cleft palate and upper lip, polydactyly, mid-gestation lethality, abnormal eye morphology. Prenatal lethality.
Ift140^{b2b1283Clo} [265]	MGI:5560263	*Ift140*	Chemically induced. Ift140 c.1138A>G, p.N380D	Mainzer-Saldino syndrome, Jeune syndrome	Kidney cysts, abnormal cardiovascular, pulmonary, and craniofacial development. Animals can be born without a spleen and possess polydactyly.
Ift20$^{tm1.1Gjp}$ (*Ift20*^{flox}) [266]	MGI:3817268	*Ift20*	Intron 1–3 replacement of *frt* flanked neomycin selection cassette followed by a *loxP* site	N/A	Dilated collecting duct (PN5). Profound renal enlargement (PN10), cyst burden and fibrosis (PN23).
Ift88^{Tg737Rpw} (*Ift88*^{orpk}, orpk, Tg737°rpk, Tg737^{orpk,} Tg737(o)(rpk), Tg737Rpw, TgN737Rpw, TgN(Imorpk)737Rpw) [267–269]	MGI:2157527	*Ift88*	Transgenic; 2.7 kb intronic deletion. Hypomorphic allele	N/A	Enlarged cystic kidneys, abnormal bile duct morphology, liver fibrosis, abnormal pancreatic morphology, and pancreatic cysts. Polydactyly and postnatal growth retardation. Mean survival PN7 days.
Invs^{inv} (inv, inv^{-}) [270,271]	MGI:1856915	*Invs*	Transgenic; Duplication/47 kb deletion. Loss of exon 3–11	NPHP	Impaired heterotaxia, resulting *in situs inversus*.
Kif3a^{tm1Gsn} (KIF3A-, *Kif3a*^{n}, *Kif3a*^{tm1Maz}) [272]	MGI:1861960	*Kif3a*	Exon 2 flanked by *loxP* sites	N/A	Enlarged cystic kidneys, renal interstitial fibrosis, and tubule atrophy. Renal failure (PN21). Postnatal growth retardation.

(*Continued*)

TABLE 10.2 (*Continued*)

Autosomal Recessive Polycystic Kidney Disease (ARPKD) and Ciliopathy Models

PKD-Associated Ciliopathy Mouse Models	MGI ID	Affected Gene	Mutant Type	Human Disease	Key Phenotypes
$Lama^{5tm3Jhm}$ ($Lama5^{neo}$)[273]	MGI:3823246	*Lama5*	Exon 15–21 flanked by *loxP* sites	N/A	Increased kidney apoptosis, renal interstitial fibrosis, dilated tubules, cystic kidneys (PN14), and kidney failure (PN21). Maximum survival PN28.
$Nek1^{kat}$ (kat,KatJ, kat(2J))[197,274]	MGI:1858030	*Nek1*	Spontaneous; *Nek1*c.791-2105del	Short rib-polydactyly syndrome type II	Progressive renal cysts. Obstructive hydrocephaly, nervous system abnormalities, abnormal craniofacial morphology, and anemia.
$Nek8^{jck}$ (jck, $Nek8^{-}$)[275–277]	MGI:1856919	*Nek8*	Spontaneous; *Nek8* p.G1348R	NPHP, Renal-hepatic-pancreatic dysplasia 2	Enlarged kidneys with cortical and medullary cysts. 25 Wks maximum survival.
$Nphp3^{pcy}$ (pcy)[204,278]	MGI:1856987	*Nphp3*	Spontaneous; *Nphp3* p.I1614S	NPHP	Kidney inflammation, interstitial fibrosis, and cysts. 10% incidence of intracranial aneurysm, craniofacial and skeletal abnormalities. 8 mo mean survival.
$Sclt1^{Tg(CAG\text{-}sb10)1Dla}$ (CAGGS-SB10, Tg(ACTB-sb10)1Dla)[279]	MGI:3612989	*Sclt1*	Transposase mutagenesis; 3 kb deletion at intron 12. No transcript post exon 13	N/A	Abnormal kidney morphology, enlarged cystic kidneys (PN8). Cleft palate, growth retardation by PN8. Maximum survival 4 Wks.
$Tsc1^{tm1Djk}$ ($Tsc1^{c}$, $Tsc1^{fl}$, $Tsc1^{flox}$, $Tsc1^{L}$, $Tsc1^{lox}$)[103]	MGI:2656240	*Tsc1*	Intron 17 insertion of floxed neomycin-TK selection cassette	TSC	Kidney epithelium hyperplasia (proximal and distal tubules). Cystic kidneys and survival of 4 Wks following doxycycline exposure *in utero*.

(Continued)

TABLE 10.2 (*Continued*)
Autosomal Recessive Polycystic Kidney Disease (ARPKD) and Ciliopathy Models

PKD-Associated Ciliopathy Rat Models	MGI ID	Affected Gene	Mutant Type	Human Disease	Key Phenotypes
Ttc21b$^{tm2c(KOMP)Wtsi}$ (*Thm1*^{flox})[277]	MGI:5587034	*Ttc21b*	Exon 4 flanked by *loxP* sites. Tamoxifen induced	NPHP, Jeubert syndrome	Cystic kidneys, elevated cAMP, decreased cilia length.
SPRD-*Anks6*/Fsn (Han:SPRD (Cy), Cy rat)[280–283]	N/A	*Anks6*	Spontaneous	NPHP	Heterozygotes develop progressive polycystic kidney disease, severe interstitial fibrosis, and renal inflammation. Males develop uremia, proteinuria, hyperlipidemia, and hypertension in advanced ages (>10 mo). Liver and pancreatic cysts are common in females (>17 mo). Lethality at 8 Wks in homozygosity.
Wistar polycystic kidney (Wpk)[284–286]	N/A	*Tmem67*	Spontaneous	MKS, JBST, BBS, COACH syndrome, NPHP, RHYNS syndrome	Massively enlarged cystic kidneys and uremia (PN21). Cysts form predominantly in the collecting duct. No liver phenotype.
Lewis polycystic kidney (LPK)[287,288]	N/A	*Nek8*	Spontaneous	NPHP, renal-hepatic-pancreatic dysplasia 2	Renal cyst formation and hypertension. No liver phenotype.
Wistar-Chi ARPK[289,290]	N/A	Unknown	Spontaneous	N/A	Slowly progressive renal cyst formation (medulla, 8 Wks), large cysts throughout the kidneys (>8 mo). No liver phenotype. Small body size, craniofacial abnormalities.

Abbreviations: Yrs: Years; mo: months; Wks: Weeks; PN: postnatal day; E: embryonic day; >: in excess of.

insight into the genetic interaction of ADPKD/ARPKD genes and the localization/function of fibrocystin, the *Pkhd1* gene product, in planar cell division, intra- and intercellular signaling, and cilia structure/function.[6,160,163,173–180]

Beyond ARPKD, an overlapping spectrum of recessively inherited renal cystic diseases resulting from mutations to a heterogeneous group of genes have been classified as ciliopathies. Ciliopathies result from defects in the formation and/or function of the primary cilia and have been the subject of many excellent reviews discussing in detail disease severity and presentation.[20,181,182] To date, mutations of more than 60 genes have been identified to cause ciliopathy-associated phenotypes in humans. Moreover, a large number of murine ciliopathy alleles have been generated, identified, and characterized. An extensive list of rodent ciliopathy models has been summarized by Norris et al.[21] Above, Table 10.2 outlines key ciliopathy models that have been used extensively among ARPKD/ADPKD researchers. Importantly, while the mechanism of cyst initiation likely differs among these ciliopathy models given their different genetic defect, models non-orthologous to ADPKD/ARPKD such as ciliopathy models, have been invaluable in understanding key cellular mechanism that contribute to cystogenesis and renal fibrosis.[21]

10.8 GENETIC BACKGROUND/STRAIN EFFECTS ON POLYCYSTIC KIDNEY DISEASE PHENOTYPES

Much of this chapter has focused on genetic alterations unique to the protein coding regions of PC1, PC2, and fibrocystin. The interpretation of studies such as these form our understanding of PKD pathophysiology and underlying pathomechanisms. Unfortunately, many of these observations have been made without acknowledging the influence rodent strain or genetic background may have on disease presentation. This problem is extremely critical to acknowledge as modifier genes have been shown to influence the functional consequence of genetic variation both in rodents and humans.[183,184]

Inbred rodent strains, classically defined by 20 or more generations of brother-sister mating, exhibit vast genetic and interphenotypic variability.[185] The Mouse Genome Informatics (MGI) Database (http://www.informatics.jax.org/) currently lists 426 inbred mouse strains and 217 inbred rat strains. One of those mouse strains, the C57BL/6J strain, is among the most widely used, largely due to its prolific breeding, mutation permissiveness, long life expectancy, and limited physiological or behavioral abnormalities (Mouse Phenome Database [https://phenome.jax.org/]). Furthermore, while considered WT, each strain can be very different phenotypically. For example, C57BL/6J are susceptible to hepatic and renal fibrosis, BALB/cJ are susceptible to hepatic fibrosis but resistant to renal fibrosis, and 129/SvJ are susceptible only to renal fibrosis.[186] Importantly, phenotypic differences among two strains such as C57BL/6J and C3H/HeOuJ can be more than a standard deviation for 43% (342/748) of the phenotypes compared in the Mouse Phenome Database.[187] Included in the list of significantly differing phenotypes between C57BL/6J and C3H/HeOuJ are kidney weight (0.39 g$\pm$0.03 g vs. 0.529 g$\pm$0.07 g total kidney weight [TKW], respectively) and TKW/ BW (1.27%$\pm$0.06% vs. 1.95%$\pm$0.15% TKW/BW, respectively). For this reason, many studies have focused on cataloging key differences among mouse strains when studying pathomechanisms. For example, inbred mouse strain variances are known to impact opiate reward learning,[188] immunological responses,[189] hippocampal function,[190] and kidney function.[191] Specific to kidney function, transcutaneous glomerular filtration rate measurements by fluorescent dye demonstrated different clearance rates in age-matched BALB/cJ, C57BL/6J, and 129/SvJ. Here, elimination of FITC-labeled sinistrin occurred the quickest in BALB/cJ, with significant differences compared with the other two strains.[191]

Beyond strain differences in renal weight, function, and fibrosis susceptibility, multiple PKD studies have emphasized the impact of genetic background, modifier genes, or quantitative trait loci (QTL) on PKD-associated phenotypes. For example, the initial publication and subsequent follow-up study of the *Pkd1*$^{tm1.1Pcha}$ (*Pkd1*$^{RC/RC}$) model highlights that PKD presentation differs among C57BL/6J *Pkd1*$^{RC/RC}$, 129/SvJ *Pkd1*$^{RC/RC}$, and BALB/cJ *Pkd1*$^{RC/RC}$ mice, with increasing disease severity respectively, independent of the shared *Pkd1* p.R3277C mutation.[4,17,88] Importantly, in neither of these studies did the *Pkd1*$^{RC/RC}$ mice present with PLD. A study by another group, using the *Pkd1*$^{RC/RC}$ mouse in a 6-month preclinical trial of fenofibrate, demonstrated a protective effect of the drug on renal and hepatic cyst development. While the intervention was convincing, it remains unclear why the *Pkd1*$^{RC/RC}$ presented with PLD, as the specifics on strain and background were not provided by the authors.[52] The potential for a secondary gene, or group of genes, modifying hepatic disease manifestation in a *Pkd1* hypomorphic setting may be responsible. If this is the case, the modifier gene or locus may have been lost or introduced during maintenance or crossing of the *Pkd1*$^{RC/RC}$ mice, an important observation for future follow-up.

Another example of strain influencing PKD-associated phenotypes is the *cpk* mouse model caused by mutations to cystin.[192,193] The model has been described in the DBA/2J,[193] C57BL/6J,[194] CD-1,[195] and BALB/cJ[196] strains. For the most part, the PKD phenotype is comparable among the strains presenting as neonatal proximal tubule dilations and adult collecting duct cysts. One key difference, though, is in the extra-renal manifestations, such as pancreatic and bile duct cysts, which are absent in C57BL/6J *cpk/cpk* mice, but present in all other strains. Similarly, both the *kat2J* mouse model,[197] caused by mutations to *Nek1*,[198] and the *bpk* mouse model,[199] caused by mutations to *Bicc1*,[200] develop variable PKD in the outbred strain ([C57BL/6J-*kat2J*/+ × CAST/Ei]F2[201] or [BALB/c-+/*bpk* × CAST/Ei]F1[202]) compared with the parental C57BL/6J or BALB/cJ strain, respectively. Genetic background effects have also been published for the *Pkhd1*^{tm1Cjwa} (*Pkhd1*$^{del2/del2}$) model, where pancreatic ductal cysts were observed in the outbred 129Sve × C57Bl/6J background but not in the C57Bl/6J or BALB/cJ backgrounds.[157] The *pcy* model,[203] caused by mutations to *Nphp3*[204] and the *Pkd1*^{tm1Djmp} (*Pkd1*$^{nl/nl}$),[119] have a markedly discordant PKD onset and severity among different strains. For example, PKD severity is drastically more severe in DBA/2FG *pcy/pcy* compared with C57Bl/6J *pcy/pcy* mice,[205] and *Pkd1*$^{nl/nl}$ mice have milder PKD with longer survival (>12 mo vs. 2 mo) if the model is in the mixed C57BL/6J and 129Ola/Hsd background versus the pure C57Bl/6J background as initially published.[206]

Collectively, these examples illustrate the importance of keeping precise records on genetic background, generation of inbreeding, and back-crossing experiments in order to be able to tease apart the phenotypic/mechanistic consequences unique to the gene of interest versus potential strain associated modifier gene/loci. This becomes especially important when generating double and triple genetically engineered rodent models across different background strains. Interestingly, murine models with dissimilar PKD-associated phenotypes across backgrounds would provide the perfect platform to identify potential modifier genes relevant to human ADPKD where large intra-/interfamilial phenotypic heterogeneity is observed. Such studies, while extremely tedious, have been performed by outcrossing inbred rodent models to a divergent strain and performing whole-genome genetic comparisons, also known as QTL mapping, among animals showing different grades of disease severity. In the case of PKD, this has been done for the *kat2J*, *bpk*, *pcy*, and Han-SPRD-cy models and suggests that genes such as *Invs, Bicc1, Col4a3,* or *Slc2la2* may modify PKD severity.[9,201,202,207,208]

10.9 CONCLUDING REMARKS

This chapter has detailed the generation, types, and use of rodent PKD models, with an emphasis on models orthologous to ADPKD. When comparing the various models, it is easily appreciable that each model presents with a unique pathology. Therefore, the choice of which model is best to use for research largely depends on the study question. For example, if the aim is to study the localization, function, and pathway interactions of PC1 or PC2, it is most important to use an orthologous ADPKD model rather than a non-orthologous ciliopathy model. In this example, severity and progression of PKD is less relevant to answering the study question. However, timing of *Pkd1* or *Pkd2* inactivation may be important to the study design, since the function of ADPKD proteins change depending on the stage of renal development, as discussed in Section 10.4.[10–12] Also, it is likely that both PC1 and PC2 have distinct functions in different tissues and possibly nephron segments (Table 10.1, Section 10.4).[108,209,210] Therefore, the expression pattern of Cre-recombinase drivers when using inducible models should be evaluated in detail, for example, using available RNA-seq and single-cell RNA-seq expression databases (e.g., Mouse Gene Expression Database [http://www.informatics.jax.org/expression.shtml], Expression Atlas [https://www.ebi.ac.uk/gxa/home], Mouse ENCODE Project [http://chromosome.sdsc.edu/ mouse/download.html], or others[211–213]).

For studies looking to understand the relationship between cystogenesis and pathophysiological or pathomechanistic features such as fibrosis, inflammation, epithelial-stromal cell interactions, metabolism, or epigenetics, it would be more suitable to choose a slowly progressive model, as introduced in Sections 10.4 and 10.5. This is because these models are thought to mirror the ADPKD patient pathophysiology better,[4,14,214] and therefore set the stage for a higher translational impact. It is also important to appreciate the difference between studying the pathomechanism impacting cyst progression apart from the pathomechanism addressing cyst initiation. Only conditional and inducible ADPKD models are suitable to study cyst initiation, because they provide the ability to precisely determine the timepoint of PC1 or PC2 loss, which is thought to be a key driver of cyst initiation in ADPKD[27,33] (Section 10.4). By comparison, models with germline mutations to *Pkd1* or *Pkd2* likely mimic patients who present with "nonvisible" microcysts at any stage of disease,[51,54] even though by histology or imaging techniques, the kidneys seem to have no cystic disease. For this reason, germline mutation models are unsuitable to study cyst initiation.

The choice of model becomes even more challenging when studying extra-renal manifestations of ADPKD such as cardiovascular or hepatic abnormalities. Very few models exist with patient relevant cardiovascular and hepatic phenotypes, as discussed throughout this chapter. For instance, only one published model develops intracranial aneurysms as observed in ADPKD patients, the *Pkd1*(TAG) mouse (Section 10.6), which genetically does not mimic the human disease.[64] This may be an artifact of focus (i.e., PKD researchers have not focused on this phenomenon in detail) or due to the fact that rodents are possibly protected against intracranial aneurysms upon PC1 or PC2 loss. Likewise, observed PLD in currently available orthologous models does not reflect human PLD, as outlined in Sections 10.4 and 10.5 of this chapter. In this case, researchers may choose to study non-orthologous ADPKD models such as the PCK rat (Section 10.7).

Performing preclinical trials further increases the stringency and thoughtfulness required in choosing the best model. For instance, while short-term treatments in rapidly progressive murine models with early disease onset are inviting from a cost and time-frame perspective, they do not imitate the clinical reality of slowly progressive disease. This is an important point, as different molecular and cellular pathways are at play during nephrogenesis versus adult renal steady-state maintenace.[215–218] For example, the developing kidney has a much higher epithelial cell proliferative index than the adult kidney.[12,13,218] This observation suggests that epithelial cell proliferation may play a much more prominent role in rapid cyst expansion of juvenile ADPKD models, compared with slow cyst progression observed in adult onset ADPKD models. Hence, compounds targeting epithelial proliferation will likely demonstrate a substantial difference in efficacy depending on the model. Beyond this, it has been well established that the pharmacodynamics of a compound differs depending on mouse strain and age.[219,220] Hence, drug biotransformation is likely distinct between a rapidly progressive and a slowly progressive ADPKD model. Similarly, the expression of a molecular target may vary with age or disease state of an animal. In this scenario, therapeutic compounds may result in better or worse antagonistic or agonist efficacy in a rapidly progressive and early-onset model, compared with a slowly progressive, adult-onset model. Consequently, many PKD researchers would agree that pathomechanistic studies and preclinical trials should be performed in at least two different models to ensure robustness and relevance to human disease. Therefore, it is with thoughtful consideration in choosing animal models, that the success of translating therapies from bench to bedside will be achieved.

Beyond model choice, there are key factors in study design that should always be considered when using rodent models. Factors such as sample size, sex, and litter variability influence the integrity of results and hence impact clinical translatability.[221–223] Sample size (number of animals) per group should be sufficiently powered to support the given observation. Dependent on PKD disease variability within a given model, it is advisable to have a sample size greater than five animals per group. Furthermore, gender and litter effects should be accounted for. Many disease models, including PKD models, present with sexually dimorphic pathology. For these reasons, it is essential to include both genders equally when performing pathomechanistic studies and preclinical trials. Also, disease variability is often smaller among litter mates compared with animals obtained from different litters. Thus, researchers often include animals from at least three or more different litters in each experimental group. Last, when it comes to analyzing results, it is important for the analyst to remain blinded as many of the PKD phenotypes can be impacted by observer bias (outlined in Section 10.3). It is also critical to apply the proper statistical method to analyze results. Often, the *t*-test assumptions, such as normal distribution of the data means, are not met in rodent research and nonparametric tests are more appropriate for analyzing groups.[224,225] This also applies for ad hoc tests when multiple groups are compared.

In summary, while optimal model choice for answering a particular PKD research question may be challenging at times, we are fortunate as a field to have such a broad repertoire of rodent models. Each model described within this chapter has provided key insights into the pathomechanisms of ADPKD and has brought us closer to an effective treatment, the first of such being Tolvaptan (Samsca, Jinarc, Jynarque). Given the complexity of genetic and cellular mechanisms at play in PKD and its associated extra-renal manifestations, it is conceivable that many more models will be developed in the future. These models will further our understanding, and hopefully address the current unknowns in PKD-related pathomechanisms including, but not limited to, the cystogenic impact of *PKD1/PKD2* allelic differences, cystic microenvironment composition, fibrosis, and genetic modifiers. Ultimately, detailed analyses of existing models, as well as the development of new models, will allow us to better understand the nuances of cyst development and progression, shedding light on why PKD presents with such heterogeneity in patients. Deeper insight into the complex pathology of PKD using rodent models will set the stage for a more personalized approach in alleviating the disease in patients.

REFERENCES

1. Lu W, Peissel B, Babakhanlou H et al. Perinatal lethality with kidney and pancreas defects in mice with a targetted *Pkd1* mutation. *Nat Genet* 1997; **17**: 179–181.
2. Wu G, Markowitz GS, Li L et al. Cardiac defects and renal failure in mice with targeted mutations in Pkd2. *Nat Genet* 2000; **24**: 75–78.
3. Wu G, D'Agati V, Cai Y et al. Somatic inactivation of Pkd2 results in polycystic kidney disease. *Cell* 1998; **93**: 177–188.
4. Hopp K, Ward CJ, Hommerding CJ et al. Functional polycystin-1 dosage governs autosomal dominant polycystic kidney disease severity. *J Clin Invest* 2012; **122**: 4257–4273.
5. Wu G, Tian X, Nishimura S et al. Trans-heterozygous *Pkd1* and *Pkd2* mutations modify expression of polycystic kidney disease. *Hum Mol Genet* 2002; **11**: 1845–1854.
6. Fedeles SV, Tian X, Gallagher AR et al. A genetic interaction network of five genes for human polycystic kidney and liver diseases defines polycystin-1 as the central determinant of cyst formation. *Nat Genet* 2011; **43**: 639–647.
7. Gresh L, Fischer E, Reimann A et al. A transcriptional network in polycystic kidney disease. *EMBO J* 2004; **23**: 1657–1668.
8. Tran U, Zakin L, Schweickert A et al. The RNA-binding protein bicaudal C regulates polycystin 2 in the kidney by antagonizing miR-17 activity. *Development* 2010; **137**: 1107–1116.

9. Guay-Woodford LM. Murine models of polycystic kidney disease: Molecular and therapeutic insights. *Am J Physiol Renal Physiol* 2003; **285**: F1034–1049.
10. Piontek K, Menezes LF, Garcia-Gonzalez MA et al. A critical developmental switch defines the kinetics of kidney cyst formation after loss of *Pkd1*. *Nat Med* 2007; **13**: 1490–1495.
11. Rogers KA, Moreno SE, Smith LA et al. Differences in the timing and magnitude of *Pkd1* gene deletion determine the severity of polycystic kidney disease in an orthologous mouse model of ADPKD. *Physiol Rep* 2016; **4**: e12846.
12. Lantinga-van Leeuwen IS, Leonhard WN, van der Wal A et al. Kidney-specific inactivation of the *Pkd1* gene induces rapid cyst formation in developing kidneys and a slow onset of disease in adult mice. *Hum Mol Genet* 2007; **16**: 3188–3196.
13. Happe H, Leonhard WN, van der Wal A et al. Toxic tubular injury in kidneys from *Pkd1*-deletion mice accelerates cystogenesis accompanied by dysregulated planar cell polarity and canonical Wnt signaling pathways. *Hum Mol Genet* 2009; **18**: 2532–2542.
14. Leonhard WN, Zandbergen M, Veraar K et al. Scattered deletion of PKD1 in kidneys causes a cystic snowball effect and recapitulates polycystic kidney disease. *J Am Soc Nephrol* 2015; **26**: 1322–1333.
15. Karihaloo A, Koraishy F, Huen SC et al. Macrophages promote cyst growth in polycystic kidney disease. *J Am Soc Nephrol* 2011; **22**: 1809–1814.
16. Swenson-Fields KI, Vivian CJ, Salah SM et al. Macrophages promote polycystic kidney disease progression. *Kidney Int* 2013; **83**: 855–864.
17. Kleczko EK, Marsh KH, Tyler LC et al. CD8(+) T cells modulate autosomal dominant polycystic kidney disease progression. *Kidney Int* 2018; **94**: 1127–1140.
18. Mrug M, Zhou J, Woo Y et al. Overexpression of innate immune response genes in a model of recessive polycystic kidney disease. *Kidney Int* 2008; **73**: 63–76.
19. Ta MH, Harris DC, Rangan GK. Role of interstitial inflammation in the pathogenesis of polycystic kidney disease. *Nephrology (Carlton)* 2013; **18**: 317–330.
20. Hildebrandt F, Benzing T, Katsanis N. Ciliopathies. *N Engl J Med* 2011; **364**: 1533–1543.
21. Norris DP, Grimes DT. Mouse models of ciliopathies: The state of the art. *Dis Model Mech* 2012; **5**: 299–312.
22. Ma M, Tian X, Igarashi P et al. Loss of cilia suppresses cyst growth in genetic models of autosomal dominant polycystic kidney disease. *Nat Genet* 2013; **45**: 1004–1012.
23. Lee SH, Somlo S. Cyst growth, polycystins, and primary cilia in autosomal dominant polycystic kidney disease. *Kidney Res Clin Pract* 2014; **33**: 73–78.
24. Patel V, Li L, Cobo-Stark P et al. Acute kidney injury and aberrant planar cell polarity induce cyst formation in mice lacking renal cilia. *Hum Mol Genet* 2008; **17**: 1578–1590.
25. Fischer E, Legue E, Doyen A et al. Defective planar cell polarity in polycystic kidney disease. *Nat Genet* 2006; **38**: 21–23.
26. Verdeguer F, Le Corre S, Fischer E et al. A mitotic transcriptional switch in polycystic kidney disease. *Nat Med* 2010; **16**: 106–110.
27. Harris PC, Torres VE. Genetic mechanisms and signaling pathways in autosomal dominant polycystic kidney disease. *J Clin Invest* 2014; **124**: 2315–2324.
28. Saigusa T, Bell PD. Molecular pathways and therapies in autosomal-dominant polycystic kidney disease. *Physiology (Bethesda)* 2015; **30**: 195–207.
29. Chapin HC, Caplan MJ. The cell biology of polycystic kidney disease. *J Cell Biol* 2010; **191**: 701–710.
30. Irazabal MV, Torres VE. Experimental therapies and ongoing clinical trials to slow down progression of ADPKD. *Curr Hypertens Rev* 2013; **9**: 44–59.
31. Patel V, Chowdhury R, Igarashi P. Advances in the pathogenesis and treatment of polycystic kidney disease. *Curr Opin Nephrol Hypertens* 2009; **18**: 99–106.
32. Chang MY, Ong AC. Mechanism-based therapeutics for autosomal dominant polycystic kidney disease: Recent progress and future prospects. *Nephron Clin Pract* 2012; **120**: c25–34; discussion c35.

33. Happe H, Peters DJ. Translational research in ADPKD: Lessons from animal models. *Nat Rev Nephrol* 2014; **10**: 587–601.
34. Ko JY, Park JH. Mouse models of polycystic kidney disease induced by defects of ciliary proteins. *BMB Rep* 2013; **46**: 73–79.
35. Wilson PD. Mouse models of polycystic kidney disease. *Curr Top Dev Biol* 2008; **84**: 311–350.
36. Menezes LF, Germino GG. Murine models of polycystic kidney disease. *Drug Discov Today Dis Mech* 2013; **10**: e153–e158.
37. Nagao S, Kugita M, Yoshihara D et al. Animal models for human polycystic kidney disease. *Exp Anim* 2012; **61**: 477–488.
38. Lee J, Rho J-i, Devkota S et al. Developing genetically engineered mouse models using engineered nucleases: Current status, challenges, and the way forward. *Drug Discov Today: Disease Models* 2016; **20**: 13–20.
39. Liao J, Cui C, Chen S et al. Generation of induced pluripotent stem cell lines from adult rat cells. *Cell Stem Cell* 2009; **4**: 11–15.
40. Ittner LM, Gotz J. Pronuclear injection for the production of transgenic mice. *Nat Protoc* 2007; **2**: 1206–1215.
41. Nie X, Arend LJ. Pkdl is required for male reproductive tract development. *Mech Dev* 2013; **130**: 567–576.
42. Bhunia AK, Piontek K, Boletta A et al. PKD1 induces p21(waf1) and regulation of the cell cycle via direct activation of the JAK-STAT signaling pathway in a process requiring PKD2. *Cell* 2002; **109**: 157–168.
43. Pennekamp P, Karcher C, Fischer A et al. The ion channel polycystin-2 is required for left-right axis determination in mice. *Curr Biol* 2002; **12**: 938–943.
44. Williams SS, Cobo-Stark P, James LR et al. Kidney cysts, pancreatic cysts, and biliary disease in a mouse model of autosomal recessive polycystic kidney disease. *Pediatr Nephrol* 2008; **23**: 733–741.
45. Freichel M, Kriebs U, Vogt D et al. Strategies and Protocols to Generate Mouse Models with Targeted Mutations to Analyze TRP Channel Functions. In: Zhu MX (ed). TRP Channels: Boca Raton (FL), 2011.
46. Perrone RD, Mouksassi MS, Romero K et al. Total kidney volume is a prognostic biomarker of renal function decline and progression to end-stage renal disease in patients with autosomal dominant polycystic kidney disease. *Kidney Int Rep* 2017; **2**: 442–450.
47. Warner G, Hein KZ, Nin V et al. Food restriction ameliorates the development of polycystic kidney disease. *J Am Soc Nephrol* 2016; **27**: 1437–1447.
48. Nowak KL, You Z, Gitomer B et al. Overweight and obesity are predictors of progression in early autosomal dominant polycystic kidney disease. *J Am Soc Nephrol* 2018; **29**: 571–578.
49. Farris AB, Alpers CE. What is the best way to measure renal fibrosis?: A pathologist's perspective. *Kidney Int Suppl (2011)* 2014; **4**: 9–15.
50. Wallace DP, Hou YP, Huang ZL et al. Tracking kidney volume in mice with polycystic kidney disease by magnetic resonance imaging. *Kidney Int* 2008; **73**: 778–781.
51. Irazabal MV, Mishra PK, Torres VE et al. Use of ultra-high field MRI in small rodent models of polycystic kidney disease for in vivo phenotyping and drug monitoring. *J Vis Exp* 2015; **100**: e52757.
52. Lakhia R, Yheskel M, Flaten A et al. PPARalpha agonist fenofibrate enhances fatty acid beta-oxidation and attenuates polycystic kidney and liver disease in mice. *Am J Physiol Renal Physiol* 2018; **314**: F122–F131.
53. Bae KT, Zhu F, Chapman AB et al. Magnetic resonance imaging evaluation of hepatic cysts in early autosomal-dominant polycystic kidney disease: The Consortium for Radiologic Imaging Studies of Polycystic Kidney Disease cohort. *Clin J Am Soc Nephrol* 2006; **1**: 64–69.

54. Kline TL, Korfiatis P, Edwards ME et al. Image texture features predict renal function decline in patients with autosomal dominant polycystic kidney disease. *Kidney Int* 2017; **92**: 1206–1216.
55. Kline TL, Edwards ME, Garg I et al. Quantitative MRI of kidneys in renal disease. *Abdom Radiol (NY)* 2018; **43**: 629–638.
56. Franke M, Baessler B, Vechtel J et al. Magnetic resonance T2 mapping and diffusion-weighted imaging for early detection of cystogenesis and response to therapy in a mouse model of polycystic kidney disease. *Kidney Int* 2017; **92**: 1544–1554.
57. Kline TL, Irazabal MV, Ebrahimi B et al. Utilizing magnetization transfer imaging to investigate tissue remodeling in a murine model of autosomal dominant polycystic kidney disease. *Magn Reson Med* 2016; **75**: 1466–1473.
58. Perrone RD, Neville J, Chapman AB et al. Therapeutic area data standards for autosomal dominant polycystic kidney disease: A report from the polycystic kidney disease outcomes consortium (PKDOC). *Am J Kidney Dis* 2015; **66**: 583–590.
59. Sadick M, Attenberger U, Kraenzlin B et al. Two non-invasive GFR-estimation methods in rat models of polycystic kidney disease: 3.0 Tesla dynamic contrast-enhanced MRI and optical imaging. *Nephrol Dial Transplant* 2011; **26**: 3101–3108.
60. Black LM, Lever JM, Traylor AM et al. Divergent effects of AKI to CKD models on inflammation and fibrosis. *Am J Physiol Renal Physiol* 2018; **315**: F1107–F1118.
61. Abdulmahdi W, Rabadi MM, Jules E et al. Kidney dysfunction in the low-birth weight murine adult: Implications of oxidative stress. *Am J Physiol Renal Physiol* 2018; **315**: F583–F594.
62. Gattone VH 2nd, Siqueira TM, Jr., Powell CR et al. Contribution of renal innervation to hypertension in rat autosomal dominant polycystic kidney disease. *Exp Biol Med (Maywood)* 2008; **233**: 952–957.
63. Yoshihara D, Kugita M, Sasaki M et al. Telmisartan ameliorates fibrocystic liver disease in an orthologous rat model of human autosomal recessive polycystic kidney disease. *PLOS ONE* 2013; **8**: e81480.
64. Kurbegovic A, Cote O, Couillard M et al. Pkd1 transgenic mice: Adult model of polycystic kidney disease with extrarenal and renal phenotypes. *Hum Mol Genet* 2010; **19**: 1174–1189.
65. Saigusa T, Dang Y, Mullick AE et al. Suppressing angiotensinogen synthesis attenuates kidney cyst formation in a Pkd1 mouse model. *FASEB J* 2016; **30**: 370–379.
66. Saigusa T, Dang Y, Bunni MA et al. Activation of the intrarenal renin-angiotensin-system in murine polycystic kidney disease. *Physiol Rep* 2015; **3**.
67. Fonseca JM, Bastos AP, Amaral AG et al. Renal cyst growth is the main determinant for hypertension and concentrating deficit in *Pkd1*-deficient mice. *Kidney Int* 2014; **85**: 1137–1150.
68. Kathem SH, Mohieldin AM, Abdul-Majeed S et al. Ciliotherapy: A novel intervention in polycystic kidney disease. *J Geriatr Cardiol* 2014; **11**: 63–73.
69. Lorenz JN. A practical guide to evaluating cardiovascular, renal, and pulmonary function in mice. *Am J Physiol Regul Integr Comp Physiol* 2002; **282**: R1565–1582.
70. Mattson DL. Long-term measurement of arterial blood pressure in conscious mice. *Am J Physiol* 1998; **274**: R564–R570.
71. Kramer K, Voss HP, Grimbergen JA et al. Telemetric monitoring of blood pressure in freely moving mice: A preliminary study. *Lab Anim* 2000; **34**: 272–280.
72. Salman IM, Sarma Kandukuri D, Harrison JL et al. Direct conscious telemetry recordings demonstrate increased renal sympathetic nerve activity in rats with chronic kidney disease. *Front Physiol* 2015; **6**: 218.
73. Zhao X, Ho D, Gao S et al. Arterial pressure monitoring in mice. *Curr Protoc Mouse Biol* 2011; **1**: 105–122.

74. Lee EJ. Cell proliferation and apoptosis in ADPKD. *Adv Exp Med Biol* 2016; **933**: 25–34.
75. Peintner L, Borner C. Role of apoptosis in the development of autosomal dominant polycystic kidney disease (ADPKD). *Cell Tissue Res* 2017; **369**: 27–39.
76. Zhou JX, Li X. Apoptosis in Polycystic Kidney Disease: From Pathogenesis to Treatment. In: Li X (ed). *Polycystic Kidney Disease*: Brisbane (AU), 2015.
77. Takakura A, Contrino L, Zhou X et al. Renal injury is a third hit promoting rapid development of adult polycystic kidney disease. *Hum Mol Genet* 2009; **18**: 2523–2531.
78. Kurbegovic A, Trudel M. Acute kidney injury induces hallmarks of polycystic kidney disease. *Am J Physiol Renal Physiol* 2016; **311**: F740–F751.
79. Puri P, Bushnell D, Schaefer CM et al. Six2creFrs2alpha knockout mice are a novel model of renal cystogenesis. *Sci Rep* 2016; **6**: 36736.
80. Gauer S, Urbschat A, Gretz N et al. Kidney injury molecule-1 is specifically expressed in cystically-transformed proximal tubules of the PKD/Mhm (cy/+) rat model of polycystic kidney disease. *Int J Mol Sci* 2016; **17**.
81. Lakhia R, Hajarnis S, Williams D et al. MicroRNA-21 aggravates cyst growth in a model of polycystic kidney disease. *J Am Soc Nephrol* 2016; **27**: 2319–2330.
82. Hajarnis S, Lakhia R, Yheskel M et al. microRNA-17 family promotes polycystic kidney disease progression through modulation of mitochondrial metabolism. *Nat Commun* 2017; **8**: 14395.
83. Parikh CR, Dahl NK, Chapman AB et al. Evaluation of urine biomarkers of kidney injury in polycystic kidney disease. *Kidney Int* 2012; **81**: 784–790.
84. Karihaloo A. Role of Inflammation in Polycystic Kidney Disease. In: Li X (ed). *Polycystic Kidney Disease*: Brisbane (AU), 2015.
85. Calvet JP. The Role of Calcium and Cyclic AMP in PKD. In: Li X (ed). *Polycystic Kidney Disease*: Brisbane (AU), 2015.
86. Torres VE, Harris PC. Strategies targeting cAMP signaling in the treatment of polycystic kidney disease. *J Am Soc Nephrol* 2014; **25**: 18–32.
87. Hopp K, Wang X, Ye H et al. Effects of hydration in rats and mice with polycystic kidney disease. *Am J Physiol Renal Physiol* 2015; **308**: F261–266.
88. Hopp K, Hommerding CJ, Wang X et al. Tolvaptan plus pasireotide shows enhanced efficacy in a PKD1 model. *J Am Soc Nephrol* 2015; **26**: 39–47.
89. Wang Q, Cobo-Stark P, Patel V et al. Adenylyl cyclase 5 deficiency reduces renal cyclic AMP and cyst growth in an orthologous mouse model of polycystic kidney disease. *Kidney Int* 2018; **93**: 403–415.
90. Boulter C, Mulroy S, Webb S et al. Cardiovascular, skeletal, and renal defects in mice with a targeted disruption of the Pkd1 gene. *Proc Natl Acad Sci USA* 2001; **98**: 12174–12179.
91. Kim K, Drummond I, Ibraghimov-Beskrovnaya O et al. Polycystin 1 is required for the structural integrity of blood vessels. *Proc Natl Acad Sci USA* 2000; **97**: 1731–1736.
92. Garcia-Gonzalez MA, Outeda P, Zhou Q et al. Pkd1 and Pkd2 are required for normal placental development. *PLOS ONE* 2010; **5**.
93. Khonsari RH, Ohazama A, Raouf R et al. Multiple postnatal craniofacial anomalies are characterized by conditional loss of polycystic kidney disease 2 (Pkd2). *Hum Mol Genet* 2013; **22**: 1873–1885.
94. Little MH. *Kidney development, disease, repair, and regeneration*. Elsevier/AP, Academic Press is an imprint of Elsevier: Amsterdam; Boston, 2016.
95. Starremans PG, Li X, Finnerty PE et al. A mouse model for polycystic kidney disease through a somatic in-frame deletion in the 5′ end of Pkd1. *Kidney Int* 2008; **73**: 1394–1405.
96. Shibazaki S, Yu Z, Nishio S et al. Cyst formation and activation of the extracellular regulated kinase pathway after kidney specific inactivation of *Pkd1*. *Hum Mol Genet* 2008; **17**: 1505–1516.

97. Shao X, Yang R, Yan M et al. Inducible expression of kallikrein in renal tubular cells protects mice against spontaneous lupus nephritis. *Arthritis Rheum* 2013; **65**: 780–791.
98. Lantinga-van Leeuwen IS, Leonhard WN, van de Wal A et al. Transgenic mice expressing tamoxifen-inducible Cre for somatic gene modification in renal epithelial cells. *Genesis* 2006; **44**: 225–232.
99. Nishio S, Tian X, Gallagher AR et al. Loss of oriented cell division does not initiate cyst formation. *J Am Soc Nephrol* 2010; **21**: 295–302.
100. Liu X, Vien T, Duan J et al. Polycystin-2 is an essential ion channel subunit in the primary cilium of the renal collecting duct epithelium. *Elife* 2018; **7**.
101. Schonig K, Schwenk F, Rajewsky K et al. Stringent doxycycline dependent control of CRE recombinase in vivo. *Nucleic Acids Res* 2002; **30**: e134.
102. Uhlmann EJ, Wong M, Baldwin RL et al. Astrocyte-specific TSC1 conditional knockout mice exhibit abnormal neuronal organization and seizures. *Ann Neurol* 2002; **52**: 285–296.
103. Traykova-Brauch M, Schonig K, Greiner O et al. An efficient and versatile system for acute and chronic modulation of renal tubular function in transgenic mice. *Nat Med* 2008; **14**: 979–984.
104. Talbot JJ, Shillingford JM, Vasanth S et al. Polycystin-1 regulates STAT activity by a dual mechanism. *Proc Natl Acad Sci USA* 2011; **108**: 7985–7990.
105. Takakura A, Nelson EA, Haque N et al. Pyrimethamine inhibits adult polycystic kidney disease by modulating STAT signaling pathways. *Hum Mol Genet* 2011; **20**: 4143–4154.
106. Belibi F, Ravichandran K, Zafar I et al. mTORC1/2 and rapamycin in female Han:SPRD rats with polycystic kidney disease. *Am J Physiol Renal Physiol* 2011; **300**: F236–244.
107. Omori S, Hida M, Fujita H et al. Extracellular signal-regulated kinase inhibition slows disease progression in mice with polycystic kidney disease. *J Am Soc Nephrol* 2006; **17**: 1604–1614.
108. Yu S, Hackmann K, Gao J et al. Essential role of cleavage of polycystin-1 at G protein-coupled receptor proteolytic site for kidney tubular structure. *Proc Natl Acad Sci USA* 2007; **104**: 18688–18693.
109. Kurbegovic A, Kim H, Xu H et al. Novel functional complexity of polycystin-1 by GPS cleavage in vivo: Role in polycystic kidney disease. *Mol Cell Biol* 2014; **34**: 3341–3353.
110. Tan M, Wettersten HI, Chu K et al. Novel inhibitors of nuclear transport cause cell cycle arrest and decrease cyst growth in ADPKD associated with decreased CDK4 levels. *Am J Physiol Renal Physiol* 2014; **307**: F1179–1186.
111. Rowe I, Chiaravalli M, Mannella V et al. Defective glucose metabolism in polycystic kidney disease identifies a new therapeutic strategy. *Nat Med* 2013; **19**: 488–493.
112. Rossetti S, Kubly VJ, Consugar MB et al. Incompletely penetrant PKD1 alleles suggest a role for gene dosage in cyst initiation in polycystic kidney disease. *Kidney Int* 2009; **75**: 848–855.
113. Vujic M, Heyer CM, Ars E et al. Incompletely penetrant PKD1 alleles mimic the renal manifestations of ARPKD. *J Am Soc Nephrol* 2010; **21**: 1097–1102.
114. Idowu J, Home T, Patel N et al. Aberrant regulation of Notch3 signaling pathway in polycystic kidney disease. *Sci Rep* 2018; **8**: 3340.
115. Gainullin VG, Hopp K, Ward CJ et al. Polycystin-1 maturation requires polycystin-2 in a dose-dependent manner. *J Clin Invest* 2015; **125**: 607–620.
116. Montford JR, Bauer C, Dobrinskikh E et al. Inhibition of 5-Lipoxygenase decreases renal fibrosis and progression of chronic kidney disease. *Am J Physiol Renal Physiol* 2019; **316**(4): F732–F742.
117. Yheskel M, Lakhia R, Cobo-Stark P et al. Anti-microRNA screen uncovers miR-17 family within miR-17~92 cluster as the primary driver of kidney cyst growth. *Sci Rep* 2019; **9**: 1920.
118. Parnell SC, Magenheimer BS, Maser RL et al. A mutation affecting polycystin-1 mediated heterotrimeric G-protein signaling causes PKD. *Hum Mol Genet* 2018; **27**: 3313–3324.

119. Lantinga-van Leeuwen IS, Dauwerse JG, Baelde HJ et al. Lowering of *Pkd1* expression is sufficient to cause polycystic kidney disease. *Hum Mol Genet* 2004; **13**: 3069–3077.
120. Jiang ST, Chiou YY, Wang E et al. Defining a link with autosomal-dominant polycystic kidney disease in mice with congenitally low expression of *Pkd1*. *Am J Pathol* 2006; **168**: 205–220.
121. Chen L, Zhou X, Fan LX et al. Macrophage migration inhibitory factor promotes cyst growth in polycystic kidney disease. *J Clin Invest* 2015; **125**: 2399–2412.
122. Huang JL, Woolf AS, Kolatsi-Joannou M et al. Vascular endothelial growth factor C for polycystic kidney diseases. *J Am Soc Nephrol* 2016; **27**: 69–77.
123. Hassane S, Claij N, Lantinga-van Leeuwen IS et al. Pathogenic sequence for dissecting aneurysm formation in a hypomorphic polycystic kidney disease 1 mouse model. *Arterioscler Thromb Vasc Biol* 2007; **27**: 2177–2183.
124. Li LX, Fan LX, Zhou JX et al. Lysine methyltransferase SMYD2 promotes cyst growth in autosomal dominant polycystic kidney disease. *J Clin Invest* 2017; **127**: 2751–2764.
125. Zhou X, Fan LX, Sweeney WE, Jr. et al. Sirtuin 1 inhibition delays cyst formation in autosomal-dominant polycystic kidney disease. *J Clin Invest* 2013; **123**: 3084–3098.
126. Zhou X, Fan LX, Peters DJ et al. Therapeutic targeting of BET bromodomain protein, Brd4, delays cyst growth in ADPKD. *Hum Mol Genet* 2015; **24**: 3982–3993.
127. Kim I, Li C, Liang D et al. Polycystin-2 expression is regulated by a PC2-binding domain in the intracellular portion of fibrocystin. *J Biol Chem* 2008; **283**: 31559–31566.
128. Doctor RB, Serkova NJ, Hasebroock KM et al. Distinct patterns of kidney and liver cyst growth in pkd2(WS25/-) mice. *Nephrol Dial Transplant* 2010; **25**: 3496–3504.
129. Stroope A, Radtke B, Huang B et al. Hepato-renal pathology in pkd2ws25/- mice, an animal model of autosomal dominant polycystic kidney disease. *Am J Pathol* 2010; **176**: 1282–1291.
130. Zafar I, Ravichandran K, Belibi FA et al. Sirolimus attenuates disease progression in an orthologous mouse model of human autosomal dominant polycystic kidney disease. *Kidney Int* 2010; **78**: 754–761.
131. Amura CR, Brodsky KS, Groff R et al. VEGF receptor inhibition blocks liver cyst growth in pkd2(WS25/-) mice. *Am J Physiol Cell Physiol* 2007; **293**: C419–428.
132. Chang MY, Parker E, El Nahas M et al. Endothelin B receptor blockade accelerates disease progression in a murine model of autosomal dominant polycystic kidney disease. *J Am Soc Nephrol* 2007; **18**: 560–569.
133. Zhang M, Srichai MB, Zhao M et al. Nonselective cyclooxygenase inhibition retards Cyst progression in a murine model of autosomal dominant polycystic kidney disease. *Int J Med Sci* 2019; **16**: 180–188.
134. Masyuk TV, Radtke BN, Stroope AJ et al. Inhibition of Cdc25A suppresses hepato-renal cystogenesis in rodent models of polycystic kidney and liver disease. *Gastroenterology* 2012; **142**: 622–633 e624.
135. Ravichandran K, Ozkok A, Wang Q et al. Antisense-mediated angiotensinogen inhibition slows polycystic kidney disease in mice with a targeted mutation in Pkd2. *Am J Physiol Renal Physiol* 2015; **308**: F349–357.
136. Masyuk TV, Radtke BN, Stroope AJ et al. Pasireotide is more effective than octreotide in reducing hepatorenal cystogenesis in rodents with polycystic kidney and liver diseases. *Hepatology* 2013; **58**: 409–421.
137. Masyuk AI, Huang BQ, Ward CJ et al. Biliary exosomes influence cholangiocyte regulatory mechanisms and proliferation through interaction with primary cilia. *Am J Physiol Gastrointest Liver Physiol* 2010; **299**: G990–999.
138. Ye H, Wang X, Sussman CR et al. Modulation of polycystic kidney disease severity by phosphodiesterase 1 and 3 subfamilies. *J Am Soc Nephrol* 2016; **27**: 1312–1320.

139. Pritchard L, Sloane-Stanley JA, Sharpe JA et al. A human PKD1 transgene generates functional polycystin-1 in mice and is associated with a cystic phenotype. *Hum Mol Genet* 2000; **9**: 2617–2627.
140. Pei Y, Watnick T, He N et al. Somatic PKD2 mutations in individual kidney and liver cysts support a "two-hit" model of cystogenesis in type 2 autosomal dominant polycystic kidney disease. *J Am Soc Nephrol* 1999; **10**: 1524–1529.
141. Watnick T, He N, Wang K et al. Mutations of PKD1 in ADPKD2 cysts suggest a pathogenic effect of trans-heterozygous mutations. *Nat Genet* 2000; **25**: 143–144.
142. Cagnazzo F, Gambacciani C, Morganti R et al. Intracranial aneurysms in patients with autosomal dominant polycystic kidney disease: Prevalence, risk of rupture, and management. A systematic review. *Acta Neurochir (Wien)* 2017; **159**: 811–821.
143. Kurbegovic A, Trudel M. Progressive development of polycystic kidney disease in the mouse model expressing Pkd1 extracellular domain. *Hum Mol Genet* 2013; **22**: 2361–2375.
144. Cai Y, Fedeles SV, Dong K et al. Altered trafficking and stability of polycystins underlie polycystic kidney disease. *J Clin Invest* 2014; **124**: 5129–5144.
145. Park EY, Sung YH, Yang MH et al. Cyst formation in kidney via B-Raf signaling in the PKD2 transgenic mice. *J Biol Chem* 2009; **284**: 7214–7222.
146. Li A, Tian X, Zhang X et al. Human polycystin-2 transgene dose-dependently rescues ADPKD phenotypes in Pkd2 mutant mice. *Am J Pathol* 2015; **185**: 2843–2860.
147. Burtey S, Riera M, Ribe E et al. Overexpression of PKD2 in the mouse is associated with renal tubulopathy. *Nephrol Dial Transplant* 2008; **23**: 1157–1165.
148. Burtey S, Riera M, Ribe E et al. Centrosome overduplication and mitotic instability in PKD2 transgenic lines. *Cell Biol Int* 2008; **32**: 1193–1198.
149. Gallagher AR, Hoffmann S, Brown N et al. A truncated polycystin-2 protein causes polycystic kidney disease and retinal degeneration in transgenic rats. *J Am Soc Nephrol* 2006; **17**: 2719–2730.
150. Sweeney WE, Avner ED. Polycystic Kidney Disease, Autosomal Recessive. In: Adam MP, Ardinger HH, Pagon RA, Wallace SE et al., (eds). *GeneReviews((R))*: Seattle (WA), 1993.
151. Willey CJ, Blais JD, Hall AK et al. Prevalence of autosomal dominant polycystic kidney disease in the European Union. *Nephrol Dial Transplant* 2017; **32**: 1356–1363.
152. Kaimori JY, Germino GG. ARPKD and ADPKD: First cousins or more distant relatives? *J Am Soc Nephrol* 2008; **19**: 416–418.
153. Harris PC, Torres VE. Polycystic kidney disease, autosomal dominant. *GeneReviews* 1993 [Updated 2011]; NBK1246.
154. Turkbey B, Ocak I, Daryanani K et al. Autosomal recessive polycystic kidney disease and congenital hepatic fibrosis (ARPKD/CHF). *Pediatr Radiol* 2009; **39**: 100–111.
155. Gallagher AR, Esquivel EL, Briere TS et al. Biliary and pancreatic dysgenesis in mice harboring a mutation in Pkhd1. *Am J Pathol* 2008; **172**: 417–429.
156. Moser M, Matthiesen S, Kirfel J et al. A mouse model for cystic biliary dysgenesis in autosomal recessive polycystic kidney disease (ARPKD). *Hepatology* 2005; **41**: 1113–1121.
157. Woollard JR, Punyashtiti R, Richardson S et al. A mouse model of autosomal recessive polycystic kidney disease with biliary duct and proximal tubule dilatation. *Kidney Int* 2007; **72**: 328–336.
158. Bakeberg JL, Tammachote R, Woollard JR et al. Epitope-tagged Pkhd1 tracks the processing, secretion, and localization of fibrocystin. *J Am Soc Nephrol* 2011; **22**: 2266–2277.
159. Shan D, Rezonzew G, Mullen S et al. Heterozygous Pkhd1(C642*) mice develop cystic liver disease and proximal tubule ectasia that mimics radiographic signs of medullary sponge kidney. *Am J Physiol Renal Physiol* 2019; **316**: F463–F472.

160. Garcia-Gonzalez MA, Menezes LF, Piontek KB et al. Genetic interaction studies link autosomal dominant and recessive polycystic kidney disease in a common pathway. *Hum Mol Genet* 2007; **16**: 1940–1950.
161. Kim I, Fu Y, Hui K et al. Fibrocystin/polyductin modulates renal tubular formation by regulating polycystin-2 expression and function. *J Am Soc Nephrol* 2008; **19**: 455–468.
162. Nagasawa Y, Matthiesen S, Onuchic LF et al. Identification and characterization of Pkhd1, the mouse orthologue of the human ARPKD gene. *J Am Soc Nephrol* 2002; **13**: 2246–2258.
163. Lager DJ, Qian Q, Bengal RJ et al. The pck rat: A new model that resembles human autosomal dominant polycystic kidney and liver disease. *Kidney Int* 2001; **59**: 126–136.
164. Mason SB, Liang Y, Sinders RM et al. Disease stage characterization of hepatorenal fibrocystic pathology in the PCK rat model of ARPKD. *Anat Rec (Hoboken)* 2010; **293**: 1279–1288.
165. Smith LA, Bukanov NO, Husson H et al. Development of polycystic kidney disease in juvenile cystic kidney mice: Insights into pathogenesis, ciliary abnormalities, and common features with human disease. *J Am Soc Nephrol* 2006; **17**: 2821–2831.
166. Cowley BD, Jr., Rupp JC, Muessel MJ et al. Gender and the effect of gonadal hormones on the progression of inherited polycystic kidney disease in rats. *Am J Kidney Dis* 1997; **29**: 265–272.
167. Harris PC, Bae KT, Rossetti S et al. Cyst number but not the rate of cystic growth is associated with the mutated gene in autosomal dominant polycystic kidney disease. *J Am Soc Nephrol* 2006; **17**: 3013–3019.
168. Heyer CM, Sundsbak JL, Abebe KZ et al. Predicted mutation strength of nontruncating *PKD1* mutations aids genotype-phenotype correlations in autosomal dominant polycystic kidney disease. *J Am Soc Nephrol* 2016; **27**: 2872–2884.
169. Cnossen WR, Drenth JP. Polycystic liver disease: An overview of pathogenesis, clinical manifestations and management. *Orphanet J Rare Dis* 2014; **9**: 69.
170. Gattone VH 2nd, Wang X, Harris PC et al. Inhibition of renal cystic disease development and progression by a vasopressin V2 receptor antagonist. *Nature Med* 2003; **9**: 1323–1326.
171. Wang X, Gattone V, Harris PC et al. Effectiveness of vasopressin V2 receptor antagonists OPC-31260 and OPC-41061 on polycystic kidney disease development in the PCK rat. *J Am Soc Nephrol* 2005; **16**: 846–851.
172. Masyuk TV, Masyuk AI, Torres VE et al. Octreotide inhibits hepatic cystogenesis in a rodent model of polycystic liver disease by reducing cholangiocyte adenosine 3′,5′-cyclic monophosphate. *Gastroenterology* 2007; **132**: 1104–1116.
173. Kaplan BS, Fay J, Shah V et al. Autosomal recessive polycystic kidney disease. *Pediatr Nephrol* 1989; **3**: 43–49.
174. Tsiokas L, Kim E, Arnould T et al. Homo- and heterodimeric interactions between the gene products of PKD1 and PKD2. *Proc Natl Acad Sci USA* 1997; **94**: 6965–6970.
175. Richards T, Modarage K, Dean C et al. Atmin modulates Pkhd1 expression and may mediate Autosomal Recessive Polycystic Kidney Disease (ARPKD) through altered non-canonical Wnt/Planar Cell Polarity (PCP) signalling. *Biochim Biophys Acta Mol Basis Dis* 2018; **1865**(2): 378–390.
176. Yang J, Zhang S, Zhou Q et al. PKHD1 gene silencing may cause cell abnormal proliferation through modulation of intracellular calcium in autosomal recessive polycystic kidney disease. *J Biochem Mol Biol* 2007; **40**: 467–474.
177. Outeda P, Menezes L, Hartung EA et al. A novel model of autosomal recessive polycystic kidney questions the role of the fibrocystin C-terminus in disease mechanism. *Kidney Int* 2017; **92**: 1130–1144.
178. Masyuk TV, Huang BQ, Ward CJ et al. Defects in cholangiocyte fibrocystin expression and ciliary structure in the PCK rat. *Gastroenterology* 2003; **125**: 1303–1310.

179. Ward CJ, Yuan D, Masyuk TV et al. Cellular and subcellular localization of the ARPKD protein; fibrocystin is expressed on primary cilia. *Hum Mol Genet* 2003; **12**: 2703–2710.
180. Besse W, Dong K, Choi J et al. Isolated polycystic liver disease genes define effectors of polycystin-1 function. *J Clin Invest* 2017; **127**: 3558.
181. Fliegauf M, Benzing T, Omran H. When cilia go bad: Cilia defects and ciliopathies. *Nat Rev Mol Cell Biol* 2007; **8**: 880–893.
182. Baker K, Beales PL. Making sense of cilia in disease: The human ciliopathies. *Am J Med Genet C Semin Med Genet* 2009; **151C**: 281–295.
183. Nadeau JH. Modifier genes in mice and humans. *Nat Rev Genet* 2001; **2**: 165–174.
184. Montagutelli X. Effect of the genetic background on the phenotype of mouse mutations. *J Am Soc Nephrol* 2000; 11 Suppl **16**: S101–S105.
185. Silver, L. M. *Mouse Genetics: Concepts and Applications*, 1995. New York, USA: Oxford University Press. ISBN : 0195075544.
186. Walkin L, Herrick SE, Summers A et al. The role of mouse strain differences in the susceptibility to fibrosis: A systematic review. *Fibrogenesis Tissue Repair* 2013; **6**: 18.
187. Barbaric I, Miller G, Dear TN. Appearances can be deceiving: Phenotypes of knockout mice. *Brief Funct Genomic Proteomic* 2007; **6**: 91–103.
188. Dockstader CL, van der Kooy D. Mouse strain differences in opiate reward learning are explained by differences in anxiety, not reward or learning. *J Neurosci* 2001; **21**: 9077–9081.
189. Sellers RS, Clifford CB, Treuting PM et al. Immunological variation between inbred laboratory mouse strains: Points to consider in phenotyping genetically immunomodified mice. *Veterinary Pathology* 2012; **49**: 32–43.
190. Paradee W, Melikian HE, Rasmussen DL et al. Fragile X mouse: Strain effects of knockout phenotype and evidence suggesting deficient amygdala function. *Neuroscience* 1999; **94**: 185–192.
191. Schock-Kusch D, Geraci S, Ermeling E et al. Reliability of transcutaneous measurement of renal function in various strains of conscious mice. *PLOS ONE* 2013; **8**: e71519.
192. Hou X, Mrug M, Yoder BK et al. Cystin, a novel cilia-associated protein, is disrupted in the cpk mouse model of polycystic kidney disease. *J Clin Invest* 2002; **109**: 533–540.
193. Fry JL, Jr., Koch WE, Jennette JC et al. A genetically determined murine model of infantile polycystic kidney disease. *J Urol* 1985; **134**: 828–833.
194. Gattone VH 2nd, Calvet JP, Cowley BD, Jr. et al. Autosomal recessive polycystic kidney disease in a murine model. A gross and microscopic description. *Lab Invest* 1988; **59**: 231–238.
195. Gattone VH 2nd, MacNaughton KA, Kraybill AL. Murine autosomal recessive polycystic kidney disease with multiorgan involvement induced by the cpk gene. *Anat Rec* 1996; **245**: 488–499.
196. Ricker JL, Gattone VH 2nd, Calvet JP et al. Development of autosomal recessive polycystic kidney disease in BALB/c-cpk/cpk mice. *J Am Soc Nephrol* 2000; **11**: 1837–1847.
197. Janaswami PM, Birkenmeier EH, Cook SA et al. Identification and genetic mapping of a new polycystic kidney disease on mouse chromosome 8. *Genomics* 1997; **40**: 101–107.
198. Upadhya P, Birkenmeier EH, Birkenmeier CS et al. Mutations in a NIMA-related kinase gene, Nek1, cause pleiotropic effects including a progressive polycystic kidney disease in mice. *Proc Natl Acad Sci USA* 2000; **97**: 217–221.
199. Nauta J, Ozawa Y, Sweeney WE, Jr. et al. Renal and biliary abnormalities in a new murine model of autosomal recessive polycystic kidney disease. *Pediatr Nephrol* 1993; **7**: 163–172.
200. Guay-Woodford LM, Bryda EC, Christine B et al. Evidence that two phenotypically distinct mouse PKD mutations, bpk and jcpk, are allelic. *Kidney Int* 1996; **50**: 1158–1165.

201. Upadhya P, Churchill G, Birkenmeier EH et al. Genetic modifiers of polycystic kidney disease in intersubspecific KAT2J mutants. *Genomics* 1999; **58**: 129–137.
202. Guay-Woodford LM, Wright CJ, Walz G et al. Quantitative trait loci modulate renal cystic disease severity in the mouse bpk model. *J Am Soc Nephrol* 2000; **11**: 1253–1260.
203. Takahashi H, Ueyama Y, Hibino T et al. A new mouse model of genetically transmitted polycystic kidney disease. *J Urol* 1986; **135**: 1280–1283.
204. Olbrich H, Fliegauf M, Hoefele J et al. Mutations in a novel gene, NPHP3, cause adolescent nephronophthisis, tapeto-retinal degeneration and hepatic fibrosis. *Nat Genet* 2003; **34**: 455–459.
205. Nagao S, Hibino T, Koyama Y et al. Strain difference in expression of the adult-type polycystic kidney disease gene, pcy, in the mouse. *Jikken Dobutsu* 1991; **40**: 45–53.
206. Happe H, van der Wal AM, Salvatori DC et al. Cyst expansion and regression in a mouse model of polycystic kidney disease. *Kidney Int* 2013; **83**: 1099–1108.
207. Bihoreau MT, Megel N, Brown JH et al. Characterization of a major modifier locus for polycystic kidney disease (Modpkdr1) in the Han:SPRD(cy/+) rat in a region conserved with a mouse modifier locus for Alport syndrome. *Hum Mol Genet* 2002; **11**: 2165–2173.
208. Woo DD, Nguyen DK, Khatibi N et al. Genetic identification of two major modifier loci of polycystic kidney disease progression in pcy mice. *J Clin Invest* 1997; **100**: 1934–1940.
209. Peters DJ, van de Wal A, Spruit L et al. Cellular localization and tissue distribution of polycystin-1. *J Pathol* 1999; **188**: 439–446.
210. Wilson PD. Polycystin: New aspects of structure, function, and regulation. *J Am Soc Nephrol* 2001; **12**: 834–845.
211. Sollner JF, Leparc G, Hildebrandt T et al. An RNA-Seq atlas of gene expression in mouse and rat normal tissues. *Sci Data* 2017; **4**: 170185.
212. Lee JW, Chou CL, Knepper MA. Deep sequencing in microdissected renal tubules identifies nephron segment-specific transcriptomes. *J Am Soc Nephrol* 2015; **26**: 2669–2677.
213. Park J, Shrestha R, Qiu C et al. Single-cell transcriptomics of the mouse kidney reveals potential cellular targets of kidney disease. *Science* 2018; **360**: 758–763.
214. Torres VE, Harris PC. Polycystic kidney disease: Genes, proteins, animal models, disease mechanisms and therapeutic opportunities. *J Intern Med* 2007; **261**: 17–31.
215. McMahon AP. Development of the mammalian kidney. *Curr Top Dev Biol* 2016; **117**: 31–64.
216. Seely JC. A brief review of kidney development, maturation, developmental abnormalities, and drug toxicity: Juvenile animal relevancy. *J Toxicol Pathol* 2017; **30**: 125–133.
217. Krause M, Rak-Raszewska A, Pietila I et al. Signaling during kidney development. *Cells* 2015; **4**: 112–132.
218. Short KM, Combes AN, Lefevre J et al. Global quantification of tissue dynamics in the developing mouse kidney. *Dev Cell* 2014; **29**: 188–202.
219. Nachshon A, Abu-Toamih Atamni HJ, Steuerman Y et al. Dissecting the effect of genetic variation on the hepatic expression of drug disposition genes across the collaborative cross mouse strains. *Front Genet* 2016; **7**: 172.
220. Prinz J, Vogt I, Adornetto G et al. A novel drug-mouse phenotypic similarity method detects molecular determinants of drug effects. *PLOS Comput Biol* 2016; **12**: e1005111.
221. Scott S, Kranz JE, Cole J et al. Design, power, and interpretation of studies in the standard murine model of ALS. *Amyotroph Lateral Scler* 2008; **9**: 4–15.
222. Wong CH, Siah KW, Lo AW. Estimation of clinical trial success rates and related parameters. *Biostatistics* 2018; **20**(2): 273–286.
223. Begley CG, Ioannidis JP. Reproducibility in science: Improving the standard for basic and preclinical research. *Circ Res* 2015; **116**: 116–126.

224. Aban IB, George B. Statistical considerations for preclinical studies. *Exp Neurol* 2015; **270**: 82–87.
225. Laajala TD, Jumppanen M, Huhtaniemi R et al. Optimized design and analysis of preclinical intervention studies in vivo. *Sci Rep* 2016; **6**: 30723.
226. Raphael KL, Strait KA, Stricklett PK et al. Inactivation of Pkd1 in principal cells causes a more severe cystic kidney disease than in intercalated cells. *Kidney Int* 2009; **75**: 626–633.
227. Yoon H, Lee DJ, Kim MH et al. Identification of genes concordantly expressed with Atoh1 during inner ear development. *Anat Cell Biol* 2011; **44**: 69–78.
228. Qiu N, Xiao Z, Cao L et al. Conditional mesenchymal disruption of pkd1 results in osteopenia and polycystic kidney disease. *PLOS ONE* 2012; **7**: e46038.
229. Kolpakova-Hart E, Nicolae C, Zhou J et al. Col2-Cre recombinase is co-expressed with endogenous type II collagen in embryonic renal epithelium and drives development of polycystic kidney disease following inactivation of ciliary genes. *Matrix Biol* 2008; **27**: 505–512.
230. Hering TM, Wirthlin L, Ravindran S et al. Changes in type II procollagen isoform expression during chondrogenesis by disruption of an alternative 5′ splice site within Col2a1 exon 2. *Matrix Biol* 2014; **36**: 51–63.
231. Kolpakova-Hart E, McBratney-Owen B, Hou B et al. Growth of cranial synchondroses and sutures requires polycystin-1. *Dev Biol* 2008; **321**: 407–419.
232. Iwano M, Plieth D, Danoff TM et al. Evidence that fibroblasts derive from epithelium during tissue fibrosis. *J Clin Invest* 2002; **110**: 341–350.
233. Paul BM, Vassmer D, Taylor A et al. Ectopic expression of Cux1 is associated with reduced p27 expression and increased apoptosis during late stage cyst progression upon inactivation of Pkd1 in collecting ducts. *Dev Dyn* 2011; **240**: 1493–1501.
234. Yu J, Carroll TJ, McMahon AP. Sonic hedgehog regulates proliferation and differentiation of mesenchymal cells in the mouse metanephric kidney. *Development* 2002; **129**: 5301–5312.
235. Piontek KB, Huso DL, Grinberg A et al. A functional floxed allele of Pkd1 that can be conditionally inactivated in vivo. *J Am Soc Nephrol* 2004; **15**: 3035–3043.
236. Shillingford JM, Piontek KB, Germino GG et al. Rapamycin ameliorates PKD resulting from conditional inactivation of Pkd1. *J Am Soc Nephrol* 2010; **21**: 489–497.
237. Fedeles SV, So JS, Shrikhande A et al. Sec63 and Xbp1 regulate IRE1alpha activity and polycystic disease severity. *J Clin Invest* 2015; **125**: 1955–1967.
238. Lang S, Benedix J, Fedeles SV et al. Different effects of Sec61alpha, Sec62 and Sec63 depletion on transport of polypeptides into the endoplasmic reticulum of mammalian cells. *J Cell Sci* 2012; **125**: 1958–1969.
239. Braren R, Hu H, Kim YH et al. Endothelial FAK is essential for vascular network stability, cell survival, and lamellipodial formation. *J Cell Biol* 2006; **172**: 151–162.
240. Li A, Fan S, Xu Y et al. Rapamycin treatment dose-dependently improves the cystic kidney in a new ADPKD mouse model via the mTORC1 and cell-cycle-associated CDK1/cyclin axis. *J Cell Mol Med* 2017; **21**: 1619–1635.
241. McCright B, Lozier J, Gridley T. Generation of new Notch2 mutant alleles. *Genesis* 2006; **44**: 29–33.
242. Tallquist MD, Soriano P. Epiblast-restricted Cre expression in MORE mice: A tool to distinguish embryonic vs. extra-embryonic gene function. *Genesis* 2000; **26**: 113–115.
243. Luquet S, Perez FA, Hnasko TS et al. NPY/AgRP neurons are essential for feeding in adult mice but can be ablated in neonates. *Science* 2005; **310**: 683–685.
244. Takakura A, Contrino L, Beck AW et al. Pkd1 inactivation induced in adulthood produces focal cystic disease. *J Am Soc Nephrol* 2008; **19**: 2351–2363.
245. Cebotaru L, Liu Q, Yanda MK et al. Inhibition of histone deacetylase 6 activity reduces cyst growth in polycystic kidney disease. *Kidney Int* 2016; **90**: 90–99.

246. Kim I, Ding T, Fu Y et al. Conditional mutation of Pkd2 causes cystogenesis and upregulates beta-catenin. *J Am Soc Nephrol* 2009; **20**: 2556–2569.
247. Spirli C, Okolicsanyi S, Fiorotto R et al. ERK1/2-dependent vascular endothelial growth factor signaling sustains cyst growth in polycystin-2 defective mice. *Gastroenterology* 2010; **138**: 360–371 e367.
248. Schrick JJ, Vogel P, Abuin A et al. ADP-ribosylation factor-like 3 is involved in kidney and photoreceptor development. *Am J Pathol* 2006; **168**: 1288–1298.
249. Veis DJ, Sorenson CM, Shutter JR et al. Bcl-2-deficient mice demonstrate fulminant lymphoid apoptosis, polycystic kidneys, and hypopigmented hair. *Cell* 1993; **75**: 229–240.
250. Sorenson CM, Rogers SA, Korsmeyer SJ et al. Fulminant metanephric apoptosis and abnormal kidney development in bcl-2-deficient mice. *Am J Physiol* 1995; **268**: F73–81.
251. Sorenson CM, Padanilam BJ, Hammerman MR. Abnormal postpartum renal development and cystogenesis in the bcl-2 (–/–) mouse. *Am J Physiol* 1996; **271**: F184–193.
252. Sorenson CM. Nuclear localization of beta-catenin and loss of apical brush border actin in cystic tubules of bcl-2 –/– mice. *Am J Physiol* 1999; **276**: F210–217.
253. Shillingford JM, Murcia NS, Larson CH et al. The mTOR pathway is regulated by polycystin-1, and its inhibition reverses renal cystogenesis in polycystic kidney disease. *Proc Natl Acad Sci USA* 2006; **103**: 5466–5471.
254. MacRae Dell K, Nemo R, Sweeney WE, Jr. et al. EGF-related growth factors in the pathogenesis of murine ARPKD. *Kidney Int* 2004; **65**: 2018–2029.
255. Veizis EI, Carlin CR, Cotton CU. Decreased amiloride-sensitive Na+ absorption in collecting duct principal cells isolated from BPK ARPKD mice. *Am J Physiol Renal Physiol* 2004; **286**: F244–254.
256. Ozawa Y, Nauta J, Sweeney WE et al. A new murine model of autosomal recessive polycystic kidney disease. *Nihon Jinzo Gakkai Shi* 1993; **35**: 349–354.
257. Mesner LD, Ray B, Hsu YH et al. Bicc1 is a genetic determinant of osteoblastogenesis and bone mineral density. *J Clin Invest* 2014; **124**: 2736–2749.
258. Cogswell C, Price SJ, Hou X et al. Positional cloning of jcpk/bpk locus of the mouse. *Mamm Genome* 2003; **14**: 242–249.
259. Chittenden L, Lu X, Cacheiro NL et al. A new mouse model for autosomal recessive polycystic kidney disease. *Genomics* 2002; **79**: 499–504.
260. Price SJ, Chittenden LR, Flaherty L et al. Characterization of the region containing the jcpk PKD gene on mouse Chromosome 10. *Cytogenet Genome Res* 2002; **98**: 61–66.
261. Flaherty L, Bryda EC, Collins D et al. New mouse model for polycystic kidney disease with both recessive and dominant gene effects. *Kidney Int* 1995; **47**: 552–558.
262. Flaherty L, Messer A, Russell LB et al. Chlorambucil-induced mutations in mice recovered in homozygotes. *Proc Natl Acad Sci USA* 1992; **89**: 2859–2863.
263. Chiu MG, Johnson TM, Woolf AS et al. Galectin-3 associates with the primary cilium and modulates cyst growth in congenital polycystic kidney disease. *Am J Pathol* 2006; **169**: 1925–1938.
264. Lu H, Galeano MCR, Ott E et al. Mutations in DZIP1L, which encodes a ciliary-transition-zone protein, cause autosomal recessive polycystic kidney disease. *Nat Genet* 2017; **49**: 1025–1034.
265. Jonassen JA, SanAgustin J, Baker SP et al. Disruption of IFT complex A causes cystic kidneys without mitotic spindle misorientation. *J Am Soc Nephrol* 2012; **23**: 641–651.
266. Jonassen JA, San Agustin J, Follit JA et al. Deletion of IFT20 in the mouse kidney causes misorientation of the mitotic spindle and cystic kidney disease. *J Cell Biol* 2008; **183**: 377–384.
267. Zhang Q, Davenport JR, Croyle MJ et al. Disruption of IFT results in both exocrine and endocrine abnormalities in the pancreas of Tg737(orpk) mutant mice. *Lab Invest* 2005; **85**: 45–64.

268. Sommardahl C, Cottrell M, Wilkinson JE et al. Phenotypic variations of orpk mutation and chromosomal localization of modifiers influencing kidney phenotype. *Physiol Genomics* 2001; **7**: 127–134.
269. Moyer JH, Lee-Tischler MJ, Kwon HY et al. Candidate gene associated with a mutation causing recessive polycystic kidney disease in mice. *Science* 1994; **264**: 1329–1333.
270. Yokoyama T, Copeland NG, Jenkins NA et al. Reversal of left-right asymmetry: A situs inversus mutation. *Science* 1993; **260**: 679–682.
271. Morgan D, Turnpenny L, Goodship J et al. Inversin, a novel gene in the vertebrate left-right axis pathway, is partially deleted in the inv mouse. *Nat Genet* 1998; **20**: 149–156.
272. Lin F, Hiesberger T, Cordes K et al. Kidney-specific inactivation of the KIF3A subunit of kinesin-II inhibits renal ciliogenesis and produces polycystic kidney disease. *Proc Natl Acad Sci USA* 2003; **100**: 5286–5291.
273. Shannon MB, Patton BL, Harvey SJ et al. A hypomorphic mutation in the mouse laminin alpha5 gene causes polycystic kidney disease. *J Am Soc Nephrol* 2006; **17**: 1913–1922.
274. Vogler C, Homan S, Pung A et al. Clinical and pathologic findings in two new allelic murine models of polycystic kidney disease. *J Am Soc Nephrol* 1999; **10**: 2534–2539.
275. Atala A, Freeman MR, Mandell J et al. Juvenile cystic kidneys (jck): A new mouse mutation which causes polycystic kidneys. *Kidney Int* 1993; **43**: 1081–1085.
276. Sun Y, Zhou J, Stayner C et al. Magnetic resonance imaging assessment of a murine model of recessive polycystic kidney disease. *Comp Med* 2002; **52**: 433–438.
277. Tran PV, Talbott GC, Turbe-Doan A et al. Downregulating hedgehog signaling reduces renal cystogenic potential of mouse models. *J Am Soc Nephrol* 2014; **25**: 2201–2212.
278. Takahashi H, Calvet JP, Dittemore-Hoover D et al. A hereditary model of slowly progressive polycystic kidney disease in the mouse. *J Am Soc Nephrol* 1991; **1**: 980–989.
279. Li J, Lu D, Liu H et al. Scltl deficiency causes cystic kidney by activating ERK and STAT3 signaling. *Hum Mol Genet* 2017; **26**: 2949–2960.
280. Bihoreau MT, Ceccherini I, Browne J et al. Location of the first genetic locus, PKDr1, controlling autosomal dominant polycystic kidney disease in Han:SPRD cy/+ rat. *Hum Mol Genet* 1997; **6**: 609–613.
281. Gretz N, Kranzlin B, Pey R et al. Rat models of autosomal dominant polycystic kidney disease. *Nephrol Dial Transplant* 1996; 11 Suppl **6**: 46–51.
282. Kaspareit-Rittinghausen J, Deerberg F, Rapp KG et al. A new rat model for polycystic kidney disease of humans. *Transplant Proc* 1990; **22**: 2582–2583.
283. Kranzlin B, Schieren G, Gretz N. Azotemia and extrarenal manifestations in old female Han:SPRD (cy/+) rats. *Kidney Int* 1997; **51**: 1160–1169.
284. Gattone VH 2nd, Tourkow BA, Trambaugh CM et al. Development of multiorgan pathology in the wpk rat model of polycystic kidney disease. *Anat Rec A Discov Mol Cell Evol Biol* 2004; **277**: 384–395.
285. Nauta J, Goedbloed MA, Herck HV et al. New rat model that phenotypically resembles autosomal recessive polycystic kidney disease. *J Am Soc Nephrol* 2000; **11**: 2272–2284.
286. Smith UM, Consugar M, Tee LJ et al. The transmembrane protein meckelin (MKS3) is mutated in Meckel-Gruber syndrome and the wpk rat. *Nat Genet* 2006; **38**: 191–196.
287. McCooke JK, Appels R, Barrero RA et al. A novel mutation causing nephronophthisis in the Lewis polycystic kidney rat localises to a conserved RCC1 domain in Nek8. *BMC Genomics* 2012; **13**: 393.
288. Ta MH, Rao P, Korgaonkar M et al. Pyrrolidine dithiocarbamate reduces the progression of total kidney volume and cyst enlargement in experimental polycystic kidney disease. *Physiol Rep* 2014; **2**.
289. Inage Z, Kikkawa Y, Minato M et al. Autosomal recessive polycystic kidney in rats. *Nephron* 1991; **59**: 637–640.
290. Ohno K, Kondo K. A mutant rat with congenital skeletal abnormalities and polycystic kidneys. *Jikken Dobutsu* 1989; **38**: 139–146.

11 Using *C. elegans* as a Model in PKD

Juan Wang and Maureen Barr

CONTENTS

11.1 INTRODUCTION

Autosomal dominant polycystic kidney disease (ADPKD) is the most common genetic cause of end-stage renal disease, with a prevalence of 1:500 to 1:1000. ADPKD is caused by mutations in the *PKD1* and *PKD2* genes, which encode polycystin-1 (PC1) and polycystin-2 (PC2), respectively. ADPKD is a systemic disorder that also manifests liver cysts, cerebral aneurysms, and cardiac valvular abnormalities. PKD is a characteristic feature observed in several ciliopathies, implicating cilia in the pathogenesis of human ADPKD. With over 30 years of research, significant insight into the disease has been achieved. However, a cure awaits until we gain a clearer understanding of molecular functions of the polycystins throughout development and adulthood and the pathogenic consequences of disabled polycystins in ADPKD.

In 1999, *C. elegans* PC1 and PC2 homologues LOV-1 and PKD-2 were identified based on their mutant phenotypes of defective male mating behaviors.[1] Transgenic reporters revealed that *lov-1* and *pkd-2* are expressed in male-specific ciliated sensory neurons and that the protein products localize to cilia (Figure 11.1). This discovery provided one of the first links between cilia and polycystic kidney disease and established the nematode *C. elegans* as a powerful animal model system for studying PKD and human ciliopathies. In 2002, Pazour and Yoder showed that the mammalian polycystins localized to cilia on renal epithelial cells.[3,4] Hence ciliary localization of the polycystins is evolutionarily conserved. These studies, along with Joel Rosenbaum's discovery of intraflagellar transport (IFT) in the algae *Chlamydomonas*[5] provided the foundation and framework for studying cilia biology in the context of human disease.[6]

In addition to their conserved ciliary localization, *C. elegans* and mammalian PC1 and PC2 localize to extracellular vesicles (EVs).[7] Submicron-sized EVs may arise via fusion of multivesicular bodies to the plasma membrane and release of intraluminal vesicles as exosomes. Alternatively, EVs may arise via budding of the plasma membrane and release as ectosomes or microvesicles. The polycystins are found in exosome-like extracellular vesicles isolated from human urine.[8,9] In *C. elegans*, green fluorescent protein (GFP)-tagged LOV-1 and PKD-2 are shed—likely as ectosomes—from cilia of living animals.[2] The fact that polycystins localize to EVs in both human and *C. elegans* suggests an undiscovered and unappreciated function for the polycystins in EVs.

C. elegans is an important model organism in both basic and biomedical research fields.[10] *C. elegans* is a 1 mm, free-living nematode that cultures on *E. coli* in Petri dishes, making for a very low-cost culture system. *C. elegans* exists primarily as a self-fertilizing hermaphrodite (with males in $<0.2\%$ of the population) and has a generation time of 3 days, which makes for quick and easy genetic analysis. *C. elegans* is transparent, which combined with genetically encoded reporters, enables visualization and manipulation of biological processes *in vivo*. With a history spanning over 50 years, "the worm" has many useful shared community resources (Table 11.1), including the NIH-funded *Caenorhabditis* Genetics Center (CGC) strain repository (https://cgc.umn.edu/) and the NIH-funded WormBase, an international consortium of biologists and computer scientists dedicated to providing the research community with accurate, current, and accessible information concerning the genetics, genomics, and biology of *C. elegans* and related nematodes (www.wormbase.org).

The goal of this chapter is to introduce basic research approaches and practical protocols used in the study of the polycystins, cilia, and PKD using *C. elegans* as a model organism. We aim to provide researchers, especially those new to the field, with sufficient information and practical descriptions to initiate research using *C. elegans* as an *in vivo* model to study human genetic diseases of cilia.

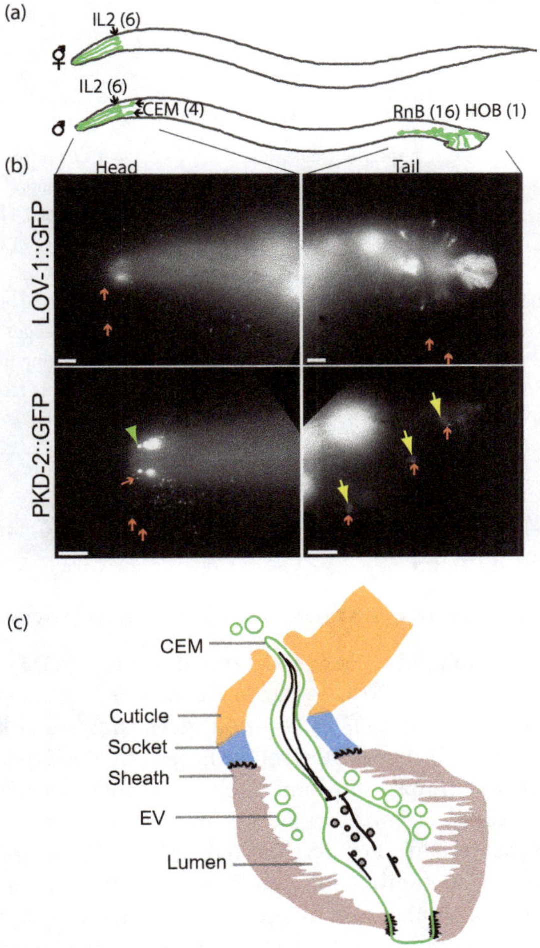

FIGURE 11.1 Extracellular vesicle releasing neuron (EVN) diagram and polycystin localization in *C. elegans*. (a) The ciliated EVNs, six in the hermaphrodite (*top*) and 27 in male (*bottom*). A subset of *C. elegans* sensory neurons release EVs into environment: six IL2 neurons in the head in both the male and hermaphrodite, four cephalic male-specific CEM neurons, 16 *Rn*B (where n = 1 to 9 but not 6 according to the ray numbering) ray *B* type neurons, and one *ho*ok *B* type HOB neuron. These 27 neurons are collectively named EVN for *EV* releasing neurons. (b) *Top*: Male head and tail images of LOV-1::GFP reporter (N-terminal extracellular domain of LOV-1 [1–991 aa] fused to GFP). In all panels, red arrows point to EVs surrounding the head and the tail. *Bottom*: Male head and tail images of PKD-2::GFP reporter. Green arrowhead points to a CEM cilium with PKD-2::GFP enriched at the tip. Yellow arrows point to the cuticular pore of the ray neurons and PKD-2::GFP release around the pore. Scale bar represents 10 μm. (c) Model based on electron tomography of the distal end of the CEM neuron and its surroundings. The glial sheath cell and socket cell form a continuous lumen surrounding the CEM neuron cilium, which is exposed to the environment directly through a cuticular opening. The polycystins EVs are stored in the extracellular lumen and released environmentally via the ciliary pore.[2] (Modified from Wang, J. et al., *Curr. Biol.* 24, 519–525.)

TABLE 11.1
Online Resources

Name	Link	Use
WormBase	https://wormbase.org	Gene information, available reagents
WormAtlas	http://www.wormatlas.org/	Anatomy, cell identification
WormBook	http://www.wormbook.org/	Biology of *C. elegans* including methods
Caenorhabditis Genetics Center (CGC)	https://cgc.umn.edu/	Providing and accepting *C. elegans* strains
National Bioresource Project for the Experimental Animal "Nematode *C. elegans*"	https://shigen.nig.ac.jp/c.elegans/	Providing mutant strains; also accepts request for gene knockouts

11.2 *C. ELEGANS* CULTURING, STRAIN CONSTRUCTION, AND RESOURCES

11.2.1 Obtaining *C. elegans* Strains and Strain Maintenance

C. elegans strains are cultured on nematode growth media (NGM) agar plates with OP50 *E. coli* as a food source at 20°C as standard conditions.[11] *C. elegans* strains may be acquired from the CGC (https://cbs.umn.edu/cgc/); strains and knockout strains may be requested from the National Bioresource Project for the Experimental Animal "Nematode *C. elegans*" (https://shigen.nig.ac.jp/c.elegans). For additional sources where one may request *C. elegans* strains, see WormBase (https://www.wormbase.org/species/c_elegans#041-10). For detailed methods of *C. elegans* culturing and strain maintenance, decontamination, synchronizing and staging, and freezing procedures for long-term storage and for recovery from the frozen state, refer to the online WormBook chapter http://www.wormbook.org/chapters/www_strainmaintain/strainmaintain.html.[12]

In *C. elegans*, the predominant sex is hermaphrodite, and males spontaneously arise only rarely (less than 0.2%). The *C. elegans* polycystins encoding genes *lov-1* and *pkd-2* are expressed and required in male-specific ciliated sensory neurons. Therefore, in all experiments in which males are tested, we use strains with either the *him-5(e1490)* or *him-8(e1489)* allele as wildtype. *C. elegans* sex is determined by X-to-autosome ratio, with hermaphrodites being XX and males being hemizygous for X. High incidence of male (Him) mutations increase the rate of X-chromosome nondisjunction.[13] *him-5(e1490)* and *him-8(e1489)* males exhibit normal mating behaviors and are used as wildtype controls.

11.2.2 Fluorescent Protein Reporter and Transgenic Worm Construction

C. elegans is a transparent, multicellular animal—a feature enabling subcellular imaging that is unprecedented in its simplicity and reproducibility. Combined with genetically encoded fluorescent reporters and time lapse imaging, the transparent worm allows real-time visualization of ciliated neuronal anatomy, morphology, protein transport, and activity in a living animal. In *C. elegans*, cilia are located at the distal-most dendritic ending of sensory neurons. (In mammalian neurons, cilia are located on the cell body.) Each sensory neuron is unique in position and the morphology of the dendrite and axon—one can use these features to identify individual neurons online at WormAtlas (http://www.wormatlas.org/neurons/Individual%20Neurons/Neuronframeset.html.) Amphid head and phasmid tail sensory neurons can also be identified by lipophilic dye filling.[14]

For detailed instructions and general considerations to make fluorescent protein reporters of your gene of interest, please refer to the WormBook chapter from the WormMethods, Gene Expression section at http://www.wormbook.org/chapters/www_reportergenefusions/reportergenefusions.html.[15]

In general, a transcriptional fluorescent reporter encodes the promoter of your favorite gene (yfg) driving the expression of GFP or another fluorescent protein (FP, FP cassettes typically include introns and the 3′ UTR of *unc-54* gene). A transcriptional GFP reporter is used to identify in what cells yfg is expressed, as GFP is soluble, cytoplasmic, diffusible, and labels the entire cell. In contrast to vertebrate genomes, the *C. elegans* genome is more compact and promoter elements are typically located 1 kb upstream of the start codon of genes.[16] If the intergenic region upstream of start codon is larger than 2 kb, a 2 kb promoter sequence may be used to drive GFP expression; if the intergenic region upstream of ATG is smaller than 2 kb, the entire intergenic region is usually included. An "X-box" motif in the promoter of yfg may indicate a ciliary gene. The DAF-19 RFX transcription factor is an evolutionarily conserved master regulator of ciliogenesis that binds X-box motifs usually located around 200 bp upstream of ATG.[17,18]

Another general consideration in generating GFP reporters is that the *C. elegans* genome makes use of secondary promoters within long introns of genes.[19] For example, the sole RFX gene, *daf-19*, encodes a set of nested isoforms that uses distinct promoters located in introns.[20–22]

To examine protein localization dynamics, trafficking, and transport, a translational GFP reporter uses the entire genomic region of yfg (promoter + exons + introns) to drive GFP expression. Alternatively, GFP may be fused to the N- or C-terminus of the cDNA and driven by the endogenous or a tissue-specific promoter. A general consideration is that N- and C-terminal fusion reporters may have different localization patterns. The gold standard is to determine whether or not the translational GFP reporter rescues mutant phenotypes of yfg, and this functional GFP reporter is considered to reflect the localization of the endogenous protein. We used this strategy to rescue *pkd-2* mutant defects with the PKD-2::GFP translational reporter and to show that PKD-2::GFP localization in living males and anti-PKD-2 antibody staining in fixed males are identical.[23]

Transgenic worms can be obtained by injecting DNA of a GFP reporter construct and of a selectable marker directly into the syncytial gonad of a young adult hermaphrodite. Transgenic animals are picked in the next generation based on the cotransformation selection scheme. The general protocol of microinjection may be found at WormBase: http://www.wormbook.org/chapters/www_transformationmicroinjection/transformationmicroinjection.html.[24]

C. elegans carries transgenic DNA as extra-chromosomal arrays, in which DNA fragments are joined end to end in varying orientation and copy number, largely determined by DNA concentration. A general rule for DNA concentration is to get a bright fluorescent signal that does not interfere with the function of yfg.

CRISPR is another efficient way to generate reporters by tagging the endogenous gene, which more likely reflects the endogenous expression pattern.[25]

11.3 FLUORESCENCE MICROSCOPY

One significant advantage of using *C. elegans* in the study of PKD and human ciliopathies is that *C. elegans* cilia-less mutants are viable in contrast to cilia-less vertebrate models that are often lethal. This allows *C. elegans* researchers to study genetic network interactions using double, triple, and n-ple mutant combinations ($n = 4$ or more). To this end, *C. elegans* has also been credited for the discovery of many new cilia genes and identifying causal genes of ciliopathies. *C. elegans* also enables study of *in vivo* protein localization and dynamics in almost any genetic background. In this regard, *C. elegans* has advanced our understanding and revealed that cilia are highly compartmentalized organelles.[26]

For protein subcellular localization and expression patterns, Z-stack images are taken, and image analysis is performed using software such as Image J. For general *C. elegans* mounting and imaging protocol, please refer to WormBook: http://www.wormbook.org/chapters/www_intromethodscellbiology/intromethodscellbiology.html.[27]

11.3.1 PKD-2::GFP Reporters and EV Quantification Protocol for Single Live Worms

Ciliary EVs are approximately 100 nm and not visible by bright field or differential interference contrast (DIC) microscopy. Hence, the visibility of EVs relies on the fluorescence intensity of the GFP reporter. Compared with other subcellular compartments within the cilium, EVs are relatively difficult to observe with the naked eye. The best way to visualize EVs is via the PKD-2::GFP reporter (PKD-2::GFP transgenic strains are available from the CGC) as a positive control and then examine other candidate EV reporters. Here, we use PKD-2::GFP EV imaging from cephalic male (CEM) ciliated sensory neurons as example.

Protocol

All strains should be kept under the same culture conditions, on reproductively growing, healthy, uncrowded plates.

1. Pick 20–30 larval stage 4 (L4) males for each strain to a seeded NGM plate 24 h ahead of imaging time.
2. The next day, prior to imaging, make one tape thick 2% agarose pad.
3. On an 18 × 18 mm No. 1 coverslip, add four separate drops of 0.5 μL 10 mM levamisole (dissolved in M9 buffer); place one or two young adult males in each drop.

4. Flip the coverslip on the agarose pad, begin imaging immediately and complete each slide within 30 min.
5. Under a 63× or 100× objective, locate the cephalic male (CEM) cilia, then focus closer to the coverslip until PKD-2::GFP containing EVs are visible (Figure 11.1).
6. Take Z-stacks of the animal from the coverslip to the cilium. Nomarski DIC images may be taken separately to orient animals and to distinguish EVs from bacterial autofluorescence.
7. EV particle intensity and numbers may be quantified by ImageJ particle analysis software.
8. For each strain, image a minimum of 20 males per day on 3 different days.
9. To quantify EV release in wildtype, mutant, and/or different conditions, process and analyze data using statistics software such as GraphPad Prism.

11.3.2 Measurement of Dendritic Movement and IFT in Male-Specific Neurons

C. elegans PKD-2::GFP is transported from the neuronal cell body along the dendrite to the ciliary base and back to the cell body at characteristic velocities (Figure 11.2).[28] Using a forward genetic screen and candidate gene approach, we identified mechanisms regulating PKD-2 ciliary localization, including roles for phosphoinositides, tubulin glutamylation and tubulin isotypes, and extracellular matrix.[14,29–32,47] To study PKD-2::GFP dendritic trafficking and ciliary localization, detailed protocols have been described by O'Hagan and Barr.[31] Here, we present a basic protocol and general considerations for the methods.

Measuring PKD-2::GFP dendritic velocity.

Protocol

All strains should be kept under the same culture conditions, on healthy, reproductively growing, uncrowded plates.

1. Pick 20–30 L4 males for each strain to a seeded NGM plate 24 h ahead of imaging time.
2. The next day, prior to imaging, make one tape thick 2% agarose pad.
3. On an 18 × 18 mm No. 1 coverslip, add four separate drops of 0.5 μL 10 mM levamisole; place one or two young adult males in each drop.
4. Flip the coverslip on the agarose pad, start imaging right away, and finish each slide within 30 min.
5. Under a 100× objective (63× objective is doable, but not ideal), focus on the cell body of CEM neuron, then switch to camera mode: moving PKD-2 dendritic puncta are not easily visible to the naked eye. The dendrite is not visible under bright field or DIC because of the small 200–300 nm diameter. PKD-2::GFP is not localized to the dendrite membrane; however, immobile PKD-2::GFP puncta provide a good landmark to focus on the dendrite.

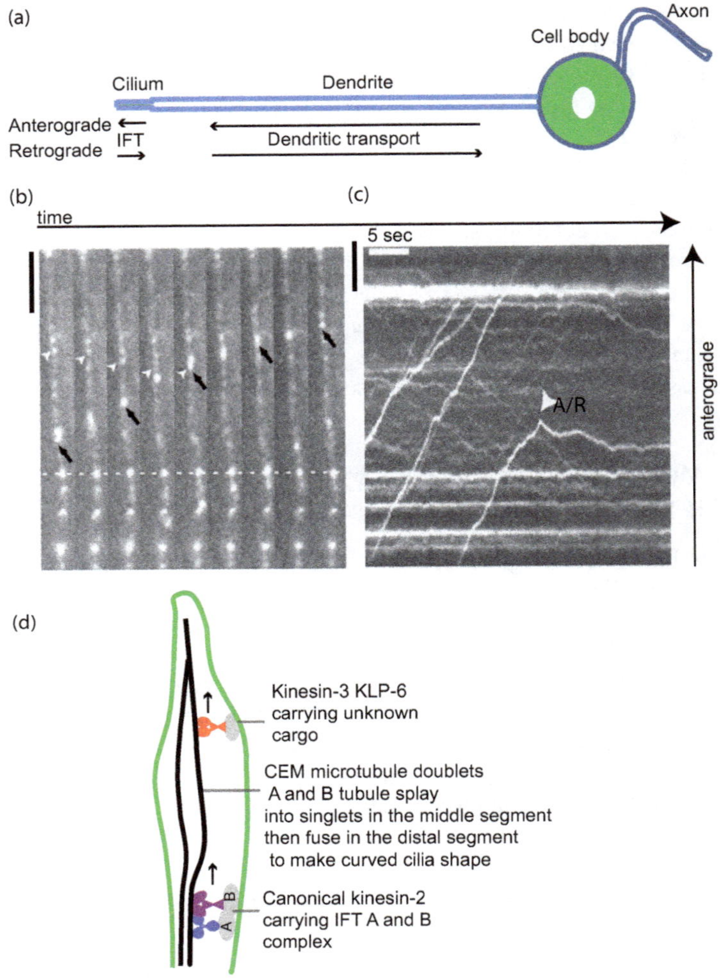

FIGURE 11.2 CEM neuronal dendritic trafficking and ciliary IFT. (a) Cartoons of CEM neuron expressing PKD-2::GFP (shown in green) and diagram of PKD-2 dendritic vesicular transport and ciliogenesis protein IFT transport in CEM neurons. (b) A montage of PKD-2::GFP particles shows both anterograde (black arrows, from the cell body to cilium) and retrograde (white arrowheads) movement. The broken white line indicates stationary particles. Scale bar: 5 µm. (c) A kymograph depicting PKD-2::GFP particle motility in the dendrite. The horizontal and vertical axes represent time and distance, respectively. Vertical scale bar: 5 µm. The white arrowhead reflects retrograde to anterograde changes. [(a) to (c) Reproduced/adapted with permission from Bae et al.[28]] (d) Diagram of CEM cilia IFT trafficking. The CEM axoneme is specialized to make curved cilium by splaying the doublet microtubule to singlet microtubules, modeled by electron tomographic TEM.[29] The CEM neuron has three kinesin motors, heterotrimetric kinesin-2, encoded by KLP-11/KLP-20/KAP-1 (in blue), homodimeric KIF17 homologue kinesin-2 OSM-3 (in purple), and a kinesin-3 protein KLP-6 (in red). Gray circle with A and B indicate IFT complex A and B. KLP-6 cargo has not been identified.

6. Turn on camera live mode. Set exposure long enough to see the moving dendritic particles on the computer screen, then adjust brightness/contrast to see the dendrite. Restrict the field of view to exclude cilia base or cell body, as the PKD-2::GFP signal in these compartments is very bright and can interfere with visualizing dendritic PKD-2::GFP signal under the same brightness/contrast settings. We use 200–300 mS exposure time and take 200 frames in continuous streaming. Generate a kymograph[31] by imaging analysis software, such as Metamorph, ImageJ, or Zen Blue from Zeiss. Measure the velocity by drawing a line along the fluorescent line on the kymograph (Figure 11.2). PKD-2::GFP moves at about 1.2 μm/sec for anterograde transport and about 0.8 μm/sec for retrograde transport.[28]

Automated quantitative kymograph analysis by software can save time and reduce human technical error.[35]

Cilia display a 9+2 or 9+0 axonemal ultrastructure structure and are constructed from the evolutionarily conserved intraflagellar transport (IFT) machinery.[36,37] However, cilia can be specialized in form and function, and *C. elegans* is a powerful system to study mechanisms underlying ciliary diversity.[14] For example, the *pkd-2* expressing cephalic male CEM cilium is highly curved and supported by a unique axonemal structure, in which nine doublet microtubules in the proximal cilium splay into 18 singlets (nine A-tubule singlets and nine B-tubule singlets) in the middle region that rejoin to generate a curved axoneme in the distal region.[2,28] A specific alpha tubulin isoform encoded by the *tba-6* gene and heterotrimeric kinesin-2 are required for axonemal curvature.[29,38] *C. elegans* amphid and phasmid cilia are constructed by canonical IFT kinesin-2 motor encoded by *klp-11*, *klp-20*, and *kap-1* and the homodimeric kinesin-2 OSM-3/KIF17.[39,40] CEM cilia are equipped with an additional kinesin-3 motor, KLP-6, which is exclusively expressed in the *C. elegans pkd-2* expressing and EV releasing neurons (Figure 11.2c).[38,41] The heterotrimeric KLP-11/KLP-20/KAP-1 motor and the homodimeric motor OSM-3 are redundantly required for PKD-2::GFP EV release, while the kinesin-3 motor KLP-6 itself is necessary for PKD-2::GFP EV release.[2] Much research is still needed to understand ciliary EV biogenesis and release. IFT analysis is a straightforward way to examine how the motors in the CEM cilia are regulated, which may relate to mechanisms of EV biogenesis and release.[7]

11.3.3 Measuring IFT

IFT genes are expressed in most if not all ciliated neurons in the worm. To visualize IFT in a single cilium, a cell-specific promoter is used to drive expression of an IFT gene fused to an FP. In CEM cilia, we use either the *pkd-2* or *klp-6* promoter to visualize IFT.

Protocol

1. Male worm culturing and mounting procedures are the same as protocol for measuring PKD-2::GFP dendritic movement and velocity.

2. Under a 100× objective (63× is objective doable, but not ideal), focus on the ciliary base of the CEM neuron, then switch to camera mode: moving IFT particle are typically not visible to the naked eye. Turn on camera live mode, set exposure long enough to see moving IFT particles on the computer screen. We use 200–300 mS exposure time and take 200 frames in continuous streaming. Generate a kymograph[31] by imaging analysis software, such as Metamorph, ImageJ, or Zen Blue from Zeiss. Measure the velocity by drawing a line along the fluorescent line on the kymograph (Figure 11.2).

   Note: A big challenge of live imaging is completely immobilizing the animal. Levamisole is used to anesthetize the worm. To reduce technical variation, always use 10 mM levamisole and take the images within about same length of time after mounting the worm. Levamisole and other drugs do affect dendritic transport.[42]

11.4 CELL SORTING AND RNAseq FOR TRANSCRIPTOME ANALYSIS

Single-cell RNAseq expands our understanding of cell biology by examining single-cell transcriptome in a functional environment and from a living animal. We developed a protocol to isolate GFP-labeled EV releasing neurons and performed transcriptome analysis by RNAseq.[43] The basic procedure is that tissues and cells are dissociated by protease treatment, FACS GFP-labeled cells are then collected for RNAseq.

Protocol Based on Kaletsky et al.[44]

1. Strain PT2519 *(myIs13 [Pklp-6::gfp + pBx] III; him-5[e1490])* contains an integrated transgenic GFP reporter that marks all EV releasing neurons; the *him-5 (e1490)* mutation is used to generate a male-enriched population. Synchronize and culture PT2519 and *him-5(e1490)* (negative control) on three 9 cm HGM plates seeded with OP50.
2. Begin protocol 45 min before scheduled sort time.
3. Wash worms into 1.5 mL clear Eppendorf tube with M9. Be careful not to collect chunks of bacteria with worms. If worm pellet exceeds 250 μL, split sample between 2 tubes and process separately.
4. Wash worms 5 times with M9 buffer; wash until all of food is removed.
5. Wash once with 500 μL lysis buffer (200 mM DTT; 0.25% SDS; 20 mM HEPES pH 8.0; 3% sucrose). Prepare 50 mL aliquots. Freeze in 1.5 mL aliquots at –20°C).
6. Add 750 μL lysis buffer, incubate at room temperature for exactly 6.5 min. Dumpy and roller strains are more sensitive to lysis conditions. In this case, incubate for exactly 6 min.
7. Wash 5 times with M9 buffer, spinning in minicentrifugation machine and working quickly. Tubes should no longer smell like DTT.

8. Add 250 μL freshly made, room temperature, 20 mg/mL Pronase (protease from *Streptomyces griseus* Sigma-Aldrich cat no: P6911-1G) in water solution to worm pellet; incubate 20 min at room temperature.
9. Pipette vigorously with P200 tip 100 times every 5 min; examine 2 μL of worms on dissecting scope after every pipetting stage to assess level of dissociation. Batches of Pronase are variable at dissociation and require different incubation times, usually between 15–25 min.
10. Prepare 2% FBS in PBS (Fetal Bovine Serum, Certified, Heat-Inactivated Invitrogen cat no: 10082-139).
11. Use PBS solution to pre-wet a 5 μM syringe filter using a 1 mL syringe. Do this between pipetting steps; store filters on ice.
12. Transfer dissociated cells to 1 mL syringe using 27 G needle, swap out needle for filter, and gently filter cells into FACS tube on ice. Wash filter with 1 mL PBS solution. Keep filtered cells on ice and in dark until sorting.
13. The filtered cells are diluted in PBS with 2% FBS and sorted using a FACSVantage SE w/ DiVa (BD Biosciences). Negative sorting gates were determined from day 1 adult *him-5* cell suspension. Fluorescently labeled events were sorted directly into tubes containing Trizol LS for subsequent RNA extraction.
14. Prepare 850 μL Trizol LS aliquots in collection tubes. Sort 150 μL cells, at a maximum. Collect about 250,000 cells directly into Trizol LS per tube.
15. RNA can be isolated using a standard Trizol/chloroform/isopropanol method, followed by DNase digestion and cleanup with the Qiagen RNEasy MinElute kit. RNA quality and quantity can be assessed using the Agilent Bioanalyzer RNA Pico chip.

We performed RNAseq by collaboration with Coleen Murphy Laboratory at Princeton University.[43,44] For specific RNAseq protocol, please consult with service provider. Specifically, in our experiments, RNA was amplified using the Nugen Ovation RNAseq v2 kit according to the manufacturer's instructions. The amplified cDNA was sheared using the Covaris E220 according to library size input requirements, and libraries were prepared using the Nugen Encore NGS Library System or the Illumina TruSeq DNA Sample Prep kit. Libraries were sequenced (single-end or paired-end) using an Illumina HiSeq 2000.

HGM (high-growth media) recipe: 3 g NaCl, 20 g Bacto Peptone, 30 g Bacto Agar; add about 975 mL ddH_2O, mix well. Add 25 mL 1 M phosphate buffer (pH = 6, refer to NGM plate from WormBook/worm culture method chapter), autoclave, and cool to 60°C. Add 4 mL 5 mg/mL cholesterol in ethanol, mix well; 1 mL 1 M $MgSO_4$, mix well; and 1 mL 1 M $CaCl_2$, mix well. Pour into 9 cm Petri dishes (about 25 mL per plate).

11.5 FUNCTIONAL READOUT OF POLYCYSTIN-EXPRESSING NEURONS: MALE BEHAVIORAL ANALYSIS

lov-1 and *pkd-2* are required for three different behaviors (mate searching, response, and vulva location) that can be separated into different genetic pathways.[43] By testing gene function in the different behaviors, we can assign genes to pathways or protein complexes/modules.

In contrast to hermaphrodites, who stay with food and reproduce, males leave available food to search for mates.[45] *lov-1* and *pkd-2* mutants are defective in sex drive as measured by a leaving assay.[46]

11.5.1 Male Mate Search Behavior: Leaving Assay

Protocol (adapted from Lipton et al.[45])

1. Prepare plastic Petri plates (9-cm diameter) with 10 mL agar medium (17 gm of agar [Difco, Detroit, MI], 2.7 gm of Bacto Peptone [Difco], 0.55 gm of Tris base [Sigma, St. Louis, MO], 0.27 gm of Tris HCl [Sigma], 2.0 gm of NaCl [Fisher Scientific, Pittsburgh, PA], and 1 mL of ethanol containing 5 mg/mL cholesterol [Sigma], per liter H_2O) and allow plates to dry (lids closed) overnight on a laboratory bench. A smaller amount of agar per plate is used (10 mL) compared with the standard amount for genetics (23 mL) because it increases the ease of scoring worm tracks on the agar surface.
2. Inoculate each plate in the center with 18 μL of *E. coli* strain OP50 grown to OD600 = 1.0 and allow plates to incubate for 12–16 hr at room temperature to establish a small circular lawn of approximately 9 mm diameter. The dryness of the agar surface and degree of bacterial growth are the most critical variables for reproducible results.
3. Pick approximately 20–30 L4 males per genotype to a freshly seeded plate 24 h ahead of assay time. *him-5(e1490)* is a positive control (males leave) while *pkd-2(sy606); him-5(e1490)* is a negative control (males are leaving assay defective—the Las phenotype).
4. To start the assay, place a male individually on the bacterial lawn. Use 20 males per genotype for a total of 20 assay plates per genotype. Keep the plates at room temperature and examine plates at set intervals (0, 1, 8, and 24 h) to determine if the male is a leaver or nonleaver. An animal is scored as a leaver if it is present outside a 3.5 cm radius circle (1 cm from the edge of the plate) or if its tracks show that the male passed beyond this circle during the preceding time interval. Tracks can often be most easily seen by observing the plate tilted at angles to observe the reflections of laboratory lights on the agar surface. Once a male is scored as a leaver, the plate is discarded.

Leaving rate is calculated as the probability of leaving per hour, P_L. Plots of log (fraction nonleavers) versus time yield, during the first 24 hours, straight lines passing through the point representing 100% nonleavers at time 0. Such linear plots reflect the constant probability per unit time that a male will leave. P_L is estimated as the hazard obtained by fitting an exponential parametric survival model to the censored data using maximum likelihood (software R; http://www.R-project.org). Under normal conditions, wildtype males leave with a P_L value of around 0.17; about half the males will have left in the first 5 hours of the assay. To control for fluctuations in conditions, the leaving rate of experimental animals should be compared to controls tested side by side on the same batch of plates on the same day. Leaving assays are done in triplicate on 3 different days.

11.5.2 Male Mate Behavior Response and Location of Vulva Efficiency Assay

C. elegans males perform a robust innate mating ritual achieved by integrating male-specific sensory-motor programming into the core neuromuscular system shared with hermaphrodites, thus representing a unique system to study genes neurons, and behavior (Barr and Garcia 2006).[33] *C. elegans* male mating behavior can be defined by five steps: (1) response to contact with the hermaphrodite by backing along her body; (2) turning around her head or tail; (3) location of the vulva: when the male encounters the vulva, he stops backward scanning; (4) insertion of spicules into the vulva; and (5) sperm transfer. *C. elegans lov-1* and *pkd-2* are required specifically for response and vulva location (*lov-1* is named for its mutant phenotype: location of vulva defective).

Response and Vulva Location Assay Protocol
This assay is done using a stereodissecting microscope.

1. The day before the assay, make a small dot about 0.5 cm diameter bacterial lawn by placing 12 μL of OP50 culture on NGM plate. Leave to dry on bench 24 hours before the assay with lid on. Pick the following genotypes from noncontaminated, noncrowded healthy growing plate to individual plates:
   i. 15 L4 males from positive control CB1490 *[him-5(e1490)]* plate
   ii. 15 L4 males from negative control plate PT9*[pkd-2(sy606); him-5(e1490)]*
   iii. 20 to 30 L4 males from test strain plate
   iv. 30 L4 *unc-31(e169)* hermaphrodites (*unc-31* uncoordinated hermaphrodites move more slowly)
2. The day of the assay: Place 20 *unc-31* hermaphrodites on a mating dot, wait 5–10 min to let the hermaphrodites spread. Usually the hermaphrodites distribute evenly on the lawn. If not, adjust by picking the crowed hermaphrodites to empty spots so that the hermaphrodites cover the bacterial lawn evenly.
3. Place a positive control male in the middle of the assay lawn using 10–50× magnification. Observe males until successful vulva location or 5 min, whichever happens first. Score the following behavioral parameters:
   Response (1–yes, 0–no); response time (sec); vulva location (1–yes, 0–no); vulva location efficiency (stops/total vulva encounters).
4. Repeat with four additional positive control males (one at a time). Expected: 90%–100% of *him-5* males should respond within 5 min time window and find the vulva within 1–2 encounters.

5. Repeat with five negative control males (one at a time). Expected: 15%–25% of *pkd-2*; *him-5* males should respond within 5 min time window and have ≥3 passes to find vulva.
6. Start with one male at a time, 2–4 males can be observed at one time with practice.
7. If positive controls and negative controls work out as expected, proceed with testing male genotype of interest. If not, check room temperature is normal, mating lawn plate is fresh, and the plate is not dry. If control animal behavior is not normal, then do not proceed with behavioral assays on that day. Perform 3 independent assays on 3 separate days.

   Be sure to note any additional behaviors (difficulty in turning, leaving the mating dot, etc.) as well.

$$\text{Response efficiency} = \text{number of males responding/number of total males tested} \times 100\%.$$

$$\text{Location of vulva efficiency of an individual male} = \text{stop/(stop + pass)} \times 100\%$$

For each assay, at least 20 males are scored per trial; perform triplicate trials to obtain statistical data. Statistical analysis can be done by multiple group comparisons by Kruskal–Wallis test for nonparametric data with Dunn's multiple comparison post-hoc test, using GraphPad Prism software.

11.6 METHODS USING *C. elegans* TO STUDY POLYCYSTIN FUNCTION IN EVs

Extracellular vesicles (EVs) are extracellular organelles that are released by cells to communicate with other cells. EVs cargo include RNA, DNA, protein, and lipids. The polycystins localize to EVs isolated from human, *C. elegans*, and the green algae *Chlamydomonas*, suggesting an evolutionarily conserved role of polycystins on both cilia and EVs.[7] Studying polycystin EV function and the relationship to ciliary localized polycystins may reveal a missing link between ciliopathies and PKD. Here, we provide detailed protocols on how to prepare EV samples from *C. elegans* culture and how to examine EV morphology by negative staining and immunostaining by gold labeling.

EVs prepared from worm cultures elicit a male tail-chasing behavior.[2] When a freely moving male encounters EVs on a bacteria lawn, he reverses, places his tail on his body, and continues to back in a circling movement. The persistent backward movement resembles response behavior. EVs isolated from mutant strains that do not release PKD-2::GFP-labeled EVs have reduced bioactivity.[2,27] These data suggest that tail-chasing behavior is a good readout of polycystin EV signaling function.

11.6.1 EV Preparation

Protocol

Worm culture. Grow *him-5* and test-strain cultures as synchronized young adults, as densely as possible, with the least amount of bacteria not yet starved.

Day 1

Use a spatula to chunk a densely populated strain from an NGM plate into 16 parts. Put one chunk on an OP50 seeded NGM plate. Prepare about 15 NGM plate for each strain. Culture at standard conditions in 20°C incubator for 3 days.

Seed HGM plates (refer to RNAseq cell preparation protocol) with 1 mL OP50 culture; spread the bacterial culture to make a continuous bacteria lawn. Prepare 15 plates for each strain.

Day 4

The NGM plates should be almost starved, with a lot of adult worms. There should be no dauers, which indicates starvation. Chunk the whole NGM plate of worm to a seeded HGM plate. Culture worms in 22°C incubator for 3 days.

Day 7

The HGM worm plate should be saturated with young adult worms, almost starved. If not, wait another day. Strains may grow at different rates. Process those strains/plates that are ready.

Harvest EVs

1. Using a sterile glass pipette, wash each plate with 1 mL M9 buffer. Collect worms into a 15 mL Falcon tube. Use 1 mL of M9 buffer to wash the worm plates again; transfer the second wash from the first plate to other plates for washing. The goal here is to collect all worms in about 15 mL M9 buffer. Finish this step as fast as possible.
2. Desktop centrifuge—3000 g for 15 minutes (room temperature) to pellet worm and bacteria—transfer the supernatant to a new tube and filter with 0.45 μm filter into a new 13.5 mL tube for the SW41 rotor. From now on, keep everything on ice. Measure and write down the worm wet weight to estimate EV production (EV quantity from wet worm weight) later. About 1 g wet worm should be collected at this step.
3. In the SW41 rotor, centrifuge at 10,000 g (7500 rpm) at 4°C for 30 min. Make sure to fill the centrifuge tubes to just 1 mm below the top, using M9 buffer to fill if supernatant from step 2 is less than 13 mL.
4. Transfer supernatant to a new 13.5 mL centrifugation tube, spin down EVs at 100,000 g (24,000 rpm for SW41) at 4°C for 70 min.
5. Wash the pellet with M9 buffer: Pour out supernatant, drain the tube on a stack of tissue paper with a Kimwipe on top. Add M9 buffer to 13.5 mL tube, spin down at 100,000 g at 4°C for 70 min. This step is to get rid of soluble components from the culture in the remaining supernatant left with the EV pellet.
6. Dissolve the pellet by adding 100 μL of M9 buffer to the centrifugation tube and place the tube on ice in 4°C fridge overnight.
7. Quantify EV production by total protein quantification. We used CBQCA Protein Quantification Assay Kit from Invitrogen (cat no. C6667, http://products.invitrogen.com/ivgn/product/C6667).

This kit detects protein concentrations from 10 ng/mL to 150 μg/mL. The dye reacts with primary amine groups on proteins in the presence of cyanide or thiols; the unreacted dye is nonfluorescent; sensitivity depends on the number of amines present and is not compatible with buffers containing amines or thiols.

Usually 5 μg EV (measured by total protein) can be obtained from 1 g wet worm in wildtype. The *klp-6* mutant gives about one-third the yield of EVs compared with wildtype. Adjust EV concentrations to the same concentration by M9 buffer and store in 4°C fridge for subsequence negative staining and EV bioactivity assays. For bioactivity assays, we usually use EV samples within 3 days. EV samples retain bioactivity up to 5 days as measured by male tail-chasing behavior.

11.6.2 Negative Staining of EV Sample to Examine EV Morphology under TEM

Protocol

1. Use glow discharge formvar/carbon coated nickel grids (200 mesh, EMS cat no. FCF200-Ni). For negative staining; any type of formvar/carbon coated grids would work. Nickel grids are used here because gold or copper grids are not compatible with the following silver enhancement immunogold labeling protocol.
2. Add 5 μL EV sample onto each grid, wait for 30 s, wick the solution with #1 Whitman filter paper from the edge of the grid, leaving a thin film of solution on the grid. Tilt the grid to get rid of remaining solution. However, do not touch the grid with filter paper, and do not let the grid dry before next step.
3. Immediately add a drop Nano-W (Methylamine Tungstate solution nanoprobes cat no. 2018: http://www.nanoprobes.com/products/Negative-Stains.html) on the grid, wait for 1 min, wick the solution with Whitman filter paper, leave as little solution on grid as possible, but do not touch the grid with filter paper. Air dry and save the grid in a clean box for TEM.

11.6.3 Immunogold Labeling of EVs

Protocol

1. Use glow discharge formvar/carbon coated nickel grids (200 mesh, EMS cat no. FCF200-Ni). For each sample; prepare at least 3 grids for repeat. Include relevant positive (antibody and antigen works for immunostaining) and negative controls (mutant that will not be labeled, no primary antibody and no secondary antibody controls).
2. Cut out a piece of parafilm, flatten the clean side on a clean surface. Lay out grids on the parafilm surface. Add 10 μL EV sample (dilute with M9 buffer to 1 ng by total protein/μL) to each grid and allow to sit at room temperature in open air for 10 min.
3. Cut out a piece of parafilm, flatten the clean side on a clean surface. Add 50 μL blocking buffer (Goat Gold Conjugates, EMS cat no. 25596) drops

on the parafilm surface. Use a piece of Whitman filter paper to wick the EV sample away and float grids on blocking solution drop. Cover grids with a dark chamber made by the bottom part of a pipette tip box. Block 30 min at room temperature.

4. Wash 2 times, 5 min each time in Aurion's incubation buffer (PBS pH 7.4, 0.1% Aurion BSA-c, 15 mM NaN3). For ordering Aurion BSA-c (10%), code 900.099—https://aurion.nl/product/aurion-bsa-c-10/.
5. Float grids on primary antibody, diluted in incubation buffer (1:50 to 1:200) for 1 h at room temperature.
6. Wash 6 times, 5 min each time in Aurion's incubation buffer.
7. Overlay grids with gold conjugated secondary antibody, diluted in incubation buffer (1:200 to 1:2000, depends on antibody) for 2 h at room temperature.
8. Wash 6 times, 5 min each time in Aurion's incubation buffer.
9. Wash 2 times, 5 min each time in PBS buffer.
10. Postfix in 4% paraformaldehyde in PBS for 5 min.
11. Wash 3 times, 5 min each time in ddH_2O.
12. Perform silver enhancement procedure. Float grids on developing solution (20 drops of enhancer + 1 drop developer, vortex mix) for 30 min at room temperature. For ordering Aurion Silver Enhancement Reagents: Aurion R-Gent SE-EM (code500.033): https://aurion.nl/product-category/silver-enhancement-reagents.
13. Wash 3 times, 5 min each time in ddH_2O.
14. Negative staining with Nano-W. (Methylamine Tungstate solution nanoprobes cat no. 2018: http://www.nanoprobes.com/products/Negative-Stains.html) on the grid, wait for 1 min, wick the solution with Whitman filter paper, leave as little solution on grid as possible, but do not touch the grid with filter paper. Air dry and save the grid in a clean box for TEM.

11.6.4 EV Functional Analysis by Tail-Chasing Assay

This protocol is adapted from previous work.[2,27] The assay is done using a stereomicroscope at 50× and a camera mounted to record behaviors.

1. Pick 20 L4 males the day before assay.
2. Prepare small 10 µL OP50 bacterial lawn within 8 h of the assay. We observed sharp declines in tail-chasing behavior if OP50 bacterial lawns were older than 8 h.
3. Use EV preps from *him-5*, and *klp-6; him-5* as positive and negative controls, respectively. Adjust EVs sample concentration to 10 ng total protein/µL. Add 10 µL EV sample or M9 buffer as control to the center of the small OP50 bacterial lawn. Wait 10 min or until the lawn is dry with Petri dish lid open.
4. Place one single young adult male in the center of the EV or M9 buffer (control) spotted lawn.
5. Start to record behavior for 5 min with a capture rate of 1 frame/second. Score reversal movement and tail-chasing (male moving backward in a circle with tail touching male's own head) events from recordings. For each assay,

at least 5 males are scored per trial, perform 4 trials to obtain statistical data. Statistics analysis can be done by multiple group comparisons by Kruskal–Wallis test for nonparametric data with Dunn's multiple comparison post hoc test, using GraphPad Prism software.

11.7 CONCLUDING REMARKS

The PKD research field, together with active participation from patients, represents a unique science community that works as a unified force from basic science to clinical research. However, an integrative understanding of polycystin molecular functions is still lacking. Is PKD an uncontrolled growth, metabolic, or aging disease? Why is ADPKD the most common genetic disease? From the evolutionary biology view, is there an advantageous trait for which heterozygote alleles of *PKD1* or *PKD2* are selected? These big picture questions require the understanding of the evolutionarily conserved functions of the polycystins across species, and can be accelerated by high-throughput research using model organisms. We see *C. elegans* is one of the most important model organisms for PKD research. We hope that many more scientists will join the field and utilize the worm as a model organism to tackle polycystins biology in the context of cilia, EVs, and a much more expanded view of the polycystins.

ACKNOWLEDGMENTS

This work is supported by NIH Grants DK059418, DK116606 to MB and P30 DK106912 to J.W. from the Kansas PKD Research and Translation Core Center. We thank Barr lab members past and present, for reviewing and discussing the manuscript and special thanks for Young Bae, Arantza Barrios, Leonard Haas, Jinghua Hu, Karla Knobel, Natalia Morsci, Robert O'Hagan, and Malan Silva for setting rigorous and reproducible standards for developing the protocols.

REFERENCES

1. Barr, M.M. and Sternberg, P.W. 1999. A polycystic kidney-disease gene homologue required for male mating behaviour in *C. elegans*. *Nature 401*, 386–389.
2. Wang, J., Silva, M., Haas, L.A., Morsci, N.S., Nguyen, K.C.Q., Hall, D.H., and Barr, M.M. 2014. *C. elegans* ciliated sensory neurons release extracellular vesicles that function in animal communication. *Curr. Biol. 24*, 519–525.
3. Pazour, G.J., San Agustin, J.T., Follit, J.A., Rosenbaum, J.L., and Witman, G.B. 2002. Polycystin-2 localizes to kidney cilia and the ciliary level is elevated in orpk mice with polycystic kidney disease. *Curr. Biol. 12*, R378–R380.
4. Yoder, B.K., Hou, X., and Guay-Woodford, L.M. 2002. The polycystic kidney disease proteins, polycystin-1, polycystin-2, polaris, and cystin, are co-localized in renal cilia. *J. Am. Soc. Nephrol. 13*, 2508–2516.
5. Kozminski, K.G., Johnson, K.A., Forscher, P., and Rosenbaum, J.L. 1993. A motility in the eukaryotic flagellum unrelated to flagellar beating. *Proc. Natl. Acad. Sci. USA. 90*, 5519–5523.
6. Satir, P. 2017. CILIA: Before and after. *Cilia 6*, 1.
7. Wang, J., and Barr, M.M. 2016. Ciliary extracellular vesicles: Txt Msg organelles. *Cell Mol. Neurobiol. 36*, 449–457.

8. Hogan, M.C. et al. 2009. Characterization of PKD protein-positive exosome-like vesicles. *J. Am. Soc. Nephrol. 20*, 278–288.
9. Pisitkun, T., Shen, R.-F., and Knepper, M.A. 2004. Identification and proteomic profiling of exosomes in human urine. *Proc. Natl. Acad. Sci. USA. 101*, 13368–13373.
10. Corsi, A.K., Wightman, B., and Chalfie, M. 2015. A transparent window into biology: A primer on *Caenorhabditis elegans. Genetics 200*, 387–407.
11. Brenner, S. 1974. The genetics of *Caenorhabditis elegans. Genetics 77*, 71–94.
12. Stiernagle, T. Maintenance of C. elegans (2006), WormBook, ed. The C. elegans Research Community, WormBook, doi/10.1895/wormbook.1.101.1
13. Hodgkin, J., Horvitz, H.R., and Brenner, S. 1979. Nondisjunction Mutants of the Nematode *Caenorhabditis elegans. Genetics 91*, 67–94.
14. Bae, Y.-K., and Barr, M.M. 2008. Sensory roles of neuronal cilia: Cilia development, morphogenesis, and function in *C. elegans. Front. Biosci. 13*, 5959–5974.
15. Boulin, T. et al. Reporter gene fusions (April 5, 2006), WormBook, ed. The *C. elegans* Research Community, WormBook, doi/10.1895/wormbook.1.106.1
16. Reinke et al. Transcriptional regulation of gene expression in *C. elegans* (June 4, 2013), WormBook, ed. The *C. elegans* Research Community, WormBook, doi/10.1895/wormbook.1.45.2
17. Efimenko, E., Bubb, K., Mak, H.Y., Holzman, T., Leroux, M.R., Ruvkun, G., Thomas, J.H., and Swoboda, P. 2005. Analysis of *xbx* genes in *C. elegans. Development 132*, 1923–1934.
18. Laurençon, A., Dubruille, R., Efimenko, E., Grenier, G., Bissett, R., Cortier, E., Rolland, V., Swoboda, P., and Durand, B. 2007. Identification of novel regulatory factor X (RFX) target genes by comparative genomics in *Drosophila* species. *Genome Biol. 8*, R195.
19. Spencer, W.C. et al. 2011 A spatial and temporal map of *C. elegans* gene expression. *Genome Res. 21*(2), 325–341.
20. Senti, G. and Swoboda, P. 2008. Distinct Isoforms of the RFX Transcription Factor DAF-19 Regulate Ciliogenesis and Maintenance of Synaptic Activity. *Mol. Biol. Cell 19*, 5517–5528.
21. De Stasio, E.A. et al. 2018. An expanded role for the RFX transcription factor DAF-19, with dual functions in ciliated and nonciliated neurons. *Genetics 208*, 1083–1097.
22. Wang, J., Schwartz, H.T., and Barr, M.M. 2010. Functional specialization of sensory cilia by an RFX transcription factor isoform. *Genetics 186*, 1295–1307.
23. Barr, M.M., DeModena, J., Braun, D., Nguyen, C.Q., Hall, D.H., and Sternberg, P.W. 2001. The *Caenorhabditis elegans* autosomal dominant polycystic kidney disease gene homologs lov-1 and pkd-2 act in the same pathway. *Curr. Biol. 11*, 1341–1346.
24. Evans, T. C., ed. Transformation and microinjection (April 6, 2006), WormBook, ed. The *C. elegans* Research Community, WormBook, doi/10.1895/wormbook.1.108.1
25. Dickinson, D.J., and Goldstein, B. 2016. CRISPR-Based Methods for *Caenorhabditis elegans* Genome Engineering. *Genetics 202*, 885–901.
26. Blacque, O.E., and Sanders, A.A. 2014. Compartments within a compartment. *Organogenesis 10*, 126–137.
27. Shaham, S., ed., WormBook: Methods in Cell Biology (January 02, 2006), WormBook, ed. The C. elegans Research Community, WormBook, doi/10.1895/wormbook.1.49.1
28. Bae, Y.-K., Qin, H., Knobel, K.M., Hu, J., Rosenbaum, J.L., and Barr, M.M. 2006. General and cell-type specific mechanisms target TRPP2/PKD-2 to cilia. *Development 133*, 3859–3870.
29. Silva, M., Morsci, N., Nguyen, K.C.Q., Rizvi, A., Rongo, C., Hall, D.H., and Barr, M.M. 2017. Cell-specific α-Tubulin isotype regulates ciliary microtubule ultrastructure, intraflagellar transport, and extracellular vesicle biology. *Curr. Biol. 27*, 968–980.
30. Bae, Y.-K., Kim, E., L'hernault, S.W., and Barr, M.M. 2009. The CIL-1 PI 5-phosphatase localizes TRP Polycystins to cilia and activates sperm in *C. elegans. Curr. Biol. 19*, 1599–1607.

31. Barr, M.M., García, L.R., and Portman, D.S. 2018. Sexual dimorphism and sex differences in *Caenorhabditis elegans* neuronal development and behavior. *Genetics 208*, 909–935.
32. O'Hagan, R., Piasecki, B.P., Silva, M., Phirke, P., Nguyen, K.C.Q., Hall, D.H., Swoboda, P., and Barr, M.M. 2011. The tubulin deglutamylase CCPP-1 regulates the function and stability of sensory cilia in *C. elegans*. *Curr. Biol. 21*, 1685–1694.
33. Barr, M.M. and Garcia, L.R. Male mating behavior (2006), WormBook, ed. The C. elegans Research Community, WormBook, doi/10.1895/wormbook.1.78.1
34. O'Hagan, R. and Barr, M.M. 2016. Kymographic analysis of transport in an individual neuronal sensory cilium in *Caenorhabditis elegans*. In *Cilia: Methods and Protocols*, P. Satir, and S.T. Christensen, eds. (New York, NY: Springer New York), pp. 107–122.
35. Mangeol, P., Prevo, B., and Peterman, E.J.G. 2016. KymographClear and KymographDirect: Two tools for the automated quantitative analysis of molecular and cellular dynamics using kymographs. *Mol. Biol. Cell 27*, 1948–1957.
36. Taschner, M. and Lorentzen, E. 2016 The intraflagellar transport machinery. *Cold Spring Harb Perspect Biol. 8*(10). pii: a028092.
37. Taschner, M., Bhogaraju, S., and Lorentzen, E. 2012. Architecture and function of IFT complex proteins in ciliogenesis. *Differentiation 83*, S12–S22.
38. Morsci, N.S. and Barr, M.M. 2011. Kinesin-3 KLP-6 Regulates intraflagellar transport in male-specific cilia of *Caenorhabditis elegans*. *Curr. Biol. 21*, 1239–1244.
39. Snow, J.J., Ou, G., Gunnarson, A.L., Walker, M.R.S., Zhou, H.M., Brust-Mascher, I., and Scholey, J.M. 2004. Two anterograde intraflagellar transport motors cooperate to build sensory cilia on *C. elegans* neurons. *Nat. Cell Biol. 6*, 1109–1113.
40. Verhey, K.J., Dishinger, J., and Kee, H.L. 2011. Kinesin motors and primary cilia. *Biochem. Soc. Trans.* 39, 1120–1125.
41. Peden, E.M. and Barr, M.M. 2005. The KLP-6 Kinesin is required for male mating behaviors and polycystin localization in *Caenorhabditis elegans*. *Curr. Biol. 15*, 394–404.
42. Mondal, S., Ahlawat, S., Rau, K., Venkataraman, V., and Koushika, S.P. 2011. Imaging in vivo neuronal transport in genetic model organisms using microfluidic devices. *Traffic 12*, 372–385.
43. Wang, J. et al. 2015. Cell-Specific Transcriptional profiling of ciliated sensory neurons reveals regulators of behavior and extracellular vesicle biogenesis. *Curr. Biol. 25*, 3232–3238.
44. Kaletsky, R., Lakhina, V., Arey, R., Williams, A., Landis, J., Ashraf, J., and Murphy, C.T. 2016. The *C. elegans* adult neuronal IIS/FOXO transcriptome reveals adult phenotype regulators. *Nature 529*, 92–96.
45. Lipton, J., Kleemann, G., Ghosh, R., Lints, R., and Emmons, S.W. 2004. Mate searching in *Caenorhabditis elegans*: A genetic model for sex drive in a simple invertebrate. *J. Neurosci. 24*, 7427–7434.
46. De Vore, D.M., Knobel, K.M., Nguyen, K.C.Q., Hall, D.H., and Barr, M.M. 2008. Sensory regulation of *C. elegans* male mate-searching behavior. *Curr. Biol. 18*, 1865–1871.
47. Barr, M.M., Vore, D.M. De, Knobel, K.M., Nguyen, K.C.Q., and Hall, D.H. 2018. Extracellular matrix regulates morphogenesis and function of ciliated sensory organs in *Caenorhabditis elegans*. *BioRxiv 376152*.

12 Approaches to Studying Polycystic Kidney Disease in Zebrafish

Jingyu Li and Ying Cao

CONTENTS

12.1 INTRODUCTION

Polycystic kidney disease (PKD) includes autosomal dominant PKD (ADPKD) and autosomal recessive PKD (ARPKD). ADPKD is one of the most common monogenic diseases in humans, with the incidence of 1 in 800 live births, and there are an estimated 12.5 million ADPKD patients worldwide. *PKD1* and *PKD2* are two causal genes for ADPKD, accounting for 85% and 15% of the disease, respectively.[1,2]

Compared with ADPKD, the incidence of ARPKD is much lower, with 1 in 20,000 live births, and the causal genes are *PKHD1* and *DZIP1L*.[3–5]

Besides ADPKD and ARPKD, there are some other genetic diseases or ciliopathies that manifest polycystic kidney syndrome. Ciliopathies refer to the diseases caused by structural or functional defects in cilia, the hair-like subcellular organelles. Polycystin-1 (PC1) and polycystin-2 (PC2), two proteins encoded by *PKD1* and *PKD2*, respectively, form complexes and are transported to the cilia.[6,7] The ciliary trafficking of the PC1 and PC2 complex are important for their *in vivo* function.[8,9] Thus, ADPKD is considered as one of the most common ciliopathies. Other ciliopathies with PKD phenotype include nephronophthisis (NPHP), Joubert syndrome and related disorders (JSRD), Meckel syndrome (MKS), Bardet–Biedl syndrome (BBS), oral-facial-digital syndrome (OFD), et al.[10] Most PKD will progress to end-stage renal disease around the age of 60. So far, there is no effective treatment except renal transplantation. Thus, it's important to understand the pathogenesis of PKD.

Zebrafish is an excellent vertebrate model to study genetic diseases and embryonic development. The unique features of zebrafish embryos are as follows: (1) Embryos develop *ex vivo*. The eggs are fertilized *ex vivo*, so the embryos develop *ex vivo* from the very beginning. (2) Embryos are transparent, so it's easy to observe and analyze the embryonic developmental processes under the microscope, especially with the help of transgenic technology. (3) The zebrafish lay large numbers of eggs at a time, which facilitates genetic screen. (4) Powerful tools, especially genetic screen, chemical screen, live imaging, and so on, are widely used in the zebrafish field.

In this chapter, we will provide an overview of the common approaches used in PKD study in zebrafish and encourage the scientists to use these tools in PKD study.

12.2 KIDNEY STRUCTURE AND DEVELOPMENT IN ZEBRAFISH

The kidneys in the zebrafish embryo are pronephros and contain only one pair of nephrons (Figure 12.1), and they are much simpler compared with mammalian kidneys that are metanephros and contain millions of nephrons. The major differences between zebrafish and mammalian kidney structure are that zebrafish pronephros lack the loop of Henle and branching morphology.[11,12] The absence of the loop of Henle is to be expected as the zebrafish is an aquatic species and therefore has no requirement to preserve water and concentrate urine. Except for the loop of Henle, other segments, including glomerulus, neck, proximal convoluted and straight tubules, distal straight and convoluted tubules, and collecting duct, are conserved in the zebrafish pronephros.[11,12]

The development processes of nephrons are similar between zebrafish and mammals. The zebrafish pronephros originates from the intermediate mesoderm, which locates along the marginal zone at the 50% gastrulation stage and next to the somites at the segmentation stage. Like metanephros, the pronephros development can be divided into four stages: (1) The pronephros specification or the commitment of intermediate mesoderm into pronephros rudiment starts around 12 hours post fertilization (hpf). (2) Tube formation occurs around 16~24 hpf. (3) Nephron patterning, during which the nephrons differentiate into glomerulus and different tubular cell types, takes place around 24~40 hpf. (4) Finally, the glomerulus forms and fuses in the middle, as well as the blood filtration starts, around 40–48 hpf.[13,14]

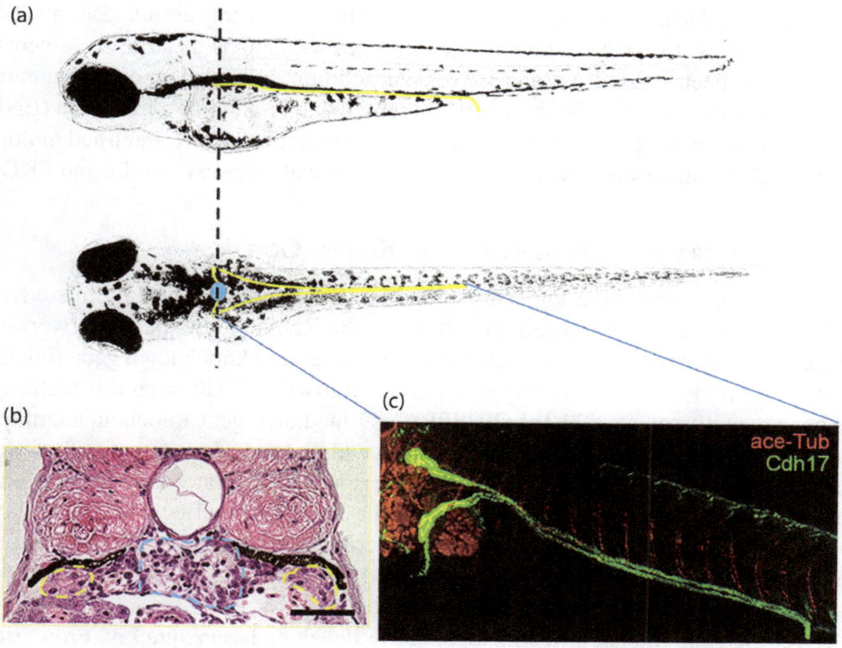

FIGURE 12.1 Zebrafish mature pronephros at 3 dpf. (a) Schematic diagram of zebrafish pronephros at 3 dpf: the upper: side view; the bottom: dorsal view. The glomerulus fuse in the middle (the blue circle in the dorsal view of the embryo), and connect to pronephric tubules, which run to the lateral to meet the ducts and fuse to the cloaca (all indicated in yellow). (b) Hematoxylin and eosin stained cross-section at the glomerulus region, as black dotted line indicated in (a), shows the glomerulus and pronephric tubules. (c) Fluorescent image from whole-mount immunostaining with antibodies against ace-Tub (cilia) and Cdh17 (pronephros), reveals the pronephric tubules and ducts.

Regardless of the simple structure and quick development of the zebrafish pronephros, the function of the kidney as well as the molecular mechanisms of the kidney development and diseases, especially PKD, are quite conserved between zebrafish and mammals. On the one hand, disease-causal genes in PKD patients also lead to PKD in zebrafish when mutated. On the other hand, PKD-causal genes first identified in zebrafish, except for primary cilia dyskinesia (PCD) genes, also lead to the clinical discovery that they also lead to PKD in humans when mutated. Thus, the zebrafish provides an excellent vertebrate model to study PKD.

12.3 GENERATING GENETIC MODELS FOR POLYCYSTIC KIDNEY DISEASES

12.3.1 Generating Disease Models by Forward Genetics

Genetic screen is one of the most powerful tools to identify genes involved in certain biological process or diseases and is one of the most significant advantages for zebrafish as a vertebrate model. Traditional genetic screens in zebrafish use chemicals or retrovirus-introduced insertions or transposon-mediated gene trap as mutagens and take three generations to establish mutant lines.[15–18] Several rounds of large-scale genetic screens have been carried out and have identified hundreds of phenotypic mutants. For PKD studies, two large-scale genetic screens, using N-ethyl-N-nitrosourea (ENU) and retrovirus-introduced insertions as mutagens, respectively, have identified multiple PKD models[14,19] and established a genome-wide association between cilia and PKD.[19]

12.3.2 Generating Disease Models by Reverse Genetics

Reverse genetics, including knockdown and knockout, are widely used in studying the function of genes. In the zebrafish field, as the RNAi technique is not working, we use morpholinos (MO), which are DNA analogues and knockdown gene function by blocking translation or splicing of the target transcripts. Although this technique has been recently replaced by the CRISPR/Cas9-mediated gene knockout technique, due to the off-target issue by the MO, it is still widely used. For example, if we want to block the maternal deposit of a gene transcript and the maternal-zygotic mutant is difficult to obtain, we use MO. To avoid the nonspecific effects of the MO, we should follow the guidelines.[20] Basically, the criteria for a reliable MO are that the phenotypes caused by MO knockdown should phenocopy the mutants and can be rescued by its mRNA overexpression.

Recently, gene knockout technologies have developed very quickly. From zinc-finger nucleases (ZFNs), transcription activator-like effectors (TALENs), to the clustered regularly interspaced short palindromic repeats (CRISPR) system, all these different methods can generate gene-specific alterations via nonhomologous end-joining (NHEJ) by ZFN-, TALEN-, or CRISPR-associated-protein 9- (Cas9-) mediated double-stranded breaks. Among all these technologies, the CRISPR/Cas9 system is the most convenient and with high mutagenesis efficiency, has thus been widely used in the zebrafish field.[21,22] Due to the high efficiency of gene knockout of the CRISPR/Cas9 system, it's possible to obtain phenotypic mutants in one generation or even right after injection, which saves 3–8 months (1–2 generation time) compared with classic genetic screen.[23] By this method, it's easy to carry out a small-scale reverse genetic screen for candidate genes. Thus, it provides a powerful tool for genetic study in the zebrafish.

Step-by-step protocols are as follows.

12.3.3 Protocols for Knockout Genes by the CRISPR/Cas9

Materials

- Tuebingen zebrafish genomic DNA
- Cas9 vector: pXT7-2NLS-Cas9 (alternatively, Cas9 protein can be ordered from company, for example, NEB, cat. no. M0646M)
- T7 mMESSAGE mMACHINE Kit: Ambion, cat. no. AM1344
- RNeasy Mini Kit: Qiagen, cat. no. 74104
- Trizol or RNAiso Plus
- 100% ethanol

- DNA gel-purification kit
- pMD19-gRNA
- rNTP mix: NEB, cat. no. 0466S, 4 mM each
- DTT 1 mM
- RNasin
- T7 RNA Polymerase: NEB, cat. no. M0251S
- Tuebingen strain wildtype zebrafish: This strain is used by Sanger for the zebrafish sequencing project

Procedure

Day 1: Target design

1. Obtain the genomic sequences of the target genes from Ensembl. Usually, we screen 10 genes per round every week.

   Note:
   - Target should be 5′-N20-NGG-3′. It could be on the top strand or bottom strand. For more information, you can check http://zifit.partners.org/ZiFiT/CSquare9Nuclease.aspx.
   - Include a restriction enzyme site in the target to facilitate mutagenesis efficiency analysis.
   - For coding gene, target should be within the two-third domain of CDS and after ATG. It would be better if the knockout can damage important domains and isoform/variant as much as possible.
   - Target should NOT be on the last exon or 5′ UTR or 3′ UTR.
   - Blast sequence on NCBI to reduce off-target.
   - Design ID primer. PCR product should be specific and about 350–600 bp.
2. Order primers.

Days 2–3: Confirm target and ID primers

1. Set up a 50 μL PCR reaction.

Component	Volume	Final Concentration
Genomic DNA	1 μL	30 ng/μL
F Primer	2 μL	4 μM
R Primer	2 μL	4 μM
10× buffer	5 μL	1×
$MgCl_2$	4 μL	
dNTP	5 μL	2.5 mM each
rTaq	0.5 μL	
ddH_2O	30.5 μL	
Amplify 35 cycles		

2. Check PCR product: Load 5 μL product for gel electrophoresis analysis with 1.5% agarose gel.

3. 25 μL for sequencing (with 5 μL primer, 4 uM) 15 μL for enzyme digestion (1 μL enzyme, 2–3 h).

Days 4–5: Order oligo

1. Design and order gRNA primers.
   Forward primer (T7gFP): T7_19~23 bp target sequence_20 bp gRNA scaffold
   IF gRNA starts with G, then the T7-gRNA forward primer (T7 g FP) is:
   5′-TAATACGACTCACTATANNNNNNNNNNNNNNNNNNNNGTTTT AGAGCTAGAAATAGC-3′
   If not start with G, then the G should be added and the T7-gRNA forward primer should be:
   5′-TAATACGACTCACTATAGNNNNNNNNNNNNNNNNNNNNGTTT T AGAGCTAGAAATAGC-3′
   gRNA scaffold Reverse Primer (gRNA RP): 20 bp gRNA scaffold
   5′–AGCACCGACTCGGTGCCACT-3′
   >T7 promotor 5′-TAATACGACTCACTATA-3′
   >gRNA scaffold sequence
   GTTTTAGAGCTAGAAATAGCAAGTTAAAATAAGGCTAGTCCGTTA TCAACTTGAAAAAGTGGCACCGAGTCGGTGCT

Day 6: Synthesize Cas9 mRNA and gRNA

1. (Optional: If you use Cas9 protein, you can skip this step.) Synthesize *Cas9* mRNA.
   - Obtain template.
     Linearization of pXT7-2NLS-Cas9 vector by XbaI (8 kb, Amp-resistant) (total volume: 100 μL, 3–5 μg DNA, 5 μL XbaI 37°C, for 4 h); purify product (dilute in 15 μL ddH_2O).
   - *In vitro* transcription with T7 mMESSAGE mMACHINE Kit.
     Set up 20 μL reactions:

Component	Volume
2× NTP/CAP	10 μL
10× reaction buffer	2 μL
Linearized template DNA	5 μL (>1 μg)
T7 RNA polymerase	2 μL
Nuclease-free water	To 20 μL
Incubate at 37°C, 2 h	

   - Gel electrophoresis: Load 0.5 μL of reaction product into wells in the gel. The expected product size is about 2 kb.
     If the mRNA is well synthesized, remove DNA template by adding 1 μL Turbo DNase, 37°C, 15 min.
   - Use RNeasy mini Kit (Qiagen) to purify Cas9 mRNA.
     Notes before starting:

Add 4 volumes of ethanol to Buffer RPE for a working solution.
- Add 80 µL RNase-free water. Add 350 µL Buffer RLT, and mix well.
- Add 250 µL ethanol to the diluted RNA, and mix well by pipetting. Do not centrifuge. Proceed immediately to the next step.
- Transfer the sample (700 µL) to an RNeasy Mini spin column placed in a 2 mL collection tube (supplied). Close the lid. Centrifuge for 15 s at ≥8000 × *g*. Discard the flow through.
- Add 500 µL buffer RPE to the RNeasy spin column. Close the lid. Centrifuge for 15 s at ≥8000 × *g* to wash the membrane. Discard the flow through.
- Add 500 µL Buffer RPE to the RNeasy spin column. Close the lid. Centrifuge for 2 min at ≥8000 × *g* to wash the membrane.
- Place the RNeasy spin column in a new 2 mL collection tube (supplied). Close the lid, and centrifuge at full speed for 1 min.
- Place the RNeasy spin column in a new 1.5 mL collection tube (supplied). Add 20 µL RNase-free water directly to the spin column membrane. Close the lid, and centrifuge for 1 min at ≥8000 × *g* to elute the RNA.

- Gel electrophoresis: Load 0.5 µL RNA in 1% agarose gel to check the RNA quality.
- Measure the concentration with nanodrop.
- Aliquot 4 µL/tube, store at −80°C.

2. Synthesize gRNA.
   - PCR gRNA template DNA (50 µL, 35 cycles).

Component	Volume	Final Concentration
pMD19-gRNA	1 µL	1–5 ng/µL
T7g FP	2 µL	4 µM
gRNA RP	2 µL	4 µM
10× buffer	5 µL	1×
$MgCl_2$	4 µL	
dNTP	5 µL	2.5 mM each
rTaq	0.5 µL	
ddH_2O	30.5 µL	

Take 5 µL PCR product for gel electrophoresis, using 1.5% agarose gel. Product should be 125 bp. Purify the product using a DNA gel-purification kit.

   - Set up 20 µL *in vitro* transcription for gRNA synthesis.

Component	Volume
gRNA template DNA	6 µL (>100 ng)
NF-water	8 µL
10× transcription buffer	2 µL
rNTP Mix	2 µL

(Continued)

DTT 1 mM	0.5 μL
RNasin	0.5 μL
T7 RNA polymerase	1 μL
Incubate at 37°C, 2 h	

Run 1 μL in 1.5% agarose gel for gel electrophoresis. Product should be 100 bp.

Add 1 μL DNase, 37°C, 15 min to remove DNA template.

- gRNA purification
  - Precool centrifuge.
  - Add 10 volume of RNAiso Plus to 1 volume of gRNA, vortex.
  - Add 40 μL chloroform. Cap the centrifuge tube and mix until the solution becomes milky.
  - Keep the solution at room temperature for 5 min. Centrifuge at 12,000 × g for 15 min at 4°C. The solution will separate into three layers: Liquid top layer (contains RNA), semisolid middle layer (mostly DNA), and bottom organic solvent layer.
  - Transfer the top liquid layer to a new centrifuge tube without touching the middle layer.
  - Measure the amount of the top layer and add an equal amount of ethanol or add up to 0.5 volume of isopropanol of the top layer. Mix together well. Keep the mixture at room temperature for 10 min.
  - Centrifuge at 12,000 × g for 10 min at 4°C to precipitate the RNA.
  - Carefully remove the supernatant; do not touch the pellet. If some isopropanol remains, that is not a problem. Add 500 μL 85% cold ethanol. Clean the precipitate by vortex. Discard supernatant. Be careful not to disturb the precipitate.
  - Add 200 μL 100% ethanol. Centrifuge at 12,000 × g for 5 min. Discard supernatant.
  - Dry the precipitate by leaving the tube open for several minutes. After the precipitate is dry, dissolve it with 15–20 μL RNase-free water.
  - Run 1 μL product for gel electrophoresis and measure the concentration using nanodrop. Store at −80°C.

Day 7: Injection

1. For Cas9 mRNA

   Recommended dose: Cas9 mRNA 300–500 ng/μL. gRNA 20–50 ng/μL. Co-inject 1.5 nL of Cas9 mRNA (300–500 ng/μL) and gRNA (gRNA 20–50 ng/μL) into the cytoplasm of 1-cell stage Tuebingen embryos. Inject at least 100 embryos for each gRNA. For gRNA targeting non-coding RNA, 200 embryos are recommended.

   - For coding RNA:

Stock	Volume	Stock Concentration
Cas9 mRNA	1 μL	600–1000 ng/μL
gRNA	1 μL	40–100 ng/μL

- For non-coding RNA (deletion):

Stock	Volume	Stock Concentration
Cas9 mRNA	0.7 μL	600–1000 ng/μL
gRNA upstream	0.7 μL	40–100 ng/μL
gRNA downstream	0.7 μL	40–100 ng/μL

2. For Cas9 protein:
The working concentration of Cas9 protein is 4 uM.

Stock	Volume	Stock Concentration
Cas9 protein	1 μL	20 μM
gRNA	4 μL	50–100 ng/μL

Day 8: Checking mutagenesis efficiency

1. Pick 15 embryos at 1~2 days postfertilization (dpf), 5 embryos/tube.
2. Add 20 μL lysis buffer, 37°C, 4 h.
3. Take 1 μL lysate, add 19 μL H_2O. Denature at 95°C, 15 min.
4. Set up a 20 μL PCR.

Component	Volume	Final Concentration
Tuebingen zebrafish genomic DNA	1 μL	30 ng/μL
F Primer	1 μL	4 μM
R Primer	1 μL	4 μM
10× buffer	2 μL	
$MgCl_2$	1.6 μL	
dNTP	2 μL	2.5 mM each
rTaq	0.2 μL	
ddH_2O	11.2 μL	
Run 35 cycles		

5. Run 4 μL in 1.5% agarose gel for gel electrophoresis.
6. (Optional): If the gRNA target sequence contains restriction enzyme site, digest with the restriction enzyme. Incubate at 37°C, for 2 h. Run a gel electrophoresis and estimate the mutagenesis efficiency according to the percentage of undigested DNA.
7. (Optional): If the gRNA target sequence does not contain restriction enzyme site, sequence the PCR product. Estimate the mutagenesis efficiency according to the sequencing result.

Day 7 and after: Observe phenotypes and establish mutant lines

1. If mutagenesis efficiency >50%, you can observe phenotypes of injected embryos directly. Otherwise, redo from the beginning.

2. (Optional): For the injected embryos that contain phenotypic ones, raise the embryos to adult, which is founder fish or F0 fish. Screen F0 to get mutants with frame shift mutations. Outcross identified F0 mutants with wildtype fish to establish the mutant line.

12.4 CHARACTERIZING PKD MUTANTS

12.4.1 General Characterization of PKD in Zebrafish

The renal cyst in zebrafish is the bubble-like structure close to the pectoral fin, which is easy to distinguish by stereoscope (Figure 12.2a). The bubble-like structure is due to the cystic glomerular. However, the tubule or duct dilation is not that visible by direct observation through a stereomicroscope and can be detected by fluorescent microscopy after immunostaining (Figure 12.2b).

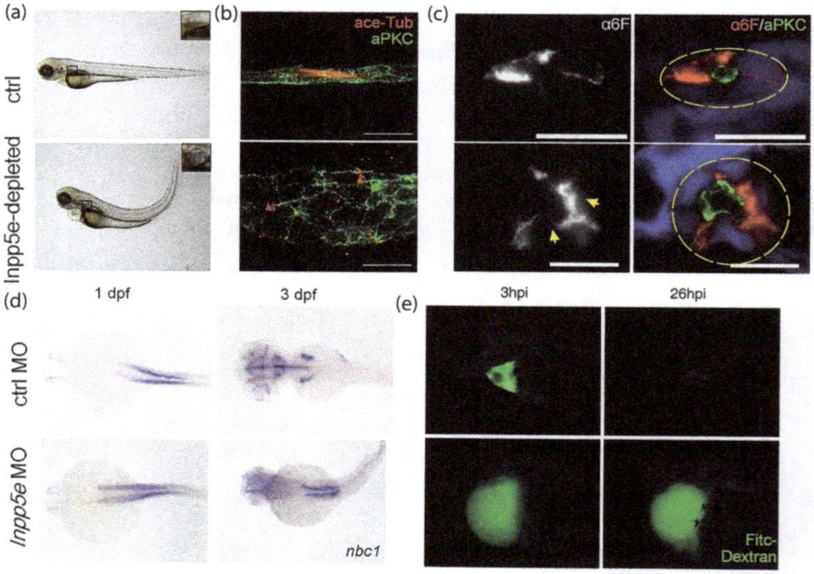

FIGURE 12.2 Examples of phenotypic and functional analysis of zebrafish PKD models. Controls on the top and inpp5e mutants (a and b) or morphants (c–e) on the bottom. (a) Dorsal body curvature and bubble-like renal cyst of *inpp5e* mutants are visible under stereoscope. (b) Fluorescent images of immunostaining with antibodies against ace-Tub (cilia) and aPKC (renal apical membrane) showed loss of cilia and dilated pronephric tubule in *inpp5e* mutants. (c) Fluorescent images of immunostaining for renal cryosections (circled by yellow dotted lines) with antibodies against a6F (Na+K+ ATPase, renal basolateral membrane) and aPKC show mislocalization of the basolateral marker Na+K+ ATPase to the apical membrane (arrow) in *inpp5e* morphants. (d) Images of *in situ* hybridization with *nbc1* probe (proximal renal tubule marker) revealed that forward migration/shift of *nbc1* from 1 to 3 dpf in wildtype, but not in *inpp5e* morphants. (e) Fluorescent images of dye clearance assay showed that majority of Fitc-Dextran was filtered and excreted in wildtype after 26 hpi, but not in inpp5e morphants. Scale bar, 10 µm. (From Xu, W. et al. *Journal of the American Society of Nephrology* 28.)

Besides renal cysts, zebrafish ADPKD models also manifest some other typical phenotypes, such as dorsal body curvature.[19,24,25] As body curvature is the most visually apparent phenotype in zebrafish *pkd2* mutants, it was used as an initial readout for a chemical modifier screen, which identified histone deacetylase (HDAC) inhibitors to ameliorate the cystic phenotype in both zebrafish and mouse ADPKD models.[26]

12.4.2 Analyzing Cell Proliferation in the Zebrafish Kidney

At the cellular level, defects of ADPKD include defects in cell proliferation, cell apoptosis, cell polarity, and cell secretion.[27] Cell proliferation is one of the most significant features of PKD in human and mouse models.[28,29] However, this seems not true in zebrafish PKD mutants, as there is no evidence that cell proliferation is increased in zebrafish PKD mutants (our unpublished data and research by Kramer-Zucker et al.[30]). Two methods have been used to estimate the proliferative cell number in the zebrafish pronephric duct (pnd). One is by direct counting the cell number change over a period of time due to the simple structure of pnd. After 24 hpf, only part of the collecting duct cells proliferate.[31] The other is by BrdU pulse treatment as in the protocol in Section 12.4.4.1 of this chapter. In general, renal cell proliferation in the zebrafish embryo is not as active as that in the mouse embryo. At 3 dpf, around 20 proliferating cells exist in the zebrafish pnd (our unpublished data).

12.4.3 Analyzing Cell Polarity and Cell Migration in the Zebrafish Kidney

Instead, cell migration or convergent extension is one of the major cellular activity in zebrafish during pronephros patterning.[31] Consistent with this, cell proliferation is not that significantly changed in zebrafish PKD models as that in mammals.

Defects in the cell polarity and cell migration are closely associated with PKD in both zebrafish and mammals. Cell polarity refers to both apical-basal polarity and planar cell polarity. Similar to other epithelial cells, renal epithelial cells have established apical-basolateral polarity with the apical side facing the lumen and basolateral connecting to the neighboring cells or cellular matrix. In both mice and zebrafish, defects in apical-basolateral polarity have been discovered in several PKD models, for example, ADPKD, JBTS (Figure 12.2c).[14,32,33] Meanwhile, disruption of proteins required for apical-basolateral polarity (e.g., Dlg5, Scribble, Crumbs) in both mice and zebrafish also leads to PKD.[34–36]

Planar cell polarity (PCP) refers to cell polarity in a tissue plane and is perpendicular to apical-basolateral polarity. PCP is regulated by noncanonical Wnt signaling (β-catenin independent) or the PCP pathway. During kidney development in both mice and zebrafish, renal tubules undergo convergent extension, a typical PCP phenomenon, which drives the tubules narrow and elongate or migrate.[31,37] Disruption of PCP is also implicated in cystic kidney formation and vice versa.[38–41]

We can use immunostaining with antibodies against polarity markers to check the cell polarity and the whole-mount *in situ* hybridization or live imaging to check the

TABLE 12.1
Markers for Analyzing Kidney Phenotypes in Zebrafish

Assay	Markers	Detection	References/Source
Immunostaining	Ace-Tub	Cilia	Sigma, cat. no. 6-11B-1
	γ–Tub	Basal bodies	Sigma, cat. no. T6557
	Na^+K^+-ATPase	Basolateral membrane of the pnd	Drummond et al.[14]/ Developmental Studies Hybridoma Bank, cat. no. α6F
	Cdh17	Basolateral membrane of the pnd	38
	aPKC	Apical membrane of the pnd	Xu et al.[33]/Santa Cruz, cat. no. SC-216
	Zo-1	Apical tight junction	Invitrogen, cat. no. 339100
	phalloidin	F-actin (apical)	Xu et al.[33]/Invitrogen, cat. no. R415
	BrdU and anti-BrdU antibody	Cell proliferation	Sigma
	TUNEL	Apoptosis	
In situ hybridization	*nbc1*	The proximal nephron	31
	trpM7	The late proximal nephron	31
	ret1	The distal nephron	31
	clcnk	The distal nephron	12
Transgenic fish	*Tg(NaK-ATPase:GFP)*	The whole nephron	31
	Tg(Cdh17:GFP)	The whole nephron	46
	Tg(ret1:GFP)	The distal nephron	31
	ET33-D10 GFP	The proximal nephron	31
	Tg(β-actin:Arl13b-GFP)	Cilia	47
	Tg(ubiquitin:arl13b-GFP)	Cilia	48

nephron segment markers to detect convergent extension or cell migration. During zebrafish kidney development, the boundary of segment marker, for example, *nbc1*, shift in a proximal-direction overtime (e.g., from 1 to 3 dpf), and this shift is disrupted in all PKD models in zebrafish, including Ift88 and Inpp5e-depleted embryos (Figure 12.2d).[31,33] Markers and transgenic lines are listed in Table 12.1.

12.4.4 Protocols for Phenotype Characterization of PKD Mutants

12.4.4.1 BrdU Pulse Treatment

Materials

- BrdU or Edu solution: 10 mM in embryo medium with 15% DMSO
- 4% formaldehyde: Freshly 1:10 diluted formalin (~40% formaldehyde) in PBST
- Pronase: 1 mg/mL in embryo medium

Procedure

1. Soaking (suitable for 6-h to 72-h-old embryos). Dechorionate embryos of desired age with pronase, and place in 10 mM BrdU or Edu solution at 6°C–8°C (or simply on top of the ice) for 20 min.
   Alternatively, inject 10 mM BrdU into the yolk (for young embryos, before segmentation) or pericardial chamber (for embryos older than 2 dpf) of embryos at the desired age.
2. Allow embryos to develop to desired age.
3. Fix in 4% formaldehyde, several hours at RT, or overnightin the cold.
4. Remove formaldehyde; wash embryos with PBST. Put embryos in fresh methanol; place at –20°C for at least an hour. Indefinite storage at –20°C is fine.
5. For BrdU detection, continue with immunostaining protocol (see Section 12.4.4.2).

12.4.4.2 Immunostaining on Whole Embryos

Materials

- PTU (Phenylthiourea, Sigma Aldrich, cat. no. P7629): 0.003% in embryo medium
- Tricaine (Sigma, E10521); 25× stock solution: 4 mg/mL, pH 7. Add 4 mL into 96 mL of embryo medium to make a working solution
- 4% formaldehyde: Freshly 1:10 diluted formalin (~40% formaldehyde) in PBST
- Dent's fixative: 80% methanol and 20% DMSO
- PBST: PBS with 0.1% Tween-20
- PBST-SD: PBS with 10% serum/1% DMSO/0.5% Tween-20
- Trypsin: 0.25% trypsin 1:20 diluted in PBST
- Antifade mounting medium: ProLong Gold Antifade Mountant (Invitrogen) or Vectashield Antifade mounting medium H-1000 (Vector Laboratories)
- Nail polish, transparent
- BABB: Benzyl alcohol: Parts benzyl benzoate 1:2, mix
- Eppendorf tubes
- Pipettes: 3 mL disposable polyethylene, e.g., BD Falcon, 357524
- Microscope slides and cover slides
- Fine-point tweezer: #5 watchmaker's tweezer

Procedure

1. Fixation of embryos

   Raise embryos to the desired stage. If embryos are older than 30 hpf, to block pigment formation, raise embryos with PTU starting at 24 hpf.

   Anesthetize embryos in tricaine and choose fixative according to the antibodies. Most antibodies work better in Dent's fixative than in formaldehyde. Test antibodies before experiments.

   Fix in Dent's fixative overnight (for embryos younger than 30 hpf, use precooled Dent's fixative and incubate at –20°C. For older embryos, use Dent's fixative at RT and incubate at RT).

Alternatively, fix embryos in 4% formaldehyde at RT for 1 h (make sure not to over fix).

2. Wash embryos with PBST 1× for 5 min at RT.
3. Dehydrate embryos through 50% methanol in PBST (5 min), then 100% methanol for 5 min.

   Replace with fresh methanol and store the tubes at −20°C (they are good for years).
4. Rehydrate embryos through 50% methanol in PBST (5 min), then PBST 5 min.
5. Permeabilization.

   For embryos fixed in Dent's, with trypsin in PBST (0.05% trypsin: PBST = 1: 20), at RT for appropriate time.

   For embryos fixed in formaldehyde, add precooled acetone and incubate at −20°C for 7 min for embryos younger than 30 hpf and 15 min for embryos older than 30 hpf. Treat embryos with trypsin in PBST (0.05% trypsin: PBST = 1: 20) at RT for 10–30 min.

   Permeabilization conditions are summarized as in the following table:

Fixative	<30 hpf	30 hpf- 5 dpf
Dent's	Trypsin, 10–20 min, RT	Trypsin, 20–40 min, RT
4% Formaldehyde	Acetone, 7 min, −20°C	Acetone, 15 min, −20°C
	Trypsin, 10–20 min, RT	Trypsin, 20–40 min, RT

   Other permeabilization condition can also be tested, such as Digitonin (1 mg/mL), Saponin (0.1%–0.5%), and Triton X-100 (0.1–0.5%).
6. Re-fix trypsin-permeabilized embryos in 4% formaldehyde at RT for 20 min.
7. Antigen retrieval by HCl (only for BrdU-treated embryos). Rinse several times in water. Wash 1 or 2 times in 2 N HCl, incubate at RT in 2 N HCl for 1 h.
8. Wash with PBST 3 times, 5 min each.
9. Block in PBST-SD for 0.5 h.
10. Incubate in first antibody (PBST+10% serum +appropriate Ab) for 2 h at RT or ON at 4°C.
11. Wash with PBST for 5 times, at least 20 min each.
12. Incubate with your favorite second antibody in PBST+2% serum from 2 h at RT or ON at 4°C. If fluorescent second antibody is used, shield samples from light with aluminum foil from this step on.
13. Wash as in step 11.
14. Observe, if fluorescent second antibody is used. For older embryos (>2 days), or deep structures (such as kidney), manually take off yolk with fine-point tweezers and then flat mount with cover slide in mounting medium. Dry at RT or 4°C for 2–4 h and seal cover slide with nail polish.

    If color reaction is used, proceed as you would with various kits, then dehydrate embryos with methanol (reverse order in step 3).

    Rinse 2× for 10 min with methanol.

    Clear embryo in BABB (1 part benzyl alcohol, 2 parts benzyl benzoate), partial dissect, and flat mount as above (BABB is toxic and will dissolve some plastics, although Eppendorf tubes seem to be fine, so handle with care)

15. Observe and take images with microscope. After imaging, store the slides at –20°C.

12.4.4.3 Immunostaining on Cryosections

Additional Materials and Equipment (not listed in Section 12.4.4.2)

- Cryostat: Leica CM1860
- Embedding medium for frozen sections, Optimal Cutting Temperature (O.C.T.) Compound (Sakura)
- Microscope Slides, Superfrost Plus Precleaned: Fisher, 12-550-15 (the choice of slides is essential to avoid losing sections during washes)
- Sucrose: 10%–30% in PBST; add 0.05% (w/v) of sodium azide and store at 4°C
- 10% sodium citrate solution: 10% sodium citrate solution in distilled water
- Cryomold: Tissue-Tek Cryomold (Sakura)
- PAP pen: Available from several manufacturers
- A humidity chamber: Alternatively, a tip box filled with tap water in the bottom compartment can serve as a homemade humidity chamber for antibody incubations.

Procedure

1. Fixation of embryos

   Raise embryos to the desired stage. If embryos are older than 30 hpf, to block pigment formation, raise embryos with PTU starting at 24 hpf.

   Anesthetize embryos in tricaine and choose fixative according to the antibodies. Most antibodies work better in Dent's fixative than in formaldehyde. Test antibodies before experiments.

   Fix in Dent's fixative (80% methanol and 20% DMSO) overnight. (For embryos younger than 30 hpf, use precooled Dent's fixative and incubate at –20°C. For older embryos, use Dent's fixative at RT and incubate at RT.)

   Alternatively, fix embryos in 4% formaldehyde at RT for 1 h (make sure not to overfix).
2. Wash embryos with PBST 1× for 5 min at RT.
3. Dehydrate embryos through 50% methanol in PBST (5 min), then 100% methanol for 5 min.

   Replace with fresh methanol and store the tubes at –20°C (they are good for years).
4. Rehydrate embryos through 50% methanol in PBST (5 min), then PBST 5 min.
5. Place embryos in 10% sucrose solution until the embryos sink, then in 20%, 30% sucrose solution sequentially.
6. Replace 30% sucrose with 50% O.C.T. compound for frozen sectioning.

   This is a very viscous substance, so use a Pasteur pipette. 50% O.C.T. helps maintain the shape of the kidney tube.
7. Transfer the embryos into the Cryomold using a Pasteur pipette.

8. Orient the embryos using a dissecting needle. Quickly cool the cryostat chambers with embryos to −80°C. O.C.T. compound will solidify and turn white in about 10 min.
9. Mount frozen blocks on a cryostat specimen holder using additional O.C.T. medium and collect sections on a prechilled slide inside the cryostat chamber set to −20°C. When brought to RT, sections will thaw and adhere to the slide.
10. Allow slides to dry at RT for 2 h. Then proceed with staining of cryosections.
11. Antigen retrieval (optional): Detection of some antigens may require special treatment such as placing slides into boiling-hot 10% sodium citrate solution for 20 min or in 1% SDS for 5 min.[42] A useful heat treatment method has also been described by Inoue and Wittbrodt.[43] After antigen retrieval, thorough washing in PBS, 10 min × 4, is recommended.
12. Using a PAP pen, create a hydrophobic grease border circling the sections. Allow the grease to dry for 2~3 min.
13. Briefly rehydrate slides by gently pipetting about 300 mL of PBST on the slide surface. This will also dissolve the frozen section medium.
14. Place slides into a humidity chamber.
15. Overlay each slide with 150 μL of the PBST-SD. The grease border will help to keep the solution on the slide but is not absolutely necessary. Incubate slides with the PBST-SD for 45 min at RT.
16. Dilute the primary antibody to appropriate concentration in PBST+10% serum.
17. Remove the PBST-SD by positioning slides vertically and collecting excess solution with a Kimwipe. Add 100 μL of primary antibody solution as in Step 15. Incubate 2 h at RT or ON at 4°C in a humidified chamber.
18. Wash slides with PBST, 3×, 5 min each.
19. Incubate with secondary antibody at RT for 2 h.
20. Wash slides with PBST, 5×, 5 min each.
21. Use a Kimwipe to remove excess PBST and apply the mounting medium. Place a coverslip over the sections and seal with nail polish.
22. Image using a confocal or conventional fluorescence microscopy.
23. This preparation may be stored at 4°C for at least a week, although the quality of the signal deteriorates over time.

12.4.4.4 Whole-Mount *In Situ* Hybridization

Materials

- 10× Dig RNA labeling mix: Roche, cat. no. 11277073910
- RNasin: RNase inhibitor from Roche or Promega
- T7 RNA polymerase: NEB
- Pronase: Sigma; make 30 mg/mL stock solution with embryo medium; dilute to 1 mg/mL with embryo medium before use
- 4% formaldehyde: Freshly diluted formalin (~40% formaldehyde) in PBST
- 10 × PBS: 80 g NaCl, 2.4 g KH_2PO_4, 2 g KCl, 14.4 g Na_2HPO_4, 900 mL ddH_2O; adjust pH to 7.4; add ddH_2O to 1 L; and autoclave

- PBST: 1 × PBS plus 0.1% Tween-20
- Proteinase K stock solution (10 mg/mL): Dissolve 100 mg proteinase K in 10 mL ddH_2O; store at –20°C
- HYB–: 50% formamide, 0.1% Tween-20, 5 × SSC; store at –20°C
- HYB+: HYB–, 50 μg/mL heparin, 500 μg/mL yeast RNA; store at –20°C
- 20 × SSC: 44.115 g sodium citrate, 87.66 g NaCl, add ddH_2O to 500 mL; autoclave
- 10 × MAB: 58.44 g NaCl, 116.07 g maleic acid, 800 mL ddH_2O, adjust pH to 7.5 (The solution is cloudy until the pH is close. Slowly add 7–8 g NaOH powder directly until the solution starts clear, then measure the pH when adding NaOH powder. When the pH is close, adjust with 5 N NaOH solution.) Add ddH_2O to 1 L; autoclave
- MABT: 1 × MAB plus 0.1% Tween-20
- 10% blocking solution: Dissolve 10 g blocking reagent (Roche) in 100 mL 1 × MAB; autoclave; aliquot and store at –20°C
- 1 × blocking solution: Mix goat serum: 10% blocking solution: 1× MABT=1:2:7 (volume) before use
- Anti-Dig-AP antibody: Anti-Digoxigenin-AP, Fab fragments antibody from sheep, Roche, cat. no. 11093274910
- Staining buffer: 100 mM Tris buffer (pH 9.5), 50 mM $MgC1_2$, 100 mM NaCI, 1 mM levamisole, 0.1% Tween-20. Make the stock solution of each component, mix before use
- BM Purple AP substrate: Roche, cat. no. 11442074001. Add 1 M levamisole 1:200 (v/v) in this product before use
- Glycerol
- *BABB*: Benzyl alcohol: Parts benzyl benzoate=1:2, mix

Procedure

Making probe

1. Cut plasmid with appropriate enzymes.
2. Set up a transcription reaction on ice:

Component	Volume
DNA template	500 ng–1 μg
10× Buffer	2 μL
10× Dig mix	2 μL
DTT	2 μL
RNasin	1 μL
T7 RNA polymerase	1 μL
RNAase-free water	to 20 μL
Incubate at 37°C for 4 h.	

3. Gel electrophoresis: Load 0.5 μL of reaction product into wells in the gel.
4. If the probe RNA is well synthesized, use RNeasy mini Kit (Qiagen) to purify the probe.

5. Gel electrophoresis: Load 0.5 μL RNA in 1% agarose gel to check the RNA quality.
6. Put product to 500 μL Hyb+ buffer to make a stock solution and store at –20°C. Try around 1:10 dilution of this stock solution.

Fixing and Permeabilization

1. Dechorionate manually or with 1 mg/mL pronase. If using pronase, transfer embryos to a glass beaker and add pronase, incubate at 28.5°C until the chorions break down (around 10–15 min). Rinse with embryo medium.
2. Fix embryos in freshly diluted 4% formaldehyde in PBST at RT for 4 h or at 4°C overnight, wash twice with PBST.
3. Dehydrate through 50% (in PBST) and 100% methanol, replace once with methanol and store at –20°C. The embryos are good for years.
4. Rehydrate through 50% methanol and PBST; wash once with PBST.
5. Permeabilize with proteinase K in PBST at RT (test each new batch first).
6. Embryos younger than bud stage are very fragile. You can either skip this step or use much less proteinase K (PK). The older the embryos, the better they can sustain this digestion (also they need this step more). Use the following conditions (all embryos are fixed in formalin):

Stage	Amount	Duration
4 SS	1 μL PK in 1 mL PBST	1 min
8–14 SS	1 μL PK in 1 mL PBST	2 min
20–25 hpf	1 μL PK in 1 mL PBST	10 min
30–34 hpf	2 μL PK in 1 mL PBST	10 min
50 hpf	5 μL PK in 1 mL PBST	10 min

7. Rinse with PBST. Refix in 4% formaldehyde at RT for 20 min.
8. Rinse with PBST, 3 × 5 min.

Hybridization

1. Add Hyb$^-$ buffer, transfer to 65°C water bath, and incubate for 5 min at 65°C–70°C.
2. Add Hyb$^+$ buffer, and incubate for one-half hour at 65°C–70°C.
3. Denature probe (diluted in Hyb$^+$ buffer) 5 min 65°C–70°C. Titrate probe amount on the first try. Usually start with 1:10 dilution.
4. Hybridize with probe overnight at 65°C–70°C.
5. Wash with preheated buffers:
   50% formamide/2× SSCT, 2 × 30 min at 65°C–70°C
   2 × SSCT, 10 min at 65°C
   0.2 × SSC/PBST, 2 × 30 min at 65°C.
6. Bring to RT shaker, 2 × 15 min PBST.

Immunoreaction

1. 0.5 h at RT in 1× blocking solution.
2. Replace with 1× blocking solution plus anti-Dig-AP antibody (1:3000). Overnight at 4°C or RT for 2 h.
3. Wash with 1× blocking solution, RT, 0.5 h.
4. Wash with MABT, 3 × 30 min at RT or overnight at 4°C.

Staining

1. 2 × 5 min in staining buffer.
2. 30 min to overnight staining with BM purple AP substrate.
3. Wash with PBST, 5 min.
4. Fix in 4% formaldehyde, 20 min
5. Wash with PBST, 5 min.

Imaging

1. Clear embryos in 50%, then 100% glycerol.
2. Use stereoscope (light from above) and digital camera to take images.
3. Data analysis.

12.4.4.5 Live Imaging Using Transgenic Fish

Live imaging is a powerful tool to understand the cell physiology *in vivo*, especially in zebrafish as the embryo is transparent and easy to label cells or subcellular organelles by transgenic tools or overexpression of fusion protein. Here, we provide a protocol to visualize pronephros migration as well as the ciliary movement. For ciliary movement recording, high-speed camera with 250 frames per second is required. The frequency of pronephric cilia movement is 20 ± 3 Hz.[30] This protocol can also be applied to observe cilia movement in other organs (e.g., neural tube and olfactory).

Step-by-step protocol is as follows.

Materials and Equipment

- Low-melting point agarose, 2% in embryo medium
- Tricaine (Sigma, E10521); 25× stock solution: 4 mg/mL, pH 7; add 4 mL into 100 mL of embryo medium to make a working solution
- PTU (Phenylthiourea, Sigma Aldrich, P7629): 0.003% in embryo medium
- BDM (2,3-butanedione monoxime, Sigma, cat. no. B0753-25G), final concentration 20 mM
- Fast-speed camera for cilia imaging
- Microscope equipped with a 40× (for renal cells imaging) or 63× (for cilia imaging) water-immersion lens
- Microscope slides

Protocol

1. At 24 hpf, treat embryos with PTU in a Petri dish to inhibit melanogenesis and incubate until the required stage.
2. Melt low-melting point agarose in a microwave oven and keep at 37°C.
3. Add Tricaine to anesthetize embryos.
4. (Optional, only for cilia imaging) To stop heartbeat and circulation, add BDM to the embryo medium for 1 min.
5. Immobilize/orient embryo on its side in 2% low-melting point agarose on the microscope slide.
6. Transfer the slide to the stage of a confocal microscope. Add a drop of water on top of the embryo before observing with water immersion lens.
7. Image. For pronephros, take image at 20–45 min intervals in a z-series stack using a 40× water inversion lens. For renal cilia, using a high-speed digital video camera to image roughly the middle part of the pronephric duct. Images of moving cilia are acquired at 250 frames per second for the duration of 1 s and analyzed in slow motion (15 frames per second) to determine the frequency of cilia movement.
8. Image process and data analysis. For pronephros, using Adobe Photoshop to adjust contrast and reassemble frames into QuickTime movies using Graphic Converter (Lemke software). For renal cilia, analyze the film in slow motion (15 frames per second) to determine the frequency of cilia movement.

12.4.5 Characterization of Renal Function in Zebrafish

The kidney plays a crucial role in filtering metabolic waste from the blood and recovering solutes to sustain body homeostasis. In human and other mammalian models, kidney function can be evaluated by a series of blood and urinary tests. High levels of urea, creatinine, and abnormal salt concentrations in blood or abnormal levels of protein, blood, bacteria, and sugar present in urine samples indicate defects in kidney function. Such tests normally require approximately 30 mL of urine or 5–10 mL of blood. It has been difficult to use these assays with the zebrafish, as it is impossible to collect sufficient blood or urine to perform the assay. Instead, by using a fluorescent dye clearance assay, we are able to monitor the filtration and excretion of the fluorescent dye from the blood via the kidney and, thus, evaluate the kidney function (Figure 12.2e).

Materials and Equipment

- Embryo medium
- Embryos
- Borosilicate glass capillaries: WPI 1B100-4
- A micropipette puller: Sutter P-97
- Microinjector: WPI PV820 Pneumatic PicoPump

- Plastic molds to make agarose-coated injection chambers[44]
- 24-well plates for tissue culture
- PTU (Phenylthiourea, Sigma Aldrich, P7629): 0.003% in embryo medium
- Tricaine (Sigma, E10521); 25× Stock Solution: 4 mg/mL, pH 7; add 4 mL into 100 mL of embryo medium to make a working solution
- Fitc-dextran (40KD) or rhodamine-dextran: 50 mg/mL in ddH_2O
- Software, ImageJ (NIH)
- Stereoscope for microinjection
- Fluorescence Stereoscope with a FITC or GFP filter

Procedure

Before injection

1. Prepare the microinjection needles the day before injection. Follow the manual instruction of the manufacturer.
2. Prepare the agarose-coated injection chambers. Place the plastic molds (teeth down) in the 100 mm Petri dish. Pour approximately 20 mL of 1% agarose in embryo medium into the dish. The plastic mold will float on the surface of the agarose. Wait until completely solidified. Wrap the Petri dish with parafilm, and store at 4°C. Warm the chambers to the room temperature before injection.

Injection

3. Cut the needle tip with a blade. Load the needle with 5 µL of Fitc-dextran at the final concentration of 10 mg/mL.
4. At 24 hpf, treat embryos with PTU in a Petri dish to inhibit melanogenesis and incubate until 72 hpf.
5. At 72 hpf, add Tricaine to anesthetize embryos.
6. Inject Fitc-dextran. Place embryos left side up on an agarose-coated injection chamber, and inject 1 nL of Fitc-dextran into the heart by piercing the pericardium.
7. Transfer injected embryos to a 24-well plate containing 1 mL of embryo medium with PTU in each well. Incubate at 28.5°C for 5 h.

Imaging and Quantification

8. Anesthetize embryos with Tricaine, and position them lateral side up. Image the entire body using a stereomicroscope equipped with a UV light source and Fitc or GFP filter.
9. Place the embryos back in assigned wells to recover.
10. At 48 h postinjection, acquire a second round of images as described above.

11. Quantify fluorescence intensity of each embryo at 5 and 48 h postinjection by specifying a 100 × 100 pixel region of interest (roi) in ImageJ. Position the heart in the center of the roi and measure the fluorescence intensity.

12.5 PATHWAY ANALYSIS: CHEMICAL MODIFIER SCREEN

It's important to analyze the pathway through which the target protein functions. Genetics and biochemical studies contribute a lot in figuring out signaling pathways by searching for phenotypically similar mutants and interacting proteins. In invertebrate models, people can also use genetic modifier screens to uncover missing pathway members, revealing novel genetic interactions. However, this genetic approach is difficult to carry out in zebrafish. With pharmacological compounds or chemicals, the modifier screen in zebrafish turns to possible and becomes a potentially powerful tool to identify the novel signaling pathway and pinpoint new drug targets.

In the PKD field, a small scale chemical modifier screen has been applied to *pkd2*^{hi4166} mutants, an ADPKD model in zebrafish, and identified HDAC inhibitors trichostatin-A (TSA) and valproic acid (VPA) as suppressors of *pkd2* mutants.[26] The effects of HDAC inhibitors in the suppression of cyst formation in zebrafish were further confirmed in a mouse ADPKD model with conditional inactivation of *Pkd1*. This result suggests that pathogenesis of ADPKD is conserved between zebrafish and mammals and the screening result from zebrafish is applicable to mammalian system.

More recently, a pharmacological screen with a GPCR compound library has identified a downstream signal of cilia in the neural tube and treatment with chemical dipivefrin, an ester prodrug for epinephrine, rescued the body curvature in primary ciliary dyskinesia mutants *zmynd10*$^{-/-}$.[45] These two studies shed light on the mechanism and possible novel treatment discoveries in PKD and the ciliopathy field.

The protocol for chemical screen in zebrafish is as follows.

12.5.1 Chemical Modifier Screen

Materials

- 12-well or 96-well plates
- Chemicals to be screened
- Embryo medium
- DMSO
- Embryos

Procedure

1. Array embryos into 12-well plates with 25 embryos per well. 12-well plates are chosen here as the embryos are from heterozygous cross and only 25% embryos are expected as homozygous mutants. For treating wildtype embryos with chemicals, use 96-well plate and 3 embryos per well instead.
2. At 50% epiboly stage, chemicals were added at a starting concentration of 10 × IC_{50} or 10 μM with 1% DMSO. Use embryos in 1% DMSO as a control at each time.

3. Analyze phenotypes, including left-right asymmetry; body curvature; and cystic kidney at 30 hpf, 2, 3, and 5 dpf, respectively. Left-right asymmetry was determined by examining the heart position in the embryos at 30 hpf.
4. Take images.
5. (Optional) Titrate: If the chemical causes severe abnormality or lethality in the embryo, decrease the concentration by 5–10 fold. On the contrary, if the chemical barely affects the embryonic development, increase the concentration by 5–10 fold.

12.6 CONCLUDING REMARKS

To understand the pathogenesis of PKD, a lot of mutants and involved genes have been identified. However, it's still a big challenge to figure out the roles and mechanisms of these gene products in cyst formation. To solve this problem, more downstream factors or signaling pathways of these gene products need to be identified. With the rapid development of sequencing, gene-editing, and imaging technologies, as well as the convenience to carry out chemical modifier screen, the zebrafish model will provide more insights into the mechanisms of PKD in the future.

ACKNOWLEDGMENTS

This project was supported by grants from the National Key Research and Development Program of China (2017YFA0104600) and the National Natural Science Foundation of China (31771597 and 31571515).

REFERENCES

1. Consortium, T.E.P.K.D. 1994. The polycystic kidney disease 1 gene encodes a 14 kb transcript and lies within a duplicated region on chromosome 16. *Cell 77*, 881–894.
2. Mochizuki, T., Wu, G., Hayashi, T., Xenophontos, S.L., Veldhuisen, B., Saris, J.J., Reynolds, D.M., Cai, Y., Gabow, P.A., Pierides, A. et al. 1996. PKD2, a gene for polycystic kidney disease that encodes an integral membrane protein. *Science (New York, NY) 272*, 1339–1342.
3. Lu, H., Galeano, M.C.R., Ott, E., Kaeslin, G., Kausalya, P.J., Kramer, C., Ortiz-Brüchle, N., Hilger, N., Metzis, V., Hiersche, M. et al. 2017. Mutations in DZIP1L, which encodes a ciliary-transition-zone protein, cause autosomal recessive polycystic kidney disease. *Nature Genetics 49*, 1025–1034.
4. Onuchic, L.F., Furu, L., Nagasawa, Y., Hou, X., Eggermann, T., Ren, Z., Bergmann, C., Senderek, J., Esquivel, E., Zeltner, R. et al. 2002. PKHD1, the polycystic kidney and hepatic disease 1 gene, encodes a novel large protein containing multiple immunoglobulin-like plexin-transcription–factor domains and parallel beta-helix 1 repeats. *The American Journal of Human Genetics 70*, 1305–1317.
5. Ward, C.J., Hogan, M.C., Rossetti, S., Walker, D., Sneddon, T., Wang, X., Kubly, V., Cunningham, J.M., Bacallao, R., Ishibashi, M. et al. 2002. The gene mutated in autosomal recessive polycystic kidney disease encodes a large, receptor-like protein. *Nature Genetics 30*, 259–269.
6. Barr, M.M., and Sternberg, P.W. 1999. A polycystic kidney-disease gene homologue required for male mating behaviour in *C. elegans*. *Nature 401*, 386–389.

7. Yoder, B.K., Hou, X., and Guay-Woodford, L.M. 2002. The polycystic kidney disease proteins, polycystin-1, polycystin-2, polaris, and cystin, are co-localized in renal cilia. *Journal of the American Society of Nephrology: JASN 13*, 2508–2516.
8. Cai, Y., Fedeles, S.V., Dong, K., Anyatonwu, G., Onoe, T., Mitobe, M., Gao, J.-D., Okuhara, D., Tian, X., Gallagher, A.-R. et al. 2014. Altered trafficking and stability of polycystins underlie polycystic kidney disease. *Journal of Clinical Investigation 124*, 5129–5144.
9. Yoshiba, S., Shiratori, H., Kuo, I.Y., Kawasumi, A., Shinohara, K., Nonaka, S., Asai, Y., Sasaki, G., Belo, J.A., Sasaki, H. et al. 2012. Cilia at the node of mouse embryos sense fluid flow for left-right determination via Pkd2. *Science (New York, NY) 338*, 226–231.
10. Harris, P.C. 2009. 2008 Homer W. Smith ward: Insights into the pathogenesis of polycystic kidney disease from gene discovery. *Journal of the American Society of Nephrology 20*, 1188–1198.
11. Wingert, R.A., and Davidson, A.J. 2008. The zebrafish pronephros: A model to study nephron segmentation. *Kidney International 73*, 1120–1127.
12. Wingert, R.A., Selleck, R., Yu, J., Song, H.-D., Chen, Z., Song, A., Zhou, Y., Thisse, B., Thisse, C., McMahon, A.P. et al. 2007. The cdx genes and retinoic acid control the positioning and segmentation of the zebrafish pronephros. *PLOS Genetics 3*, 1922–1938.
13. Drummond, I. 2003. Making a zebrafish kidney: A tale of two tubes. *Trends in Cell Biology 13*, 357–365.
14. Drummond, I.A., Majumdar, A., Hentschel, H., Elger, M., Solnica-Krezel, L., Schier, A.F., Neuhauss, S.C., Stemple, D.L., Zwartkruis, F., Rangini, Z. et al. 1998. Early development of the zebrafish pronephros and analysis of mutations affecting pronephric function. *Development (Cambridge, England) 125*, 4655–4667.
15. Driever, W., Solnica-Krezel, L., Schier, A.F., Neuhauss, S.C., Malicki, J., Stemple, D.L., Stainier, D.Y., Zwartkruis, F., Abdelilah, S., Rangini, Z. et al. 1996. A genetic screen for mutations affecting embryogenesis in zebrafish. *Development 123*, 37–46.
16. Gaiano, N., Amsterdam, A., Kawakami, K., Allende, M., Becker, T., and Hopkins, N. 1996. Insertional mutagenesis and rapid cloning of essential genes in zebrafish. *Nature 383*, 829–832.
17. Haffter, P., Granato, M., Brand, M., Mullins, M.C., Hammerschmidt, M., Kane, D.A., Odenthal, J., van Eeden, F.J., Jiang, Y.J., Heisenberg, C.P. et al. 1996. The identification of genes with unique and essential functions in the development of the zebrafish, *Danio rerio. Development 123*, 1–36.
18. Kawakami, K., Takeda, H., Kawakami, N., Kobayashi, M., Matsuda, N., and Mishina, M. 2004. A transposon-mediated gene trap approach identifies developmentally regulated genes in zebrafish. *Developmental Cell 7*, 133–144.
19. Sun, Z., Amsterdam, A., Pazour, G.J., Cole, D.G., Miller, M.S., and Hopkins, N. 2004. A genetic screen in zebrafish identifies cilia genes as a principal cause of cystic kidney. *Development 131*, 4085–4093.
20. Stainier, D.Y.R., Raz, E., Lawson, N.D., Ekker, S.C., Burdine, R.D., Eisen, J.S., Ingham, P.W., Schulte-Merker, S., Yelon, D., Weinstein, B.M. et al. 2017. Guidelines for morpholino use in zebrafish. *PLOS Genetics 13*, e1007000.
21. Chang, N., Sun, C., Gao, L., Zhu, D., Xu, X., Zhu, X., Xiong, J.W., and Xi, J.J. 2013. Genome editing with RNA-guided Cas9 nuclease in zebrafish embryos. *Cell Research 23*, 465–472.
22. Hruscha, A., Krawitz, P., Rechenberg, A., Heinrich, V., Hecht, J., Haass, C., and Schmid, B. 2013. Efficient CRISPR/Cas9 genome editing with low off-target effects in zebrafish. *Development 140*, 4982–4987.
23. Varshney, G.K., Pei, W., LaFave, M.C., Idol, J., Xu, L., Gallardo, V., Carrington, B., Bishop, K., Jones, M., Li, M. et al. 2015. High-throughput gene targeting and phenotyping in zebrafish using CRISPR/Cas9. *Genome Research 25*, 1030–1042.

24. Mangos, S., Lam, P.Y., Zhao, A., Liu, Y., Mudumana, S., Vasilyev, A., Liu, A., and Drummond, I.A. 2010. The ADPKD genes *pkd1*a/b and *pkd2* regulate extracellular matrix formation. *Disease Models & Mechanisms 3*, 354–365.
25. Schottenfeld, J., Sullivan-Brown, J., and Burdine, R.D. 2007. Zebrafish curly up encodes a Pkd2 ortholog that restricts left-side-specific expression of southpaw. *Development 134*, 1605–1615.
26. Cao, Y., Semanchik, N., Lee, S.H., Somlo, S., Barbano, P.E., Coifman, R., and Sun, Z. 2009. Chemical modifier screen identifies HDAC inhibitors as suppressors of PKD models. *Proceedings of the National Academy of Sciences of the United States of America 106*.
27. Sutter, M., and Germino, G.G. 2003. Autosomal dominant polycystic kidney disease: Molecular genetics and pathophysiology. *Journal of Laboratory and Clinical Medicine 141*, 91–101.
28. MacRae Dell, K., Nemo, R., Sweeney, W.E., Jr., and Avner, E.D. 2004. EGF-related growth factors in the pathogenesis of murine ARPKD. *Kidney International 65*, 2018–2029.
29. Nadasdy, T., Laszik, Z., Lajoie, G., Blick, K.E., Wheeler, D.E., and Silva, F.G. 1995. Proliferative activity of cyst epithelium in human renal cystic diseases. *Journal of the American Society of Nephrology 5*, 1462–1468.
30. Kramer-Zucker, A.G., Olale, F., Haycraft, C.J., Yoder, B.K., Schier, A.F., and Drummond, I.A. 2005. Cilia-driven fluid flow in the zebrafish pronephros, brain and Kupffer's vesicle is required for normal organogenesis. *Development (Cambridge, England) 132*, 1907–1921.
31. Vasilyev, A., Liu, Y., Mudumana, S., Mangos, S., Lam, P.-Y., Majumdar, A., Zhao, J., Poon, K.-L., Kondrychyn, I., Korzh, V. et al. 2009. Collective cell migration drives morphogenesis of the kidney nephron. *PLOS Biology 7*, e9.
32. Wilson, P.D., Sherwood, A.C., Palla, K., Du, J., Watson, R., and Norman, J.T. 1991. Reversed polarity of Na(+)-K(+)-ATPase: Mislocation to apical plasma membranes in polycystic kidney disease epithelia. *The American Journal of Physiology 260*, F420–430.
33. Xu, W., Jin, M., Hu, R., Wang, H., Zhang, F., Yuan, S., and Cao, Y. 2017. The Joubert syndrome protein Inpp5e controls ciliogenesis by regulating phosphoinositides at the apical membrane. *Journal of the American Society of Nephrology 28*, 118–129.
34. Nechiporuk, T., Fernandez, T.E., and Vasioukhin, V. 2007. Failure of epithelial tube maintenance causes hydrocephalus and renal cysts in Dlg5-/- mice. *Developmental Cell 13*, 338–350.
35. Omori, Y., and Malicki, J. 2006. *oko meduzy* and related *crumbs* genes are determinants of apical cell features in the vertebrate embryo. *Current biology: CB 16*, 945–957.
36. Skouloudaki, K., Puetz, M., Simons, M., Courbard, J.R., Boehlke, C., Hartleben, B., Engel, C., Moeller, M.J., Englert, C., Bollig, F. et al. 2009. Scribble participates in Hippo signaling and is required for normal zebrafish pronephros development. *Proceedings of the National Academy of Sciences of the United States of America 106*, 8579–8584.
37. Karner, C.M., Chirumamilla, R., Aoki, S., Igarashi, P., Wallingford, J.B., and Carroll, T.J. 2009. Wnt9b signaling regulates planar cell polarity and kidney tubule morphogenesis. *Nat Genet 41*, 793–799.
38. Cao, Y., Park, A., and Sun, Z. 2010. Intraflagellar transport proteins are essential for cilia formation and for planar cell polarity. *J Am Soc Nephrol 21*, 1326–1333.
39. Fischer, E., Legue, E., Doyen, A., Nato, F., Nicolas, J.F., Torres, V., Yaniv, M., and Pontoglio, M. 2006. Defective planar cell polarity in polycystic kidney disease. *Nature Genetics 38*, 21–23.
40. Saburi, S., Hester, I., Fischer, E., Pontoglio, M., Eremina, V., Gessler, M., Quaggin, S.E., Harrison, R., Mount, R., and McNeill, H. 2008. Loss of Fat4 disrupts PCP signaling and oriented cell division and leads to cystic kidney disease. *Nature Genetics 40*, 1010–1015.

41. Simons, M., Gloy, J., Ganner, A., Bullerkotte, A., Bashkurov, M., Kronig, C., Schermer, B., Benzing, T., Cabello, O.A., Jenny, A. et al. 2005. Inversin, the gene product mutated in nephronophthisis type II, functions as a molecular switch between Wnt signaling pathways. *Nature Genetics 37*, 537–543.
42. Brzica, H., Breljak, D., Ljubojevic, M., Balen, D., Micek, V., Anzai, N., and Sabolic, I. 2009. Optimal methods of antigen retrieval for organic anion transporters in cryosections of the rat kidney. *Arhiv za higijenu rada i toksikologiju 60*, 7–17.
43. Inoue, D., and Wittbrodt, J. 2011. One for all--a highly efficient and versatile method for fluorescent immunostaining in fish embryos. *PLOS ONE 6*, e19713.
44. Westerfield, M. 2000. *The Zebrafish Book: A Guide for the Laboratory Use of Zebrafish (Danio rerio)*, Eugene, OR: University of Oregon Press.
45. Zhang, X., Jia, S., Chen, Z., Chong, Y.L., Xie, H., Feng, D., Wu, X., Song, D.Z., Roy, S., and Zhao, C. 2018. Cilia-driven cerebrospinal fluid flow directs expression of urotensin neuropeptides to straighten the vertebrate body axis. *Nature Genetics 50*, 1666–1673.
46. Zhou, W., Boucher, R.C., Bollig, F., Englert, C., and Hildebrandt, F. 2010. Characterization of mesonephric development and regeneration using transgenic zebrafish. *American Journal of Physiology Renal Physiology 299*, F1040–1047.
47. Borovina, A., Superina, S., Voskas, D., and Ciruna, B. 2010. Vangl2 directs the posterior tilting and asymmetric localization of motile primary cilia. *Nature Cell Biology 12*, 407–412.
48. Austin-Tse, C., Halbritter, J., Zariwala, M.A., Gilberti, R.M., Gee, H.Y., Hellman, N., Pathak, N., Liu, Y., Panizzi, J.R., Patel-King, R.S. et al. 2013. Zebrafish ciliopathy screen plus human mutational analysis identifies C21orf59 and CCDC65 defects as causing primary ciliary dyskinesia. *American Journal of Human Genetics 93*, 672–686.

13 Investigation of DNA Methylation in Autosomal Dominant Polycystic Kidney Disease

Ewud Agborbesong and Xiaogang Li

CONTENTS

13.1 INTRODUCTION

Autosomal dominant polycystic kidney disease (ADPKD) is the most common inherited kidney disorder, with an incidence of 1 in 500 to 1 in 1000 individuals.[1–4] Characterized by the progression of fluid-filled cysts, this disease results in the dramatic enlargement of the kidneys; with roughly 50% of the patients progressing to end-stage renal disease (ESRD).[5,6] Cyst initiation and progression in ADPKD is a complex process, attributed to underlying germline mutations in the *PKD1* and *PKD2* genes, which encode polycystin-1 (PC1) and polycystin-2 (PC2), respectively.[3,4] Despite the identification of these major causative genes, the mechanisms of cyst formation remain unclear. Moreover, reports suggest that the landscape of PKD mutations alone is insufficient[7,8] to explain the pervasive gene expression changes and alterations to cellular function observed in ADPKD patients.[9]

In recent years, studies have provided evidence for the role of epigenetic alterations in the pathogenesis of ADPKD. Epigenetics is defined as a heritable change in the pattern of gene expression that is not mediated by alterations in the DNA sequence.[10,11] Epigenetic mechanisms include but are not limited to chromatin folding, packaging of DNA around nucleosomes, histone tail modifications, and DNA methylation.[12] More so, there is increasing recognition for the influence of regulatory, small RNAs and non-coding RNAs on gene transcription as a mechanism of epigenetic gene regulation.[13,14] It is important to mention that a single or combination of these epigenetic modifications could influence gene transcription at any given time. Among these, DNA methylation is the most common epigenetic modification, playing a major role in the regulation of transcriptional activity.[10,11]

While the DNA sequence is more or less permanent, in ADPKD, reports indicate that via DNA methylation, gene expression and cellular functions[15] are subject to differential regulation. For example, it was found that the *PKD1* gene and other genes in cystic renal cells related to ion transport and cell adhesion are hypermethylated in their gene body region, correlating with a downregulation of their expression.[15] Furthermore, treatment with DNA-methylation inhibitors significantly repressed cyst growth concomitant with a clear increase in the level of *Pkd1* expression.[15] These findings implicate DNA methylation–mediated gene silencing as a key mechanism underlying cystogenesis. In addition, it also suggests that the DNA-methylation status can be used as a novel epigenetic biomarker and therapeutic target in ADPKD. Currently, the diagnosis of ADPKD is principally based on the detection of kidney cysts via ultrasound imaging,[4,16] which has limited sensitivity in patients with a mild-disease phenotype.[16] Interestingly, in cancer studies, DNA-methylation marks appear early, are tissue-specific,[17] and can be used for disease diagnosis.[18,19] Therefore, we could predict that DNA methylation on specific gene(s) could be used as potential biomarkers in ADPKD, for renal cyst development, altered renal function, and ultimately disease progression.[20]

In this chapter, we mainly focus on the roles of DNA methylation in the human genome and the methods used to screen for DNA methylation, including information about methodologies in epigenome-wide and gene-specific screening, respectively. Note that these methods are often proprietary and/or kit based, involving newer technologies that are constantly evolving. We will discuss these methods along with advantages and disadvantages of their use. Overall, this chapter provides an overview of DNA methylation, insights on sample preparation, and necessary resources to select and design appropriate methods for DNA-methylation screening experiments for ADPKD research.

13.2 DNA METHYLATION

DNA methylation is the first identified and the major epigenetic factor to influence gene expression. It involves the covalent transfer of a methyl group from S-adenyl methionine (SAM) to the C-5 position of the cytosine of DNA to form 5-methylcytosine (5mC).[21,22] The majority of DNA methylation occurs on cytosines that precede a guanine nucleotide in areas of the genome commonly referred to as CpG sites.[21] In general, methylation of DNA on promoter regions is a gene-silencing mechanism that blocks transcription factor binding by packaging the chromatin or recruiting co-repressors.[23]

DNA methylation is catalyzed by a family of DNA methyltransferases (DNMTs): DNMT1, DNMT3a, DNMT3b, and DNMT3L,[21,22] which differ in structure and function. Apart from DNMT2, which has very weak activity toward DNA methylation,[24] all DNMTs comprise a C-terminal catalytic domain and an N-terminal regulatory domain. DNMT3a and DNMT3b are extremely similar in structure and function, capable of methylating both naked and hemi-methylated DNA. For this reason, they are referred to as *de novo* DNMTs. Localized at the replication fork, the ubiquitously expressed DNMT1 displays a preference for hemi-methylated DNA, copying methylation patterns from parental DNA established by the DNMT3 family (DNMT 3a and DNMT 3b) onto newly synthesized DNA during DNA replication[21,22] and DNA repair.[25] As such, DNMT1 functions to maintain DNA methylation in our genome. Like DNMT1, DNMT3a is also ubiquitously expressed and mice lacking DNMT3a die at about 4 weeks of age. DNMT3b on the other hand is poorly expressed in the majority of differentiated tissues[26] and knockout of DNMT3b induces embryonic lethality.[27] These animal studies suggest that DNMT3a is required for normal cellular differentiation, while DNMT3b is required during early development. The final member of the DNMT family, DNMT3L, lacks the catalytic domain present in the other enzymes. However, it acts as a cofactor to DNMT3a and DNMT3b.[28–30] Mouse studies show that DNMT3L is expressed during gametogenesis and is required for the establishment of maternal genomic imprinting[31,32] and mice lacking DNMT3L die early during development.[32]

In general, DNA methylation plays an important role during development and disease progression. Under normal conditions, the pattern of DNA methylation changes as a result of a dynamic process involving both *de novo* DNA methylation and demethylation.[21] As a consequence, stable and unique DNA-methylation patterns develop and regulate tissue-specific gene transcription in differentiated cells. On the other hand, in disease conditions, these methylation patterns are liable to change; either preceding the disease or occurring as a consequence of it. In kidney diseases, for example, site-specific DNA methylation changes have been detected in patients with chronic kidney diseases (CKD)[33] and diabetic nephropathy.[34] Since DNA-methylation patterns are known to be fairly stable and unique in differentiated cells, profiling the methylation pattern of genes could extend our understanding of the pathophysiology of ADPKD.

13.3 SAMPLE PROCESSING FOR DNA-METHYLATION ANALYSIS

The DNA-methylation pattern of genes could be examined in different sample types, such as formalin-fixed paraffin-embedded (FFPE) or fresh-frozen tissues,[35] but the most easily available sources are body fluids, including blood and urine.[36] The types of samples used in the assays of DNA methylation are critical factors in the quality of data obtained. Likewise, the preparation of DNA (both quality and quantity) is equally important. To obtain the latter, particular care is required when processing

DNA samples for methylation assays. In this section, we will briefly discuss the application of cell-free DNA (cfDNA) and kidney biopsy/tissue for DNA-methylation studies. In addition, we will highlight possible advantages and shortcomings for each sample choice, providing recommendations for optimal sample processing.

13.3.1 Cell-Free DNA (cfDNA)

Cell-free DNA (cfDNA) in body fluids is an analyte which originates from apoptosis of nucleated cells in healthy individuals. In diseased conditions, such as tumors, cfDNA could be generated from apoptosis, necrosis, and direct secretion.[37,38] cfDNAs are quite diverse, varying in size between 180 and 10,000 bp.[39] With a half-life of about 16 minutes, cfDNA can be complexed with cellular or noncellular components, such as glycoproteins, resulting in an increase in its stability.[40]

In recent years, cfDNAs have gained great interest in basic and clinical research, in which they are used to determine the states of diseases. Numerous studies have demonstrated the potential of genetic and epigenetic alterations in cfDNA as biomarkers.[41] However, there seems to be no consensus on the isolation methods. The methods reported so far vary from in-house protocols to the use of commercially available kits such as magnetic beads or silica-gel membrane technology.[42] It is important to mention that since cfDNA fragments are usually short, the choice of commercial DNA isolation kits that are able to isolate low-molecular weight DNA fragments should be taken into consideration.[43] Regardless of the method used, there is sufficient information to suggest that the factors (1) choice of specimen type (blood or urine); (2) processing time; (3) centrifugation time and speed, and number of spins; (4) cfDNA isolation method; and (5) cfDNA quantification method, influence the yield and quality, and promote optimal and reproducible analysis of cfDNA.[44]

DNA-methylation analyses usually require bisulfite conversion, thus the need for sufficiently high concentrations of DNA. However, with the usually low amounts of free-circulating DNA, appropriate measures need to be taken to maximize DNA yield and quality. In the subsequent section, we briefly discuss sample processing of circulating cell-free DNA (ccfDNA) from blood/plasma and urine, with recommendations for optimal sample recovery.

13.3.1.1 Circulating Cell-Free DNA (ccfDNA)

Circulating cell-free DNA (ccfDNA) is DNA that is found in the bloodstream and can be captured as a biological sample, such as blood or serum, for diverse analysis. In healthy individuals, the quantity of ccfDNA is typically low,[37] ranging from 1.8 to 44 ng/mL[45] but these levels can increase following exercise.[45,46] In diseased conditions, such as tumors, the levels of ccfDNA could increase depending on the stage of the disease.[47] In addition, the amount of ccfDNA also depends on the activity of DNase enzymes in the circulatory system. In cancer studies for example, it has been shown that the increase in ccfDNA concentration is related to a decrease in DNase I enzyme activity.[48]

Several diverse extraction methods for ccfDNA exit, most of which are kit based. Of these, the QIAamp Circulating Nucleic Acid Kit (Qiagen) appears to be the most widely used, and comparable studies have shown that this kit led to the highest yield of total ccfDNA.[49] This kit provides variable steps for ccfDNA isolation from plasma or serum, depending on the volume of start-up sample. However, because of the huge individual to individual variance in the yield, it is recommended to use as much plasma or serum as possible. In the case of large sample size, the duration of ccfDNA isolation can be reduced by the use of automated ccfDNA extraction methods. To ensure maximum yield and quality of ccfDNA, the following recommendations could be applied.

First and foremost, the choice of specimen is to be considered. Studies have found ccfDNA isolated from serum to have significantly higher integrity than that from plasma,[50,51] a phenomenon attributed to the presence of high-molecular-weight DNA from hematopoietic cells in serum.[50] Next, the processing time between blood collection and centrifugation has been shown to have an effect on ccfDNA, with ccfDNA levels increasing over time prior to centrifugation of EDTA-stabilized blood.[52] As a result, specialized tubes containing a formaldehyde-free preservative reagent that inhibits the degradation of ccfDNA by nucleases, such as Cell-Free DNA BCT (Streck), are now made available to permit the storage of whole blood at room temperature for several days prior to centrifugation.[53] At this stage, the efficacy of removal of contaminating cellular DNA from plasma is dependent on the blood volume, the centrifugation speed, and time. For volumes of blood 7 mL or less, centrifugation at 800 × g for 10 minutes is sufficient.[54] However, for samples above 7 mL, a second centrifugation (16,000 × g for minutes) of plasma is required.[54,55] In addition to centrifugation speed and time, pipetting efficacy is also important to avoid collecting the dead or dying cells from the buffy coat as these may dilute your ccfDNA. Finally, once the DNA is obtained, it has to be quantified. Studies comparing quantitation methods report a correlation in total ccfDNA levels between qPCR and fluorometry.[56]

13.3.1.2 Urinary Circulating Cell-Free DNA (uccfDNA)

Urinary circulating cell-free DNA (uccfDNA) originates from cell-free DNA in blood that crosses the kidney barriers, or from cells shed into urine from the genitourinary tract.[57] Given its passage through the kidney, where cells associated with kidney function are located, urine represents a logical source to obtain DNA used for DNA-methylation research in kidney diseases.[58] The advantage of using urine over blood is its noninvasive sampling method without the need for medical staff. In addition, unlike serum or plasma collections, which require specialized equipment, the collection of urine only requires a sterile container.

In contrast to blood ccfDNA, there is insufficient data on the quantity and variability of ccfDNA in urine. Fortunately, larger volumes of start-up sample can be obtained and therefore maximize yield. Similar to blood ccfDNA, urine ccfDNA could be isolated using either commercial kits or classical laboratory techniques.[59,60] However, since urine collection does not require specialized tubes, it is advisable to process the sample as soon as possible to avoid contamination from dead cells. In the case of urine processing, samples are centrifuged twice to remove cellular pellet, first at 200 × g for 10 minutes, followed by 1800 × g for 10 minutes. Note that the cellular pellet contains genomic DNA, which could also be used for DNA-methylation assays. From the supernatant, uccfDNA can be isolated by a variety of methods such as the Norgen-based ccfDNA isolation kit and PerkinElmer-based isolation kit, according to the manufacturers' instructions. As for the quantification of uccfDNA, the most commonly used approaches include the spectrophotometric and fluorimetric methods.[43]

As mentioned above, the cellular pellet obtained from urine centrifugation can be used as a source for genomic DNA. To extract the DNA, a Qiagen QIAmp DNA Mini Kit or a comparable genomic DNA isolation method can be used.[58] After the extraction process, the DNA is quantified by using either a spectrophotometer such as the NanoDrop, fluorescent dyes, or by agarose gel electrophoresis. It is important to note that the purity and integrity of the DNA sample is essential for future assays. Therefore, taking into account the advantages and limitations of each of these quantification techniques, it might be helpful to use a combination of two. Once quantification of DNA is completed, the sample is ready for immediate use or can be stored at −20°C for analysis later.

13.3.2 Kidney Tissue/Biopsy

Unlike working with blood and urine samples, using kidney tissue is advantageous in that a larger yield of genomic DNA can be obtained. However, sample collection is invasive and not easy to get. In mouse studies, an appropriate method is used to euthanize the mouse, followed by perfusion of the kidney to get rid of excess blood. Once the kidneys are harvested, the tissue sample should either be processed fresh or flash frozen and stored at −80°C until needed. When working with frozen tissue, tissue should be removed from the freezer and allowed to thaw on ice if necessary. With an appropriate method, such as the phenol/chloroform extraction technique, genomic DNA is extracted from the tissue, quantified, and either used immediately or stored at −20°C.

In humans, formalin-fixed, paraffin-embedded (FFPE) tissue samples, routinely collected for histopathological diagnosis, represent a major resource for clinical studies. In many instances, these samples are the only available source of DNA for epigenetic studies such as DNA methylation.[35,61] However, working with FFPE tissue in DNA-methylation analysis is challenging for a variety of reasons. Though formalin fixation does not alter the methylation status of the genome,[62] this process leads to DNA damage as a result of fragmentation. Depending on the type of methylation assay used, the already fragile DNA could incur further damage; making methylation analysis on FFPE samples even more challenging.[35] Furthermore, the fixation process creates DNA-protein cross-links. Because the quality of DNA extracted affects downstream assays, it is important to break the DNA-protein cross-links and efficiently lyse all proteins.[35,61] A number of commercial kits for the extraction of DNA from FFPE exist; for example, QIAamp DNA FFPE Tissue Kit (Qiagen #56404), with MinElute columns[35] and Puregene Core kit A (Qiagen).[61] It should be noted that in both studies, though the authors adopted their respective protocols, they added slight modifications: longer incubation in proteinase K and RNase, respectively, to improve DNA quality.

In sum, the examination of DNA-methylation status can be performed from different sample types. However, for each sample type, there are significant differences in the amount of purified DNA between studies, a factor which is greatly influenced in part by the amount of start-up sample and by the isolation method. So far, there exist different DNA isolation methods, a wide variety of which are commercial-kit based. Since DNA-methylation analyses usually require bisulfite conversion, a process that can be harsh to DNA, it is therefore very important to adhere to the manufacturer's protocol to optimize quality and quantity, and to minimize variability in recovery of DNA between studies.

13.4 DNA-METHYLATION ANALYSIS TECHNIQUES

There is no one right method for analyzing DNA methylation. However, the appropriate approach for any DNA-methylation analysis will depend upon the goal of the study. As indicated in Figure 13.1, there are currently various methods available for profiling specific genes and genome-wide DNA-methylation patterns; most of which rely on an initial genomic fractionation step. These methods can be classified into three main categories: (1) those based on the bisulfite conversion of genomic DNA,[63] (2) those based

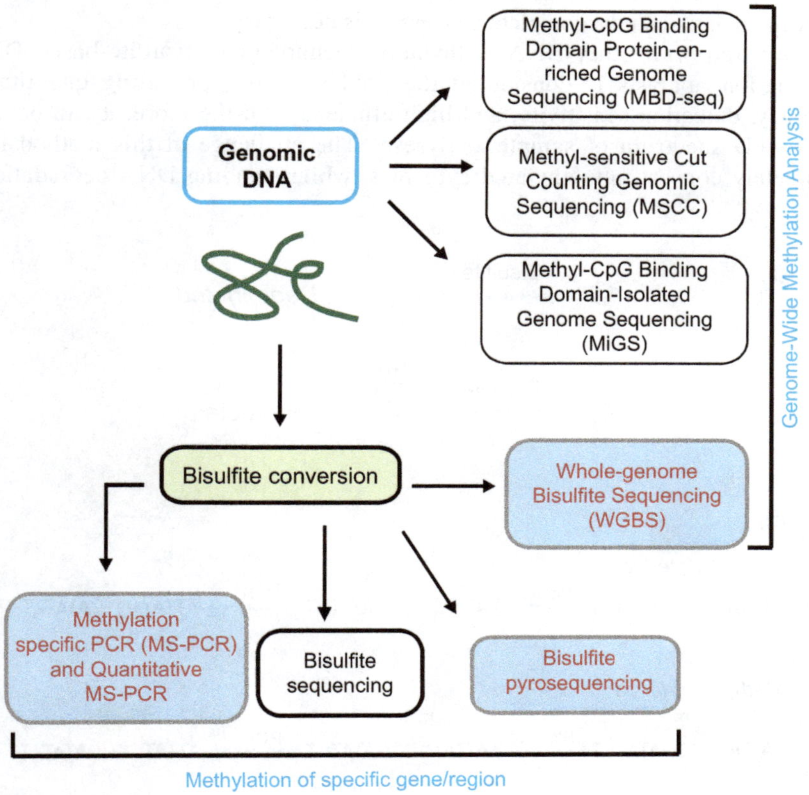

FIGURE 13.1 DNA-methylation analysis, an overview of methods available for profiling specific gene regions and genome-wide DNA-methylation patterns.

on the use of methylation-sensitive restriction endonucleases,[64] and (3) those based on the enrichment of methylated DNA by capturing with antibodies against 5-methylcytosine[65] or methyl-DNA binding protein domain (MBD).[66] Comparison studies show that methods that rely on bisulfite conversion are more specific and, in some cases, can detect changes at single-nucleotide resolution.[67,68] For this reason, we will mainly focus on the process of bisulfite conversion and some of its related techniques.

13.4.1 Bisulfite Conversion

Bisulfite conversion is a popular technique used to study DNA methylation. This technique involves the bisulfite treatment of DNA, leading to the deamination and conversion of unmethylated cytosines to uracils, while leaving the methylated cytosines (5mC) and 5-hydroxymethylcytosines (5hmC) intact. Subsequently, the uracils are converted to thymidines during a PCR reaction by DNA polymerase[63,69] (Figure 13.2). This selective deamination process of bisulfite conversion provides the basis for using different applications, such as PCR or sequencing, to differentiate between methylated and unmethylated cytosine residues, offering up to single-nucleotide-resolution information about the methylated areas of DNA. For a successful DNA-methylation study, complete conversion is necessary.

Compared with other DNA-methylation technologies, bisulfite-based DNA-methylation analysis is considered the gold standard, providing quantitative accuracy, detection sensitivity, and high efficiency. Furthermore, it can be used for a wide spectrum of sample analyses.[69] The challenge of this method is to completely convert unmethylated cytosines, while avoiding DNA degradation.[70]

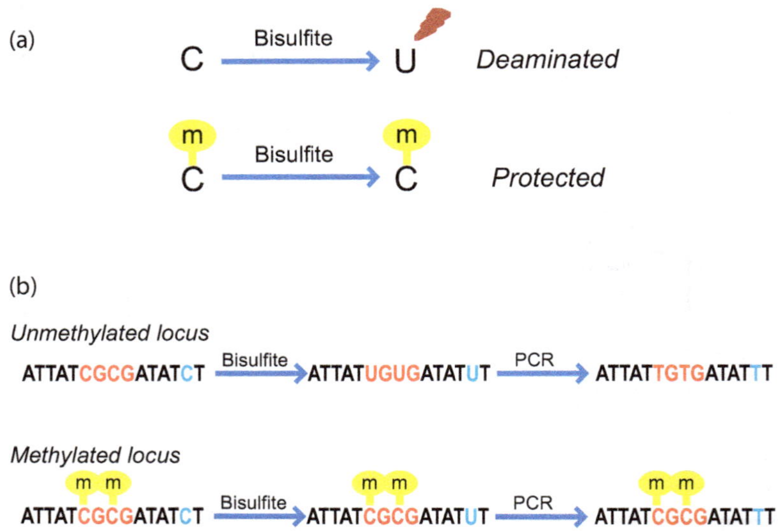

FIGURE 13.2 The application of bisulfite conversion of genomic DNA in studying DNA methylation. (a) Bisulfite treatment converts unmethylated cytosine to uracil. (b) DNA Polymerase substitutes dU for dT after bisulfite conversion.

Because of its typically harsh chemical reaction, involving acidic conditions and high temperatures, the bisulfite conversion process usually generates fragile DNA fragments of different sizes, which could in turn affect the quality of results obtained from the DNA-methylation analysis. To adjust this balance (conversion conditions versus DNA degradation), several commercially available kits are addressing these issues.[69,70] Until all the parameters involved in the bisulfite conversion are fine-tuned, the downstream application can help determine which bisulfite conversion method would be ideal.

Popular downstream methods for analyzing DNA methylation following bisulfite conversion include methylation-specific PCR (MS-PCR), bisulfite sequencing, and methylation-based microarrays. Here, we will discuss the applications of MS-PCR, quantitative real-time methylation-specific PCR (qMS-PCR), pyrosequencing. and whole-genome bisulfite sequencing (WGBS), respectively.

13.4.2 Methylation-Specific PCR (MS-PCR)

MS-PCR is the most convenient method for studying the DNA-methylation status of promoter regions of individual genes. This highly specific and sensitive method uses PCR to amplify bisulfite-modified DNA on specific loci. For proper discrimination between cytosine and thymine bases derived from methylated and unmethylated cytosines following bisulfite conversion, two pairs of primers are designed for the amplification step: one pair specific for the methylated DNA and the other for the unmethylated DNA.[71] Note that the primers for detecting the methylation of alleles should include at least two CpG sites, preferentially at the 3′-end of the sequence, an adequate number of non-CpG cytosines and have similar annealing temperatures. Furthermore, the same CpG sites need to be incorporated into the primers for both the methylated (M pair) and unmethylated (U pair) DNA. For example:

Forward primer sequence in the M pair: 5′ATTAGTTTCGTTTAAGGTTCGA3′
Forward primer sequence in the U pair: 5′ATTAGTTTTGTTTAAGGTTTGA3′

In addition, both methylated and unmethylated DNA would produce amplicons of different sizes. Note that two qPCR reactions are performed, one for each sample, and the relative methylation is calculated based on the difference of their Ct values.[67] Proper optimization of the PCR reaction will provide detection of methylated CpG sites at single-nucleotide resolution.

Forward primer sequence in the M pair: 5′ATTAGTTTCGTTTAAGGTTCGA3′
Forward primer sequence in the U pair: 5′ATTAGTTTTGTTTAAGGTTTGA3′

It is important to point out that 1) standard MS-PCR cannot quantify methylation patterns when both methylated and unmethylated alleles are present in the DNA sample, as such, complete bisulfite conversion is required;[69] and 2) the use of low-quality DNA, limits MS-PCR amplification since bisulfite conversion results in DNA fragmentation.[62] These limitations indicate the necessity for more advanced techniques such as quantitative real-time MS-PCR (qMS-PCR), and pyrosequencing.

13.4.3 Quantitative Real-Time MS-PCR (qMS-PCR)

Quantitative real-time MSP (qMS-PCR) is based on standard MSP. Similar to standard MS-PCR, sequence discrimination occurs during PCR amplification. The primers for the methylated or unmethylated CpG sites are designed to overlap potential sites of DNA methylation.[72] However, unlike standard MS-PCR, qMS-PCR is highly sensitive and specific, able to quantify methylated alleles using fluorescence-based quantitative PCR (qPCR) technology. As a result of this, up to a single-nucleotide resolution may be achieved. In addition, qMS-PCR, it is capable of detecting low-frequency, hypermethylated alleles. The high sensitivity and specificity of this technique, makes, qMS-PCR ideal for the detection of DNA-methylation biomarkers.

Described below is an established protocol for qMS-PCR, adapted from MS-PCR sequencing of urinary cells.[73]

1. DNA is extracted from the desired sample using any comparable method.
2. For methylation analysis, 1–2 μg DNA is treated with sodium bisulfite using the CpGenome kit (Chemicon), or comparable method, according to the manufacturer's instructions. The bisulfite-treated DNA is then used as a template for a quantitative based real-time methylation-specific PCR (qMS-PCR).
3. PCR is performed to probe for the gene promoter region(s) of interest.

   Note: Bisulfite-specific primers to actin should be used as the internal reference gene, and *SssI* (a CpG-specific methylase) treated DNA, also known as a universally methylated DNA, as positive control.

4. qMS-PCR Accuprime Taq polymerase (Invitrogen) should be used if working with the ABI Prism 7000 Sequences Detection System. Alternatively, the fluorescent probe can be adapted based on the detection system available.
5. For a 10 μL PCR volume, add 1 μL of Accuprime buffer, 0.25 μL Accuprime Taq polymerase, and 5 pmol/L forward and reverse primers, 5 pmol/L probe, 0.2 μL Rox reference dye, and water. Sixty cycles of denaturation (95°C for 45 s), annealing (specific primer temperature for 2 min), and extension (72°C for 1 min).

13.4.4 Pyrosequencing

Pyrosequencing is a high-throughput "sequencing by synthesis" technique that can be used to quantify DNA methylation.[74] Prior to sequencing, bisulfite conversion of the DNA is essential. Using PCR, the DNA is amplified and tagged with a biotinylated primer. The PCR product is then mixed with streptavidin beads, and due to biotin's high affinity for streptavidin, they bind to form a complex. The DNA-bound beads are purified, isolated, and then dispensed into pyrosequencing plates that contain sequencing primers. The pyrosequencing reaction takes place in a pyrosequencer in which a mixture of the DNA template, the sequencing primer, and the enzymes DNA polymerase, ATP sulfurylase, luciferase, and apyrase are mixed, together with the substrates adenosine 5′-phosphosulfate (APS) and luciferin.[75]

Pyrosequencing relies on light generation after nucleotides are incorporated in a growing chain of DNA. When the first one of the four deoxynucleotides (dNTPs) is added to the sequencing reaction, DNA polymerase catalyzes its incorporation into the DNA strand, in case there is complementarity. With each incorporation event, a phosphodiester bond is formed between the dNTPs, releasing pyrophosphate (PPi). Unincorporated nucleotides are degraded by apyrase before the next nucleotide dispensation occurs. In the presence of APS, ATP sulfurylase utilizes the PPi to produce ATP, which is used to drive the conversion of luciferin to oxyluciferin by luciferase. The intensity of light produced and detected by this reaction is proportional to the amount of ATP used and reflects the amount of nucleotide incorporated at specified sequences surrounding CpG sites; this is translated as a peak in a pyrogram, with the height of each peak representing the number of nucleotides incorporated. From these pyrograms, methylation percentages can be calculated.[74–76]

Pyrosequencing is useful for DNA methylation for several reasons. Primarily, it is easy to use and the results obtained are of high quality and quantitative.[76] In addition, it is quite sensitive, providing accurate reads with each run[77] and has the ability to detect small differences in methylation (down to 5%).[67] This makes it a good technique for heterogeneous samples where only a fraction of cells have differentially methylated genes of interest. Though the reagents required are fairly expensive, all reactions use the same reagents with only primers varying between different assays. Finally, with the creation of software, such as PyroMark Assay Design by Qiagen, primer design is fairly simple with inbuilt algorithms to reduce the amount of complications, such as dimerization, during runs.

Though pyrosequencing is useful, it also has its shortcomings. First, there are variations in the number of reads obtained for each sample, often influenced by the quality and/or secondary structure of DNA.[78] Second, there is a lack of resolution in homopolymer regions, as identical nucleotide incorporation in a sequence can be blurred across various nucleotide steps.[78] Last, the sensitivity of pyrosequencing can lead to false signals due to errors or perceived failed bisulfite conversions, typically due to mechanical errors or low template availability. Therefore, high-quality primer design and proper template amplification is crucial for each assay.[12]

13.4.5 Whole-Genome Bisulfite Sequencing (WGBS)

WGBS is the most comprehensive of all existing methods, permitting the genome-wide evaluation of DNA methylation at single-base resolution.[68,79] For this reason, it is considered the "gold standard" for profiling DNA methylation.[68] The most considerable limitations are the cost and difficulties in analysis of the next generation sequencing (NGS) data.[67] Until recently, another limitation of WGBS was the considerable amount of DNA required for the analysis. However, there exist modified protocols capable of postponing the adaptor ligation step until after bisulfite treatment. This allows WGBS to be performed routinely from ~30 ng or even as little as 125 pg of DNA.[80] Below is a brief description of an established protocol adapted from studies in tissues.[68]

1. Genomic DNA is isolated from kidney tissues and renal cells, using the MasterPure DNA Purification Kit (Epicentre), or comparable

method, according to the manufacturer's instructions. This kit employs a nonenzymatic approach to cell lysis, followed by protein precipitation and subsequent nucleic acid isolation.

2. Extracted DNA is resuspended in TE buffer (1× buffer: 10 mM Tris-HCl, 1 mM EDTA at pH 8.0) and quantified by fluorometry.
3. Genomic DNA is converted by bisulfite treatment. Briefly, 50–100 ng of purified genomic DNA is treated with Zymo Lightning Conversion Reagent in a thermal cycle for 8 min at 98°C, followed by 60 min at 54°C.

   Note: Bisulfite treatment is known to fragment DNA.

4. Bisulfite-treated DNA is purified on a spin column, and used to prepare a sequencing library using the EpiGnome Kit (Epicentre). In this procedure, bisulfite-treated single-stranded DNA is randomly primed using a polymerase able to read uracil nucleotides to synthesize DNA containing a specific sequence tag.
5. The generated libraries are diluted and loaded onto the cBot DNA Cluster Generation System. After cluster generation is complete, the flow cell is transferred to the HiSeq 2500 System for sequencing using 75 bp paired-end reads.

The advantages of WGBS include (1) it typically covers over 90% of the CpG sites in the genome in an unbiased representation and (2) it allows the identification of non-CG methylation as well as partially methylated domains (PMDs). On the down side, analyzing sequencing data is difficult as developing bioinformatics techniques for processing WGBS data remain a critical challenge. Furthermore, WGBS is expensive to run, with the library prep requiring relatively large and high quantities of DNA.[81] In situations where differential methylation occurs in only a small fraction of the genome, an alternative analysis, called reduced representation bisulfite sequencing (RRBS), could be used. In this method, only a fraction of the genome is sequenced.

13.5 CONCLUDING REMARKS

ADPKD is increasingly becoming a great epidemiologic concern, accounting for 2%–8% of end-stage renal disease worldwide. Despite the considerable advances in research, the pathophysiologic pathways involved in the progression of ADPKD remain elusive. In recent years, multiple studies in ADPKD have shown that DNA methylation is involved in the disease progression. Therefore, to better understand the pathophysiology of ADPKD, it is important to identify the changes in DNA methylation in disease conditions compared with healthy individuals. Application of these newly developed high-throughput technologies to profile the DNA-methylation status in ADPKD patients will be instrumental in finding hidden sources of variation in the disease and therapeutic selection. Targeting abnormal DNA methylation with a de-methylating agent(s) is not only an innovative concept but also the next trend in PKD therapy.

Moreover, though total kidney volume (TKV) is an appropriate surrogate marker for ADPKD disease progression, patients with unilateral, segmental, asymmetric, lopsided, bilateral with unilateral atrophy, and bilateral with bilateral atrophy would not be suitable for TKV measurements, as kidney volume has not been shown to predict change in kidney function in these presentations.[66] The variable disease course of ADPKD makes it important to develop new biomarkers that can predict disease progression from a patient perspective and to select patients for reno-protective treatment. Compared with the other kinds of biomarkers, DNA-methylation alterations may occur in advance of the alterations of mRNA and protein levels in ADPKD, and thus might be a better early potential biomarker for better patient prognosis and design of clinical trials.

ACKNOWLEDGMENTS

X. Li acknowledges support from National Institutes of Health Grant R01 DK084097, the PKD Foundation and the Kansas Research and Translation Core Center (P30 DK106912) as well as the Mayo Translation PKD Center (P30 DK090728).

REFERENCES

1. Rangan, G.K. et al., Recent advances in autosomal-dominant polycystic kidney disease. *Intern Med J*, 2016. **46**(8): p. 883–892.
2. Ferreira, F.M., E.H. Watanabe, and L.F. Onuchic, Polycystins and molecular basis of autosomal dominant polycystic kidney disease, in *Polycystic Kidney Disease*, X. Li, Editor. 2015: Brisbane (AU).
3. Grantham, J.J., Clinical practice. Autosomal dominant polycystic kidney disease. *N Engl J Med*, 2008. **359**(14): p. 1477–1485.
4. Torres, V.E., P.C. Harris, and Y. Pirson, Autosomal dominant polycystic kidney disease. *Lancet*, 2007. **369**(9569): p. 1287–1301.
5. Karihaloo, A., Role of inflammation in polycystic kidney disease, in *Polycystic Kidney Disease*, X. Li, Editor. 2015: Brisbane (AU).
6. Mochizuki, T., K. Tsuchiya, and K. Nitta, Autosomal dominant polycystic kidney disease: Recent advances in pathogenesis and potential therapies. *Clin Exp Nephrol*, 2013. **17**(3): p. 317–326.
7. Antignac, C. et al., The future of polycystic kidney disease research—As seen by the 12 Kaplan Awardees. *J Am Soc Nephrol*, 2015. **26**(9): p. 2081–2095.
8. Leonhard, W.N., H. Happe, and D.J. Peters, Variable cyst development in autosomal dominant polycystic kidney disease: The biologic context. *J Am Soc Nephrol*, 2016. **27**(12): p. 3530–3538.
9. de Almeida, R.M. et al., Transcriptome analysis reveals manifold mechanisms of cyst development in ADPKD. *Hum Genomics*, 2016. **10**(1): p. 37.
10. Chu, A.Y. et al., Epigenome-wide association studies identify DNA methylation associated with kidney function. *Nat Commun*, 2017. **8**(1): p. 1286.
11. Reddy, M.A. and R. Natarajan, Epigenetics in diabetic kidney disease. *J Am Soc Nephrol*, 2011. **22**(12): p. 2182–2185.
12. Sant, K.E., M.S. Nahar, and D.C. Dolinoy, DNA methylation screening and analysis. *Methods Mol Biol*, 2012. **889**: p. 385–406.
13. Matzke, M.A. and J.A. Birchler, RNAi-mediated pathways in the nucleus. *Nat Rev Genet*, 2005. **6**(1): p. 24–35.
14. Holoch, D. and D. Moazed, RNA-mediated epigenetic regulation of gene expression. *Nat Rev Genet*, 2015. **16**(2): p. 71–84.

15. Woo, Y.M. et al., Genome-wide methylation profiling of ADPKD identified epigenetically regulated genes associated with renal cyst development. *Hum Genet*, 2014. **133**(3): p. 281–297.
16. Pei, Y. et al., Unified criteria for ultrasonographic diagnosis of ADPKD. *J Am Soc Nephrol*, 2009. **20**(1): p. 205–212.
17. Illingworth, R. et al., A novel CpG island set identifies tissue-specific methylation at developmental gene loci. *PLoS Biol*, 2008. **6**(1): p. e22.
18. Chung, W. et al., Detection of bladder cancer using novel DNA methylation biomarkers in urine sediments. *Cancer Epidemiology Biomarkers & Prevention*, 2011. **20**(7): p. 1483–1491.
19. Reinert, T., Methylation markers for urine-based detection of bladder cancer: The next generation of urinary markers for diagnosis and surveillance of bladder cancer. *Adv Urol*, 2012. **2012**: p. 503271.
20. Woo, Y.M. et al., Epigenetic silencing of the MUPCDH gene as a possible prognostic biomarker for cyst growth in ADPKD. *Sci Rep*, 2015. **5**: p. 15238.
21. Moore, L.D., T. Le, and G. Fan, DNA methylation and its basic function. *Neuropsychopharmacology*, 2013. **38**(1): p. 23–38.
22. Jin, B., Y. Li, and K.D. Robertson, DNA methylation: Superior or subordinate in the epigenetic hierarchy? *Genes Cancer*, 2011. **2**(6): p. 607–617.
23. Beckerman, P., Y.A. Ko, and K. Susztak, Epigenetics: A new way to look at kidney diseases. *Nephrol Dial Transplant*, 2014. **29**(10): p. 1821–1827.
24. Goll, M.G. et al., Methylation of tRNAAsp by the DNA methyltransferase homolog Dnmt2. *Science*, 2006. **311**(5759): p. 395–398.
25. Mortusewicz, O. et al., Recruitment of DNA methyltransferase I to DNA repair sites. *Proc Natl Acad Sci USA*, 2005. **102**(25): p. 8905–8909.
26. Xie, S. et al., Cloning, expression and chromosome locations of the human DNMT3 gene family. *Gene*, 1999. **236**(1): p. 87–95.
27. Okano, M. et al., DNA methyltransferases Dnmt3a and Dnmt3b are essential for de novo methylation and mammalian development. *Cell*, 1999. **99**(3): p. 247–257.
28. Kareta, M.S. et al., Reconstitution and mechanism of the stimulation of de novo methylation by human DNMT3L. *J Biol Chem*, 2006. **281**(36): p. 25893–25902.
29. Suetake, I. et al., DNMT3L stimulates the DNA methylation activity of Dnmt3a and Dnmt3b through a direct interaction. *J Biol Chem*, 2004. **279**(26): p. 27816–27823.
30. Jia, D. et al., Structure of Dnmt3a bound to Dnmt3L suggests a model for de novo DNA methylation. *Nature*, 2007. **449**(7159): p. 248–251.
31. Bourc'his, D. et al., Dnmt3L and the establishment of maternal genomic imprints. *Science*, 2001. **294**(5551): p. 2536–2539.
32. Hata, K. et al., Dnmt3L cooperates with the Dnmt3 family of de novo DNA methyltransferases to establish maternal imprints in mice. *Development*, 2002. **129**(8): p. 1983–1993.
33. Ko, Y.A. et al., Cytosine methylation changes in enhancer regions of core pro-fibrotic genes characterize kidney fibrosis development. *Genome Biol*, 2013. **14**(10): p. R108.
34. Bell, C.G. et al., Genome-wide DNA methylation analysis for diabetic nephropathy in type 1 diabetes mellitus. *BMC Med Genomics*, 2010. **3**: p. 33.
35. Ludgate, J.L. et al., A streamlined method for analysing genome-wide DNA methylation patterns from low amounts of FFPE DNA. *BMC Med Genomics*, 2017. **10**(1): p. 54.
36. Diaz, L.A., Jr. and A. Bardelli, Liquid biopsies: Genotyping circulating tumor DNA. *J Clin Oncol*, 2014. **32**(6): p. 579–586.
37. Stroun, M. et al., About the possible origin and mechanism of circulating DNA – Apoptosis and active DNA release. *Clinica Chimica Acta*, 2001. **313**(1–2): p. 139–142.

38. Jahr, S. et al., DNA fragments in the blood plasma of cancer patients: Quantitations and evidence for their origin from apoptotic and necrotic cells. *Cancer Research*, 2001. **61**(4): p. 1659–1665.
39. van der Vaart, M. and P.J. Pretorius, Circulating DNA. Its origin and fluctuation. *Ann N Y Acad Sci*, 2008. **1137**: p. 18–26.
40. Thierry, A.R. et al., Origin and quantification of circulating DNA in mice with human colorectal cancer xenografts. *Nucleic Acids Res*, 2010. **38**(18): p. 6159–6175.
41. Schwarzenbach, H., D.S.B. Hoon, and K. Pantel, Cell-free nucleic acids as biomarkers in cancer patients. *Nature Reviews Cancer*, 2011. **11**(6): p. 426–437.
42. Xue, X.Y. et al., Optimizing the yield and utility of circulating cell-free DNA from plasma and serum. *Clinica Chimica Acta*, 2009. **404**(2): p. 100–104.
43. Ralla, B. et al., Nucleic acid-based biomarkers in body fluids of patients with urologic malignancies. *Critical Reviews in Clinical Laboratory Sciences*, 2014. **51**(4): p. A2–231.
44. Trigg, R.M. et al., Factors that influence quality and yield of circulating-free DNA: A systematic review of the methodology literature. *Heliyon*, 2018. **4**(7).
45. Malentacchi, F. et al., Influence of storage conditions and extraction methods on the quantity and quality of circulating cell-free DNA (ccfDNA): The SPIDIA-DNAplas External Quality Assessment experience. *Clinical Chemistry and Laboratory Medicine*, 2015. **53**(12): p. 1935–1942.
46. Fatouros, I.G. et al., Time of sampling is crucial for measurement of cell-free plasma DNA following acute aseptic inflammation induced by exercise. *Clinical Biochemistry*, 2010. **43**(16–17): p. 1368–1370.
47. Jung, K., M. Fleischhacker, and A. Rabien, Cell-free DNA in the blood as a solid tumor biomarker-A critical appraisal of the literature. *Clinica Chimica Acta*, 2010. **411**(21–22): p. 1611–1624.
48. Patel, P.S. et al., Evaluation of serum alkaline DNase activity in treatment monitoring of head and neck cancer patients. *Tumour Biol*, 2000. **21**(2): p. 82–89.
49. Page, K. et al., Influence of plasma processing on recovery and analysis of circulating nucleic acids. *PLOS ONE*, 2013. **8**(10).
50. Chan, K.C.A. et al., Effects of preanalytical factors on the molecular size of cell-free DNA in blood. *Clinical Chemistry*, 2005. **51**(4): p. 781–784.
51. Holdenrieder, S. et al., DNA integrity in plasma and serum of patients with malignant and benign diseases. *Circulating Nucleic Acids in Plasma and Serum V*, 2008. **1137**: p. 162–170.
52. Page, K. et al., Detection of HER2 amplification in circulating free DNA in patients with breast cancer. *British Journal of Cancer*, 2011. **104**(8): p. 1342–1348.
53. Bartak, B.K. et al., Colorectal adenoma and cancer detection based on altered methylation pattern of SFRP1, SFRP2, SDC2, and PRIMA1 in plasma samples. *Epigenetics*, 2017. **12**(9): p. 751–763.
54. Swinkels, D.W. et al., Effects of blood-processing protocols on cell-free DNA quantification in plasma. *Clinical Chemistry*, 2003. **49**(3): p. 525–526.
55. Chiu, R.W.K. et al., Effects of blood-processing protocols on fetal and total DNA quantification in maternal plasma. *Clinical Chemistry*, 2001. **47**(9): p. 1607–1613.
56. Ramachandran, K. et al., Free circulating DNA as a biomarker of prostate cancer: Comparison of quantitation methods. *Anticancer Research*, 2013. **33**(10): p. 4521–4529.
57. Lichtenstein, A.V. et al., Circulating nucleic acids and apoptosis. *Ann N Y Acad Sci*, 2001. **945**: p. 239–249.
58. Lecamwasam, A. et al., DNA methylation profiling of genomic DNA isolated from urine in diabetic chronic kidney disease: A pilot study. *PLoS One*, 2018. **13**(2): p. e0190280.
59. Salvi, S. et al., The potential use of urine cell free DNA as a marker for cancer. *Expert Review of Molecular Diagnostics*, 2016. **16**(12): p. 1283–1290.

60. Su, Y.H. et al., Human urine contains small, 150 to 250 nucleotide-sized, soluble DNA derived from the circulation and may be useful in the detection of colorectal cancer. *J Mol Diagn*, 2004. **6**(2): p. 101–107.
61. Jasmine, F. et al., Interpretation of genome-wide infinium methylation data from ligated DNA in formalin-fixed, paraffin-embedded paired tumor and normal tissue. *BMC Res Notes*, 2012. **5**: p. 117.
62. Kitazawa, S., R. Kitazawa, and S. Maeda, Identification of methylated cytosine from archival formalin-fixed paraffin-embedded specimens. *Lab Invest*, 2000. **80**(2): p. 275–276.
63. Grunau, C., S.J. Clark, and A. Rosenthal, Bisulfite genomic sequencing: Systematic investigation of critical experimental parameters. *Nucleic Acids Res*, 2001. **29**(13): p. E65–E65.
64. Meissner, A. et al., Genome-scale DNA methylation maps of pluripotent and differentiated cells. *Nature*, 2008. **454**(7205): p. 766–770.
65. Weber, M. et al., Chromosome-wide and promoter-specific analyses identify sites of differential DNA methylation in normal and transformed human cells. *Nat Genet*, 2005. **37**(8): p. 853–862.
66. Brinkman, A.B. et al., Whole-genome DNA methylation profiling using MethylCap-seq. *Methods*, 2010. **52**(3): p. 232–236.
67. Kurdyukov, S. and M. Bullock, DNA Methylation analysis: Choosing the right method. *Biology (Basel)*, 2016. **5**(1).
68. Wreczycka, K. et al., Strategies for analyzing bisulfite sequencing data. *J Biotechnol*, 2017. **261**: p. 105–115.
69. Li, Y. and T.O. Tollefsbol, DNA methylation detection: Bisulfite genomic sequencing analysis. *Methods Mol Biol*, 2011. **791**: p. 11–21.
70. Leontiou, C.A. et al., Bisulfite conversion of DNA: Performance comparison of different kits and methylation quantitation of epigenetic biomarkers that have the potential to be used in non-invasive prenatal testing. *PLOS ONE*, 2015. **10**(8): p. e0135058.
71. Herman, J.G. et al., Methylation-specific PCR: A novel PCR assay for methylation status of CpG islands. *Proc Natl Acad Sci USA*, 1996. **93**(18): p. 9821–9826.
72. Eads, C.A. et al., MethyLight: A high-throughput assay to measure DNA methylation. *Nucleic Acids Research*, 2000. **28**(8).
73. Roupret, M. et al., Molecular detection of localized prostate cancer using quantitative methylation-specific PCR on urinary cells obtained following prostate massage. *Clin Cancer Res*, 2007. **13**(6): p. 1720–1725.
74. Colyer, H.A. et al., Detection and analysis of DNA methylation by pyrosequencing. *Methods Mol Biol*, 2012. **863**: p. 281–292.
75. Tost, J. and I.G. Gut, DNA methylation analysis by pyrosequencing. *Nat Protoc*, 2007. **2**(9): p. 2265–2275.
76. Tost, J. and I.G. Gut, Analysis of gene-specific DNA methylation patterns by pyrosequencing technology. *Methods Mol Biol*, 2007. **373**: p. 89–102.
77. Mahapatra, S. et al., Global methylation profiling for risk prediction of prostate cancer. *Clin Cancer Res*, 2012. **18**(10): p. 2882–2895.
78. Potapova, A. et al., Systematic cross-validation of 454 sequencing and pyrosequencing for the exact quantification of DNA methylation patterns with single CpG resolution. *BMC Biotechnology*, 2011. **11**.
79. Yong, W.S., F.M. Hsu, and P.Y. Chen, Profiling genome-wide DNA methylation. *Epigenetics Chromatin*, 2016. **9**: p. 26.
80. Miura, F. and T. Ito, Highly sensitive targeted methylome sequencing by post-bisulfite adaptor tagging. *DNA Res*, 2015. **22**(1): p. 13–18.
81. Stirzaker, C. et al., Mining cancer methylomes: Prospects and challenges. *Trends Genet*, 2014. **30**(2): p. 75–84.

14 Molecular Diagnosis of Autosomal Dominant Polycystic Kidney Disease

Matthew Lanktree, Amirreza Haghighi, Xueweng Song, and York Pei

CONTENTS

14.1 INTRODUCTION

Autosomal dominant polycystic kidney disease (ADPKD) is the most common hereditary kidney disease worldwide, equally affecting all racial groups, with an estimated cumulative lifetime risk of approximately 1 in 1000.[1] In ADPKD, deficiency of polycystin-1 or polycystin-2 leads to the progressive development and growth of kidney cysts and subsequent increase in total kidney volume, regional ischemia, tubular obstruction, and eventually reduced kidney function and end-stage renal disease (ESRD).[2] Symptoms of ADPKD can include hypertension, urinary concentrating deficits leading to polyuria and nocturia, mass effects leading to chronic discomfort, and acute pain events due to kidney stones, cyst rupture, or infections.[3] Extra-renal manifestations of ADPKD can include polycystic liver disease, colonic diverticular disease, hernias, heart valve defects, and intracerebral and aortic aneurysms.[3] However, considerable variation in disease severity exists, with some patients experiencing significant symptoms as early as childhood, with other patients not suffering consequences until late in life.[4] The current gold standard for assessing disease severity and providing kidney disease prognosis is the Mayo Clinic imaging classification that uses magnetic resonance measured height- and age-adjusted total kidney volume (TKV).[5] However, genetic heterogeneity underlies the phenotypic variability, with over a thousand different mutations now reported, and identification of the responsible mutation can provide concrete evidence of ADPKD, even very early in the disease course.[1]

14.2 DIAGNOSIS AND INDICATIONS FOR GENETIC TESTING IN ADPKD

The diagnosis of ADPKD is based upon the presence of numerous bilateral kidney cysts with renal enlargement, in the context of a positive family history of kidney cysts and ESRD.[6] Given autosomal-dominant inheritance, a positive parental history of ADPKD indicates a pretest probability of 50%. However, up to one-quarter of patients clinically recognized to have ADPKD have no apparent family history.[7] No apparent family history may be due to development of disease by a *de novo* mutation; but just as frequently, the lack of history is due to unavailability of familial medical records, unrecognized mild disease, poor access to medical care in the parental generation, or mortality from another cause.[7]

Ultrasound examination of the kidney is the modality employed for initial screening. Understanding the progressive nature of ADPKD is important for diagnosis, leading to age-dependent imaging criteria. In the context of a positive family history of polycystic kidney disease, between age 18 and 39, the presence of at least three visible cysts is diagnostic of ADPKD.[8] Due to the increasing prevalence of simple cysts with age, in a patient over age 60, four cysts must be visible in each kidney. Genetic testing can be useful to provide diagnostic certainty, often in the context of no apparent family history or equivocal imaging findings. A requirement for disease exclusion at a young age, such as in living donor transplant assessments or prenatal or preimplantation diagnostics, are additional indications for genetic testing. Scenarios for which genetic testing is being increasingly used in a clinical context include early and severe disease, risk stratification to identify "high-risk" patients for treatment with therapies associated with a therapeutic burden, marked intrafamilial discordance in disease severity, atypical imaging findings and discrepancy between imaging findings and decline in renal function, and suspicion of a phenocopy (i.e., autosomal dominant tubulointerstitial kidney disease) or syndromic (i.e., nephronophthisis) form of polycystic kidney disease.[9,10]

Genetic testing in a presymptomatic state can be associated with significant emotional and psychological stress, and inclusion of a genetic counselor in the care of patients is recommended. All patients seeking genetic testing for ADPKD should be counselled about the potential for genetic discrimination. Children of patients with ADPKD will have a 50% risk of also having ADPKD, but as there are no approved treatments in the pediatric population, the current recommendation is to avoid screening for ADPKD in children beyond measurement of blood pressure.

14.3 DISEASE GENES IN ADPKD

ADPKD is primarily due to mutations in either *PKD1* or *PKD2*. Mutations in *PKD1*, localized to chromosome 16p13.3, are associated with more-severe disease compared with mutations in *PKD2*, found at chromosome 4q21-22 (Table 14.1). Depending on the ascertainment scheme for identifying ADPKD patients, the relative prevalence of the different mutation types changes. For example, cohorts collected with severe disease have a greater relative prevalence of *PKD1* mutations,[11] while cohorts based on clinical ascertainment have a relatively greater prevalence of *PKD2* mutations.[4] In the clinically ascertained Toronto Genetic Epidemiology Study of Polycystic Kidney Disease (TGESP), 70% of genetically resolved families had a *PKD1* mutation (Figure 14.1).[4] Mutations in *PKD1* are about 3 to 4 times more common in *PKD2* because of its larger genomic size.[1] *PKD1* is a very large gene, spanning over 50 kilobases and 46 exons, producing a 12,909 base-pair mRNA transcript and a 4303 amino acid, 500 kilodalton, polycystin-1 protein. By contrast, *PKD2* spans 70 kilobases, but has only 15 exons producing a 2907 base pair mRNA transcript and a 968 amino acid, 110 kilodalton, polycystin-2 protein.[12] Polycystin-1 and polycystin-2 interact via their 3′ cytoplasmic tails to form the polycystin complex, which localizes to the renal tubular epithelial cell cilia. The exact pathogenic mechanism of cystogenesis resulting from decreased polycystin complex on the cilia remains unknown, but a loss of intracellular inhibitory signaling and increased proliferation results.[13]

TABLE 14.1
Comparison of Sequencing Options in Molecular Diagnostics of ADPKD

	Long-Range PCR	Exome or Gene Panel	Whole Genome
Timing of target selection	Primer design	Capture	Computationally postsequencing
Capture of GC rich content	With troubleshooting	Low	Improved
Allelic dropout	Yes	No	No
Copy number variation detection	No	Possible	Likely
Lab intensity	High	Automated	Automated
Bioinformatic resources	Low	High	Highest
Additional cyst or modifier genes	No	If captured	Yes
Promoters, enhancers, intronic sequence, transcription factor binding sites	No	If captured, generally no	Yes
Detection of somatic mosaicism	No	Potentially if suitable read depth	Potentially if suitable read depth
Cost of research sequencing	$$$	$	$$$

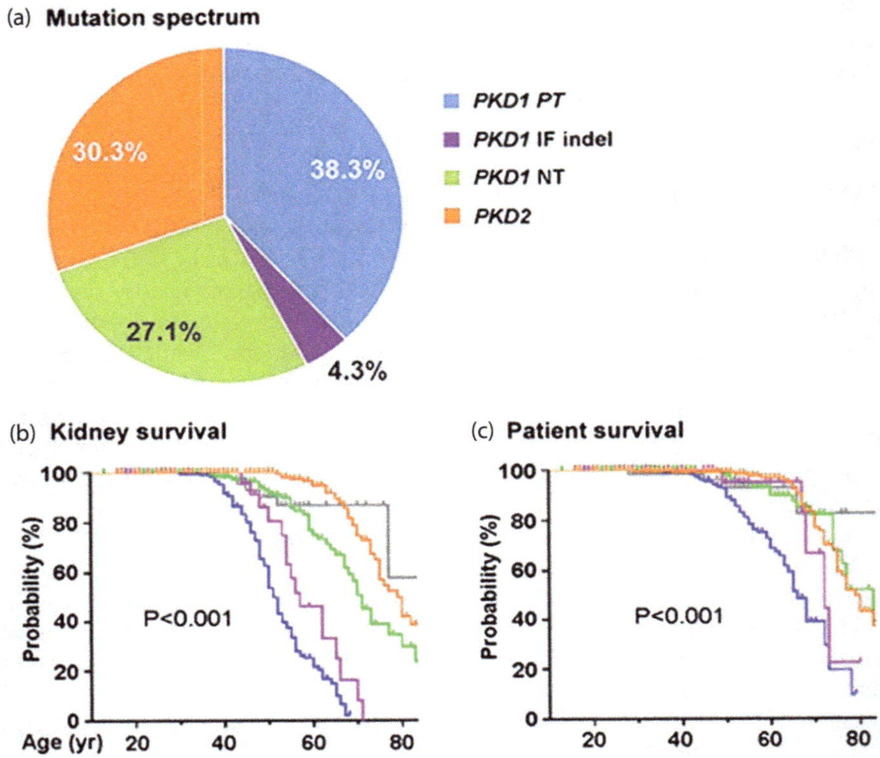

FIGURE 14.1 Genetics of ADPKD. (a) Breakdown of *PKD1* protein truncating (PT), in-frame insertion deletion (IF indel), nontruncating (NT), and *PKD2* mutations in genetically resolved cases in the Toronto Genetic Epidemiology Study of Polycystic Kidney Disease (TGESP) cohort. Mutation class is associated with kidney (b) and overall (c) survival in TGESP. (Adapted from Hwang, Y.-H. et al. *J Am Soc Nephrol* 27, 1862 and 1864, 2016.)

Mutation analysis of *PKD2* and the single copy region of *PKD1* (exon 34–46) is fairly straightforward. Genomic analysis of the remainder of *PKD1*, including the first (5′) 33 exons, is complicated by the presence of six nearby pseudogenes (*PKD1P1-P6*) with greater than 98% sequence identity. Incorrectly sequencing pseudogene sequence can lead to both false-positive and false-negative genotype calls, and strategies to avoid this are described below.[12] More recently, exome sequencing of families with no mutation detected was employed, resulting in two additional genes having been found to be mutated and causing ADPKD: glucosidase II alpha subunit (*GANAB*)[14] and DNA J heat shock protein family member B11 (*DNAJB11*).[15] *GANAB* was a strong candidate gene for causing kidney cysts because of its function as an endoplasmic reticulum (ER) resident protein required for the maturation and trafficking of membrane proteins including polycystin-1. Mutations in *PRKCSH*, which encodes the beta subunit of the same glucosidase II protein, cause autosomal dominant polycystic liver disease (ADPLD).[16]

The phenotype in patients with *GANAB* and *DNAJB11* mutations are different than patients with *PKD1* or *PKD2*, with *GANAB* mutations causing more pronounced polycystic liver disease with a mild kidney phenotype including fewer larger cysts[14] and *DNAJB11* causing a fibrotic cystic kidney phenotype not associated with kidney enlargement.[15] Together, *GANAB* and *DNAJB11* account for less than 1% of ADPKD patients.

14.4 MUTATION TYPE PREDICTS DISEASE SEVERITY IN ADPKD

Locus heterogeneity underlies a proportion of variability in disease presentation: patients with *PKD1* mutations develop more severe disease with greater cyst burden, larger TKV, and younger age at ESRD than those with *PKD2* mutations.[4,16,17] Allelic heterogeneity also contributes: patients with protein-truncating mutations have more severe disease than those with nontruncating missense mutations.[4] Truncating mutations include nonsense mutations, frameshift insertions or deletions, and canonical splice mutations or large deletions resulting in omission of large portions of the transcript. Nontruncating mutations can include small in-frame insertions or deletions or missense mutations. However, substantial variable expressivity in the ADPKD phenotype is common and family members carrying the same main-effect mutation can have extremely discordant phenotypes.

14.5 SEQUENCING METHODOLOGY APPLIED TO ADPKD

Since the 1970s, DNA sequencing by the Sanger method has been regarded as the gold standard for mutation analysis in molecular diagnostics. Because of the large degree of genetic heterogeneity, with more than 1250 and 200 pathogenic mutations in *PKD1* and *PKD2*, respectively, sequencing of the whole genes are required to confidently identify a responsible mutation. Several methods have been developed to screen for *PKD1* mutations, including denaturing gradient gel electrophoresis (DGGE),[18] denaturing high-performance liquid chromatography (DHPLC),[19] and direct sequencing of long-range PCR (LR-PCR) products.[20,21] However, LR-PCR with primers directed toward rare mismatch sites that distinguish *PKD1* from the pseudogenes, followed by nested PCR and Sanger sequencing, is the predominant method used by most diagnostic laboratories.[12] Four workflows underpin strategies for sequencing *PKD1*: long-range PCR followed by either capillary sequencing or next generation sequencing, DNA capture followed by next generation sequencing, or whole-genome sequencing (Figure 14.2). The core steps in preparing DNA for PCR-based targeted resequencing include (1) amplification and pooling of PCR amplicons equimolarly, (2) fragmenting the pooled DNA to a desired length, (3) ligating sequencing adapters to the ends of target fragments, and (4) quantitating the final library product for sequencing. For our research studies, our laboratory has implemented a *PKD1*-targeted sequencing protocol using the LR-PCR primers developed by Rossetti et al.[22] with some minor modifications and sequencing on the Illumina HiSeq2500.

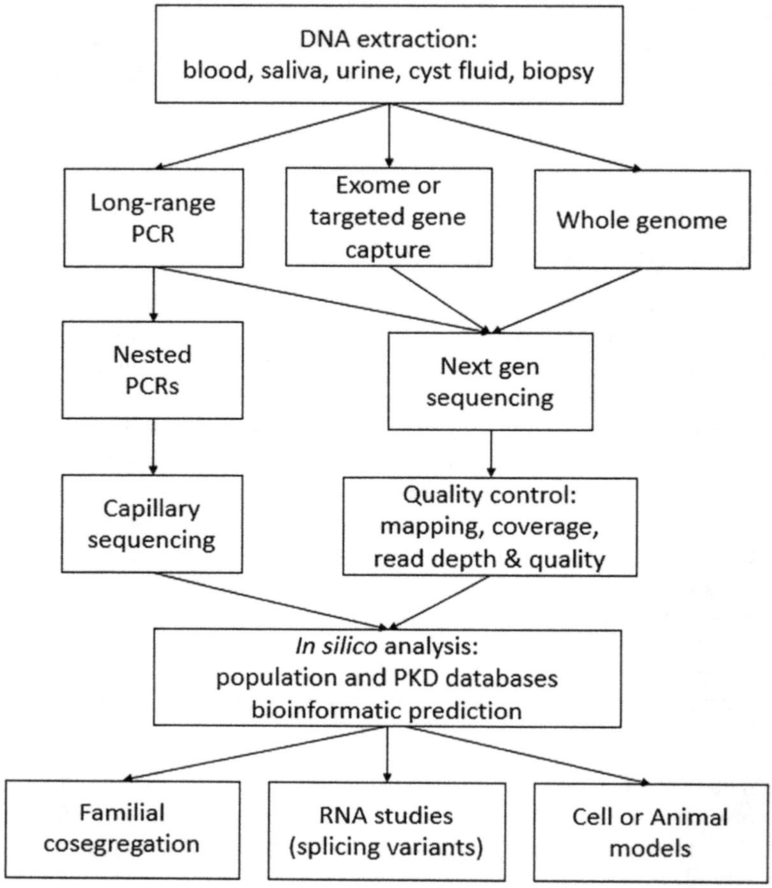

FIGURE 14.2 Analytical approach for molecular diagnostics in ADPKD.

14.5.1 Long-Range PCR Followed by Capillary or Next-Generation Sequencing

Both blood leukocyte and saliva DNA can be used, but leukocytes are preferred. Blood samples are collected in lavender-top ethylenediaminetetraacetic acid (EDTA) Vacutainer tubes (BD Diagnostics), and processed using Flexigene DNA Kit (Qiagen). Saliva samples are collected using Oragene DNA Self Collection Kits (DNA Genotek) according to the manufacturer's instructions, and processed using the prepIT•L2P DNA Purification Kit (DNA Genotek). Purified DNA is quantified using NanoDrop Spectrophotometers (Thermo Fisher Scientific) according to the manufacturer's protocol. Seven LR-PCR reactions with PKD1 locus-specific primers (Table 14.2). The 7 amplicons cover *PKD1* all 46 exons with total length of ~35,000 bp.

The choice of DNA polymerase and PCR conditions is critical for successful PCR amplification as some enzyme mixes are more robust than others. Although *PKD1*

TABLE 14.2
List of Primer Sequences Used for 7 *PKD1* LR-PCR

PKD1 LR-PCR Fragments	Forward Primers (5′–3′)	Reverse Primers (5′–3′)	Genomic Coordinates (GRCh37/hg19)	Amplicon Size (bp)
Exon 1	CGCAGCCTTACCATCCACCT	TCATCGCCCCTTCCTAAGCA	Chr16: 2,187,307-2,185,030	2278
Exon 2–15	CCAGCTCTCTGTCTACTCACCTCCGCATC	CCCTGTCCCTCCAGGCAGTCCAGCTGTAGG	Chr16: 2,171,636-2,160,151	11,486
Exon 13–15	TGGAGGGAGGGACGCCAATC	GTCAACGTGGGCCTCCAAGT	Chr16: 2,163,036-2,158,646	4391
Exon 15–21	AGCGCAACTACTTGGAGGCCC	GGAGCCCAGGCTGGAGGCTCA	Chr16: 2,158,674-2,155,209	3466
Exon 22–34	CCGTGTAGAGAGGAGGGCGTGTGCAAGGA	CACACCTGAGCATAGGAGGGCATGGCAGAG	Chr16: 2,154,714-2,146,347	8368
Exon 35–41	GTGGGCGATGGGTTTATCAGCA	GGGCTGTGGAAGCCGCCTA	Chr16: 2,144,471-2,141,678	2794
Exon 42–46	CAGCCAGGAGCCCACCCTCAC	AAGTGCTGAAGCCCACAGACAGACAGA	Chr16: 2,141,737-2,139,571	2167

gene-specific LR-PCR primers have been carefully designed to differentiate the true *PKD1* gene from its six pseudogenes, it remains difficult to efficiently amplify all fragments under standard PCR conditions because of the high GC content and sequence complexity.[20,22] KOD Xtreme Hot Start DNA Polymerase (EMD Millipore) is an optimized PCR enzyme for the amplification of long (up to 24 kb), GC-rich (up to 90% GC content) and crude (minimally processed) DNA templates. By using KOD Xtreme, we could successfully amplify the extreme GC richness of *PKD1* exon 1 fragment (2278 bp), and the longest *PKD1* fragments (exon 2–15 (11,486 bp)) with minimal optimizations, which routinely failed when other commercial LR-PCR polymerases for GC-rich templates were used. The detailed PCR amplification conditions for the seven LR-PCR fragments are described in Table 14.3. Using the LR-PCR products as the input DNA, primers directed to flank each exon can be used for capillary sequencing with input DNA free from pseudogene sequence. Alternatively, the LR-PCR products can be advanced for high-throughput sequencing.

Recent advances in next-generation sequencing (NGS) are transforming molecular diagnosis of ADPKD. Since Rossetti et al. reported the first application of NGS in ADPKD following a nested LR-PCR,[22] this approach has been adapted for mutation screening in ADPKD by laboratories worldwide.[23–25] Compared with broader approaches, such as whole-genome sequencing (WGS) or whole-exome sequencing (WES), targeted resequencing enriches the regions of interest for subsequent sequencing and produces much higher coverage and read depth and is well suited for detecting both ADPKD germline mutations and somatic mosaicism. Samples can be multiplex by barcoding technology increasing throughput and reducing cost per sample. Barcoding entails ligating a short oligonucleotide to each DNA fragment before mixing samples, allowing for tracing of each DNA read to a particular sample (Table 14.4).

The size of the pooled amplicons in the library is a key parameter for NGS library construction. Library construction involves random fragmentation of pooled DNA (mechanical or enzymatic) followed by the ligation of sequencing adapters for amplification. Mechanical shearing requires 2–5 μg of pooled LR-PCR DNA for library preparation. Illumina's Nextera tagmentation technology combines simultaneous fragmentation of DNA and adapter sequence ligation in a single reaction mediated by a transposase loaded with adapter oligos. This technique produces high-quality libraries using standard lab equipment and is ideal for precious samples with limited quantity available. The Illumina Nextera XT DNA Library Preparation kit has dual-barcoding options, allowing the user to pool up to 384 samples together. In our laboratory, we construct sequencing libraries using the Nextera XT kit and 96 samples with a starting input of 1 ng pooled *PKD1* amplicons. The 7 *PKD1* LR-PCR fragments from each patient are pooled together at an equal molar ratio and quantified using the Quant-iT PicoGreen dsDNA Assay Kit (Invitrogen) to ensure equal coverage and to achieve optimal tagmentation. A limited-cycle PCR reaction using primers containing adapter sequences amplifies the tagmented DNA and adds sequencing primer sequences and index sequences on both ends of the DNA. Amplified libraries are purified by AMPure XP beads (Beckman Coulter Life Sciences) to eliminate PCR primers and very short library fragments. The purified libraries from 96 patients are analyzed using the DNA high-sensitivity chip on an Agilent 2100 Bioanalyzer

TABLE 14.3
Reaction Mixture and PCR Conditions

PKD1 LR-PCR Fragments	DNA Polymerase	PCR Conditions
Exon 1 & Exon 42–46	KOD Xtreme	Initial step of 94°C for 2 min, followed by a step-down protocol: 5 cycles at 98°C for 10 s, 74°C for 2.5 min, 5 cycles at 98°C for 10 s, 72°C for 2.5 min, 5 cycles at 98°C for 10 s, 70°C for 2.5 min, 5 cycles at 98°C for 10 s, 68°C for 2.5 min, then followed by 20 cycles of 98°C for 10 s, 66°C for 30 sec, and 68°C for 2 min, with a final extension step of 68°C for 5 min.
Exon 2–15	KOD Xtreme	Initial step of 94°C for 2 min, followed by a step-down protocol: 5 cycles at 98°C for 10 s, 74°C for 10 min, 5 cycles at 98°C for 10 s, 72°C for 10 min, 5 cycles at 98°C for 10 s, 70°C for 10 min, then followed by 20 cycles of 98°C for 10 s, 68°C for 10 min, with a final extension step of 68°C for 10 min.
Exon 13–15	KOD Xtreme	Initial step of 94°C for 2 min, followed by a step-down protocol: 5 cycles at 98°C for 10 s, 74°C for 3 min, 5 cycles at 98°C for 10 s, 72°C for 3 min, 5 cycles at 98°C for 10 s, 70°C for 3 min, then followed by 25 cycles of 98°C for 10 s, 68°C for 3 min, with a final extension step of 68°C for 5 min.
Exon 15–21	KOD Xtreme	Initial step of 94°C for 2 min, followed by a step-down protocol: 5 cycles at 98°C for 10 s, 74°C for 3 min, 5 cycles at 98°C for 10 s, 72°C for 3 min, then followed by 30 cycles of 98°C for 10 s, 70°C for 3 min, with a final extension step of 70°C for 5 min.
Exon 22–34	LongAmp Taq	Initial step of 94°C for 3 min, followed by a step-down protocol: 4 cycles at 94°C for 30 s, 72°C for 6.5 min, 10 cycles at 94°C for 30 s, 70°C for 6.5 min, then followed by 20 cycles of 94°C for 30 s, 70°C for 3 min with increasing 10 s per cycle, with a final extension step of 70°C for 10 min.
Exon 35–41	LongAmp Taq	Initial step of 94°C for 3 min, followed by a step-down protocol: 2 cycles at 94°C for 30 s, 72°C for 3 min, then 34 cycles at 94°C for 30 s, 71°C for 3 min, with a final extension step of 71°C for 5 min.

Note: Composition of reaction mixture of volume of 25 μL: KOD Xtreme (EMD Millipore): 50 ng of genomic DNA, 400 μmol/L dNTPs, 0.3 μmol/L of each primer, manufacturer's supplied buffer, and 0.5 U of enzyme.

LongAmp Taq (New England Biolab): 100 ng of genomic DNA, 300 μmol/L dNTPs, 0.4 μmol/L of each primer, manufacture's supplied buffer, and 2.5 U of enzyme.

system (Agilent Technologies) for quality control, quantified using qPCR, and pooled together at equimolar amounts. The final library pool is then diluted, loaded onto a flow cell of HiSeq2500, subjected to cluster formation, and sequenced using a paired-end 125-bp cycle protocol at the Center for Applied Genomics (TCAG; SickKids, Toronto, ON, Canada) according to the manufacturer's instructions.

After sequencing, FASTQ files from the sequencer undergo quality control, and are demultiplexed by assigning to the proper sample using the unique oligonucleotide

TABLE 14.4
Differential Diagnosis of Diseases Including Kidney Cysts

	Causative Gene	Renal Features	Extra-Renal Features
Simple cysts		One to few variable size	None
Acquired cystic kidney disease		Associated with advanced CKD, small kidneys	None
Medullary sponge kidney		Calcification, stones, no progression	None
ARPKD	*PKHD1* *DZIP1L*	Large, echogenic, cystic kidneys diagnosed *in utero*	Congenital hepatic fibrosis, Caroli disease, hepatosplenomegaly
Tuberous sclerosis complex	*TSC1* *TSC2*	Angiomyolipomas, cysts	Cutaneous angiofibromas, retinal hamartomas, cardiac rhabdomyomas, lymphangioleiomyomatosis, neurological lesions
ADTKD	*UMOD* *REN* *MUC1*	Cysts, kidneys normal in size or small, progressive CKD	Hyperuricemia and gout, hypotension, anemia, hyperkalemia
HNF1beta-associated disease	*HNF1B*	Cysts, CKD, renal malformations (e.g., multicystic dysplasia, renal hypoplasia)	MODY, hypomagnesemia, hypocalciuria, hyperuricemia with gout, mental retardation
ADPLD	*PRKCSH* *SEC63* *GANAB* *SEC61B* *ALG8* *LRP5*	No to few renal cysts, no progression to ESRD	Diffuse cystic involvement of the liver
Von Hippel-Lindau disease	*VHL*	Cysts, clear cell carcinomas of the kidney	Retinal hemangiomas, cerebellar and spinal hemangioblastomas, pheochromocytoma
Nephronophthisis	*NPHP1-13*	Corticomedullary cysts with normal-sized kidneys, urinary concentrating and sodium reabsorption defect, progressive CKD	Retinitis pigmentosa, cerebellar vermis aplasia, hepatic fibrosis, skeletal dysplasia
Oral-facial-digital syndrome	*OFD1*	Polycystic kidneys	Oral, facial, dental, digital, and central nervous system anomalies
HANAC	*COL4A1*	Cysts, hematuria, decreased GFR	Muscle cramps, mild cerebral small vessel disease, retinal arteriolar tortuosity, intracranial aneurysms

(Continued)

TABLE 14.4 (*Continued*)
Differential Diagnosis of Diseases Including Kidney Cysts

	Causative Gene	Renal Features	Extra-Renal Features
HIPKD	*PMM2*	Antenatal or childhood onset polycystic kidney disease, CKD	Liver cysts, infantile hyperinsulinemic hypoglycemia
CAKUT	Many genes	Renal hypoplasia, dysplasia, agenesis, multicystic dysplasia; anomalies may affect contralateral kidney	

Abbreviations: ADTKD: autosomal dominant tubulointerstitial kidney disease; ADPLD: autosomal dominant polycystic liver disease; ARPKD: autosomal recessive polycystic kidney disease; CAKUT: congenital anomalies of the kidney and urinary tract; CKD: chronic kidney disease; ESRD: end-stage renal disease, GFR: glomerular filtration rate; HANAC: hereditary angiopathy, neuropathy, aneurysms, and muscle cramps; HIPKD: hyperinsulinemia with hypoglycemia and polycystic kidney disease; MODY: maturity-onset diabetes of the young.

barcode. The trimmed raw sequence is aligned to the human reference genome (hg19, NCBI build GRCh37) and *PKD1* targeted region using the Burrows-Wheeler Aligner BWA-MEM alignment algorithm (BWA-0.7.12).[26] BWA sequence alignments are converted into an analysis-ready binary alignment (BAM) file using SAMtools, and PCR duplicate reads are marked using Picard tools–1.123. Local realignment and base recalibration are performed using the Genome Analysis Tool Kit (GATK 3.6).[27] Using the BAM file as input, single nucleotide variations and small insertion or deletions (InDels) are detected simultaneously using GATK HalotypeCaller 3.6, which produces a variant call format (VCF) file containing all the observed variation. For detecting mosaic or somatic variants, both HalotypeCaller 3.6 and FreeBayes caller v0.9.20-8-gfef284a (https://github.com/ekg/freebayes/) are employed. Freebayes has a tunable allele frequency setting, and we set the alternate allele fraction $\geq$5% for maximum sensitivity. To exclude false-positive calls, all variants are visually inspected on the Golden Helix Genome Browser (Golden Helix, Bozeman, Montana, USA), which vallows for observation of the variants at the level of the individual read. Poly-T, -C, -A, -G stretches, GC-rich areas and InDel regions may influence the mapping qualities or variant calls, creating false-positive calls. For assessment of mosaic or somatic variants with low alternate allele fraction ($\leq$5%), the recurrent variants observed in multiple unrelated samples are considered sequencing artifacts and are excluded.

14.5.2 DNA Capture for Exome or Gene Panel Sequencing

There is considerable concern about the possibility of both false-positive and false-negative genotype calls, but particularly missed mutations, due to the duplicated regions of *PKD1*.[23,28] Several recent publications have shown promising results using custom capture gene panel strategy and reported that exonic regions of PKD1 gene are well covered.[24,29,30] The DNA capture gene panel approach provides the possibility of sequencing of a smaller fraction of the genome across a larger number of patients. This approach can also deliver very high coverage and read depth, allowing the identification of very rare variants and mosaicism.[12] Targeted resequencing using DNA capture methods (hybridization capture-based target enrichment) enriches genomic DNA fragments from the target regions (i.e., multiple preselected genes with a combined size up to 50 Mb) hybridized to custom-designed oligonucleotide probes of 50–120 base pairs.[31] This approach has the potential to provide an "all-in-one" and cost-effective platform with high sensitivity and specificity to detect all types of genetic variants, from single nucleotide variants to copy number variants, addresses the molecular diagnosis of typical ADPKD as well as atypical PKD (e.g., mosaicism, complex genetics, and non-ADPKD cystic disease), and can include additional genes in cystogenesis (e.g., *GANAB, DNAJB11, PKHD1, PRKCSH, LRP5*, etc.).

The hybridization capture-based target enrichment approach employs oligonucleotide probes to capture target sequences in an NGS library.[31] The principle is based on the hybridization of prepared DNA fragments complementary to the regions of interest to a standard shotgun sequencing library made from genomic DNA. Hybridization of the target regions can occur either on a solid surface or in solution. Solid-phase methods use a microarray, where the complementary probes are affixed to a glass slide for the hybrid capture reaction. In the solution-based or fluid-phase method, prior to NGS, pools of biotinylated oligonucleotide probes are hybridized to a sequencing library in solution. Following hybridization, the biotinylated probes are pulled down using streptavidin-coated magnetic beads to achieve libraries highly enriched for the target regions, and nontarget sequences are washed away. There is also the possibility to incorporate molecular barcodes in the DNA library, either to multiplex multiple samples in the same experiment, or to improve sequencing error detection and improve base calling accuracy for variants with a low variant allele frequency (i.e., somatic variants).[32] The amount of DNA required for library preparation varies based on different protocols and kits. Usually 1–3 μg DNA input is needed but specific kits designed for samples with low availability, enable libraries even from 10 ng DNA input. Hybrid capture-based target enrichment is commercialized mostly by Agilent Technologies (SureSelect) and Roche-NimbleGen (SeqCap/SeqCap EZ), and they offer products including library preparation and target enrichment kits, catalog and custom probes, software solutions, sample quality control, and automation platforms.

A growing number of studies using DNA capture gene panel approaches have been successfully applied to identify genetic variants of kidney diseases including ADPKD. Bullich et al. applied a DNA capture method using an extensive 140 kidney disease gene panel to study the Spanish cohort of 421 patients with inherited cystic (including ADPKD) and glomerular diseases.[30] Using a custom NimbleGen SeqCap EZ Choice Library (NimbleGen; Roche, Madison, WI), all exons and exon-intron boundaries (plus 20 base pairs at each end) of 140 genes were captured for a final targeted region of 1.05 Mb. The mean depth ranged from 774 to 1776 across the 140 genes in the panel and 852 and 1366 across *PKD1* and *PKD2* genes. Mutations were identified in 99% of the validation cohort (116 patients), while a genetic cause was identified in 78% of the cystic (n = 207), and 62% of the glomerular (n = 98), cases.[30] Specifically, among the patients in diagnostic cohort, the mutations were found in 88% (89 out of 102) of the cohort including ADPKD (n = 87), autosomal dominant polycystic liver disease (ADPLD) (n = 1), and oral-facial-digital syndrome (OFD) (n = 1). Three patients with ADPKD were due to copy number variants.[30] Two other studies of targeted resequencing in ADPKD using DNA capture have been recently published. Trujillano et al. performed resequencing of a panel of target genes including *PKD1* and *PKD2* enriched by in-solution hybridization using a custom-designed probe set (NimbleGen SeqCap EZ Choice, Roche, Basel, Switzerland).[24] Using this approach, they identified pathogenic *PKD1* and *PKD2* mutations in 35/36 (97.2%) of previously genetically resolved and 10/12 (83.3%) of previously ungenotyped patients. The identified mutations include 24 mutations within the *PKD1* duplicated region and one large gene deletion each in *PKD1* and *PKD2*. Despite these promising results, ~5% of the *PKD1* and *PKD2* coding regions had less then 20× of read depth and three very GC-rich exons (i.e., *PKD1* exons 1 and 42; *PKD2* exon 1) were not captured.[24] Similarly, Eisenberger et al. performed resequencing on a panel of target genes including *PKD1* and *PKD2* using a custom-designed probe set (NimbleGen SeqCap EZ choice, Roche, Basel, Switzerland) with 12 times more probes assigned to GC-rich regions, a process known as rebalancing.[29] They reported a 99% positive diagnostic rate based on 681/683 of Sanger-verified sequencing variants (including pathogenic mutations) from 55 patient samples. In general, this study provided better coverage and read depth for the GC-rich regions of *PKD1* and *PKD2* than the Trujillano et al. study. However, 1%–1.5% of the *PKD1* coding regions had <20× of read depth and some very high-GC exons were not well captured, again posing a small risk of false-negative calls. All these studies included probes that allowed for parallel capture of the duplicated regions in the six pseudogenes (PKD1P1-P6) for sequencing error identification. To compensate, accurate mapping of longer 150 bp paired-end reads to the duplicated region of *PKD1* and its pseudogenes was critical using sophisticated bioinformatics analyses; repeated analyses using different alternative/reference allele ratio cutoffs (i.e., down to 8%); and follow-up validation using locus-specific LR-PCR and Sanger sequencing for variant confirmation were also needed.

In our lab, we developed a cystic kidney disease gene panel composed of 49 genes associated with multicystic kidney or liver disease using Agilent's SureSelect capture technology (Agilent Biosystems, Santa Clara, California, USA). The probes were designed using Agilent SureDesign, a web-based design portal to capture the regions of interest including all protein coding regions and intron-exon boundaries (±200 bp) of our selected 49 genes. Inclusion of intronic sequence improves coverage of putative and canonical splice sites as well as inclusion of polymorphic intronic sequence for haplotype-based analyses. High-quality genomic DNA is sent to The Centre for Applied Genomics (Sick Children's Hospital, Toronto, Canada) for exome library preparation using Agilent SureSelect XT HS custom capture Library Preparation using the Agilent Bravo Automation System and paired-end sequencing on an Illumina HiSeq 2500 platform. In brief, 500 ng of genomic DNA was fragmented to 200 bp, on average, using a Covaris LE220 instrument. Sheared DNA was end-repaired and the 3′ ends adenylated prior to ligation of molecular-barcoded adapters with overhang-T. The genomic library was amplified by PCR using 9 cycles and hybridized with biotinylated probes that targeted the region of interest; the enriched targeted libraries were amplified by an additional 12 cycles of PCR. Exomic libraries were validated on a Bioanalyzer 2100 DNA High Sensitivity chip (Agilent Technologies) for size and by qPCR using the Kapa Library Quantification Illumina/ABI Prism Kit protocol (KAPA Biosystems) for quantities. Exome libraries were pooled and sequenced with the TruSeq SBS sequencing chemistry on a HiSeq 4000 platform following Illumina's recommended protocol, to generate paired-end data of 150 bases.

Illumina provided software bcl2fastq was used to convert the per-cycle BCL base call files generated by the Illumina Sequencing systems to standard primary sequencing output in FASTQ format. During the conversion step, demultiplexing of samples was performed by sorting reads according to the molecular barcode and removing the associated oligonucleotide sequence from each read. The program fastQC was used to assess the quality of the raw data. Prior to alignment, the reads were processed using Agilent Genomics NextGen Toolkit (AgeNT) v4.0.1. SureCall Trimmer was used to remove adapters and to trim low-quality bases. Preprocessed reads were aligned to the reference genome (GRCh37/hg19) using the BWA-mem alignment tool (v0.7.8). Files generated in SAM format by BWA for each sample were analyzed using the Agilent software SureCall (v. 4.0.1). VCF files generated by SureCall run were annotated using a custom pipeline developed in-house.

In summary, a gene panel approach provides the possibility of sequencing of a smaller fraction of the genome, targeted to regions of interest including *PKD1* and *PKD2*, across a large number of patients. Our method is a cost- and time-effective tool for initial screening of patients with cystic kidney diseases. The possibility of identifying large gene rearrangements, copy number variants, and mosaic mutations without additional testing further underlines the advantages of this approach. However, our approach, along with other exome or gene panel–based sequencing approaches, remains limited by (1) less versatility for unique probe design due to high DNA sequence identity to the pseudogenes, (2) lower capture efficiency in GC-rich regions, (3) a high proportion of "off-target sequence capture" from the pseudogene regions, (4) reduced ability to detect mosaicism due to the low alternate-to-reference allele ratio resulting from contaminating pseudogene sequences, and (5) difficulty identifying copy number variation (CNV) when the breakpoint is not found in targeted sequence.

14.5.3 Whole-Genome Sequencing

Whole-genome sequencing (WGS) provides the most comprehensive coverage of all forms of genetic variation within exons, introns, and intergenic regions across the entire genome. WGS provides higher diagnostic yield than DNA capture methods and can detect all types of disease-causing variants including single-nucleotide variants, small indels, and CNVs using a single test.[33] Variants within all potentially cystogenic genes are included in a single sequencing experiment. WGS is a less labor-intensive method from a laboratory perspective, using only DNA extraction and library preparation to sequence the entire genome. However, a shorter laboratory workflow comes at the expense of increased computational requirements, as bioinformatics is used to obtain the sequence of interest instead of biochemical capture. WGS does avoid capture bias that arises from targeted sequencing and provides more uniform coverage, which appears to overcome the *PKD1* pseudogene issues.[34] The consistent coverage, the inclusion of all sequence allowing for identification of intergenic breakpoints, and bioinformatic advancements, improves the detection CNVs and structural variants and also decreases the minimum average depth required to achieve 99% variant detection sensitivity (~30× for WGS compared with ~80–100× for WES).[34,35]

Mallawaarachchi et al. reported the first use of WGS to diagnose ADPKD.[34] Using Illumina HiSeq X and 150 base pair paired-end reads, their experience with WGS provided a positive diagnostic rate of 86% in 24 out of 28 patients with ADPKD.[34] Mean depth coverage for all coding exons of *PKD1* and *PKD2* was similar (~30× read depths) in this study and on average, WGS provided 97.3 and 98.5% mean coverage of the *PKD1* and *PKD2* coding regions, respectively, with at least 15× read depth. However, the read depth was reduced for GC-rich exons 1, 42, and 43 of *PKD1* and exon 1 of *PKD2*. They found 17 disease-causing variants in *PKD1*, and 71% of these variants (n = 12) were within the duplicated region (exons 1–33). The 17 detected *PKD1* variants consisted of splice site (n = 2), frameshift (n = 7), nonsense (n = 6), and missense (n = 1) mutations as well as a large deletion (2199 bp: deleting exons 31–34).

In our laboratory, 500 ng to 1 ug of genomic DNA is submitted to the core sequencing facility at the Centre for Applied Genomics (Sick Children's Hospital, Toronto, Canada) for genomic library preparation and whole-genome sequencing. DNA samples are quantified using Qubit High Sensitivity Assay and sample purity checked by NanoDrop OD260/280 ratio. One-hundred nanograms of DNA is used as input material for library preparation with the Illumina TruSeq Nano DNA Library Prep Kit following the manufacturer's protocol. In brief, DNA is fragmented to 350 bp using sonication on a Covaris LE220 instrument; fragmented DNA is end-repaired, A-tailed and TruSeq Illumina adapters with overhang-T are added; and libraries are validated for size and absence of primer dimers using a Bioanalyzer DNA High Sensitivity chip and quantified by qPCR using Kapa Library Quantification Illumina/ABI Prism Kit protocol (KAPA Biosystems). Validated libraries are pooled in equimolar quantities and paired-end sequenced on an Illumina HiSeq X platform following Illumina's recommended protocol to generate paired-end reads of 150-bases in length.

Illumina provided software bcl2fastq (v2.20) is used to convert the per-cycle BCL base call files to standard sequencing output in FASTQ format. Reads are aligned to the reference human genome (GRCh37) using BWA mem (v0.7.12). Duplicate reads are marked using Picard Tools (v2.5.0). Indel realignment, base quality score recalibration and germline variant detection using HaplotypeCaller are performed using GATK 3.7 following recommended best practices. Variants are annotated using an ANNOVAR-based pipeline. ERDS (v1.1) and CNVnator (v0.3.2) call CNVs and a custom annotation and prioritization pipeline is used define rare CNVs.[35] Manta (v0.29.6) is used to call structural variants.

14.6 EVALUATION OF PUTATIVE PATHOGENIC VARIANTS

Regardless of the sequencing technology employed, putative variants in cystogenic genes will be identified that will require evaluation for pathogenicity. Quality control starts with examination of the electropherogram in capillary sequencing, or coverage, read depth, and sequencing quality scores in next generation sequencing technology. A BLAST search should be performed to determine if the variant sequence is found within pseudogene sequence, and thus likely due to a sequencing error. Next, identification of presence and frequency of the mutation in published control and case populations, cosegregation of the variant through family members when available, and bioinformatic prediction is required. Finally, creation of cell- or animal-based models can be considered.

14.6.1 Allele Frequency in Population Databases (ExAC, gnomAD, BRAVO)

Publicly available databases of whole-genome and whole-exome sequenced participants are now available. The Exome Aggregation Consortium (ExAC), which subsequently rolled into the Genome Aggregation Database (gnomAD), is an international coalition of investigators and housed at the Broad Institute of Harvard and MIT, which now includes 125,748 exome and 15,708 genome sequences (gnomad.broadinstitute.org).[36] BRAVO includes 62,784 whole-genome sequences, and is shared by the National Heart, Lung, and Blood Institute (NHLBI) Trans-Omics for Precision Medicine (TOP-MED) program (bravo.sph.umich.edu/freeze5/hg38). Both resources provide population frequencies of all variants observed in participants sequenced in various adult-onset, disease-specific, and population sequencing genetic studies. Participants are checked to be unrelated, and those with a pediatric-onset disease in themselves or a first-degree relative were excluded. While participants with ADPKD are likely present in the sequenced population, identification of the population allele frequency of a putative mutation identified in an ADPKD case is the first step of analysis. As the observed prevalence of ADPKD is about 1 in 1000, and over 1000 different mutations have been observed causing ADPKD with no single variant representing a significant proportion of cases, a variant causative of ADPKD cannot have a population allele frequency greater than 0.0001 (or 0.01%).

14.6.2 ADPKD Mutation Databases

Following assessment of a putative variant in the general population, databases that contain observed causative mutations should be queried. The Mayo PKD database (pkdb.mayo.edu) contains published and unpublished reports of mutations in *PKD1* and *PKD2*. Each mutation is assigned a clinical significance of either "Definitely Pathogenic," "Highly Likely Pathogenic," "Likely Pathogenic," "Likely Hypomorphic," "Likely Neutral," or "Indeterminant." Presence in the Mayo PKD database with a "Definitely" or "Highly Likely" pathogenic classification is very suggestive that the putative mutation is indeed causative of the ADPKD. However, we recently reported that the cumulative prevalence of "Likely Pathogenic" variants in Mayo PKD database is higher (1 in 576) than the prevalence of ADPKD identified through epidemiological studies, suggesting either under-recognition of mild disease or incomplete penetrance of the mutations.[1] Disease-specific mutation databases are continuing to improve in quality with careful curation and the increasing quantity of sequence data availability, but as they are based on sequencing of cases only, they are likely to overestimate the function of rare variants. Additional data on the putative variants can be found in ClinVar (www.ncbi.nlm.nih.gov/clinvar/) and the human gene mutation database (HGMD; http://www.hgmd.cf.ac.uk).

14.6.3 Bioinformatic Prediction

Truncating mutations (nonsense, frameshift, large deletions, or canonical splice site mutations) are generally considered pathogenic in ADPKD genes, but bioinformatic prediction algorithms can be helpful for assessing missense and noncanonical splicing variants.[37] At least 15 different bioinformatic algorithms have been developed to assist in the prediction of the pathogenicity of rare variants. Popular tools include SIFT, PolyPhen2, MutationTaster, MutationAssessor, CADD, VEST, LRT, FATHMM, MetaSVM, METALR, GERP++, DANN, Eigen score, Max Entropy Scan (MES), and Human Splice Finder (HSF). Bioinformatic prediction algorithms generally work by examining evolutionary conservation of the mutated amino acid across many species, physiochemical properties of the induced amino acid change, database annotations, observed allele frequencies, and potential protein structural changes. Many algorithms provide quantitative outputs with user-defined thresholds for calling variants "pathogenic." Newer algorithms (such as REVEL, VEST3, MetaLR, MetaSVM, Condel, Mcap, Eigen, and CADD) appear to have more reliable performance and are robust to underlying inheritance pattern and technical artifacts.[38] The area under the receiver operator curve (AUC) for most algorithms compared to ClinVar assigned pathogenicity is about 0.8, with sensitivity and specificity ranging around 50%–90%.[39] Ensuring agreement of multiple bioinformatics algorithms would be expected to increase specificity, but also reduces sensitivity. Overall, bioinformatics algorithms tend to overcall pathogenicity, and we recently found bioinformatic-predicted pathogenic variants in *PKD1* and *PKD2* had a cumulative prevalence of 1 in 222 in BRAVO and gnomad, above the epidemiologically observed prevalence of ADPKD.[1] ANNOVAR (http://annovar.openbioinformatics.org/en/latest/) is an efficient tool that utilizes many bioinformatic prediction tools and annotates each variant.[37]

14.6.4 Familial Cosegregation

After identification of a putative variant in a proband, targeted sequencing of the identified mutation to verify its presence in affected family members and absence in unaffected family members can improve confidence that the variant is truly causative. Sequencing of a previously identified familial variant is faster and easier than sequencing the whole gene. As each first-degree family member (i.e., sibling, parent, or child) would have a 50% chance of carrying a putative variant, five family members are required to reach a statistical significance level of 0.05. More distantly related affected family members (i.e., cousins) have a greater effect on your confidence the variant is indeed causal, as they share a smaller proportion of their genome (12.5%).

14.6.5 Cell, Organoid, and Animal Models

Functional assays evaluating the effect of the mutation on the protein product are the gold standard, and clustered regularly interspaced short palindromic repeat (CRISPR) systems may improve access to such tests in the future. Mutations have been inserted into cultured renal tubular epithelial cells,[40] stem cell derived organoids,[41] and mouse and rat models.[42] While these systems provide the opportunity to learn more about ADPKD pathophysiology, development of models with all of >1000 mutations observed in ADPKD will be problematic without the development of more high-throughput techniques.

14.7 DIFFERENTIAL DIAGNOSIS FOR PATIENTS WITH NO MUTATION DETECTED

After sequencing approximately 10%–20% of patients will remain with no mutation detected, due to either missing a *PKD1* or *PKD2* variant, or the presentation is the result of different cystic disease. Up to 5% of ADPKD is due to a large heterozygous deletion or copy number variation that are missed with sequencing. Small deletions, either in-frame or causing frameshifts, are easily recognized in the sequencing results. However, the breakpoints for large deletions are often located in an intron and not sequenced, and the single remaining copy of the gene provides the template DNA for sequencing, with no mutation detected. Such deletions may be identified using whole-genome sequencing if copy number variation analysis is specifically performed that evaluates the read depth across the genome and attempts to search for break points. Multiplex ligation dependent probe amplification (MLPA) is the method of choice to specifically test for large duplications or deletions in *PKD1* or *PKD2*.[25] Probes targeted to each exon and to control regions of the genome are amplified and quantified relative to each other. A relative deficiency of one or more exon probes provides evidence of a deletion. An additional explanation for not identifying a responsible mutation is somatic mosaicism, where multiple cell populations with different genotypes occur within a single person, and the mutation isn't present in the cells selected for genotyping.[7,43] Intronic promotor or atypical splicing variants may affect the quantity transcript or functional polycystin-1 protein and go unrecognized. Mutations in *GANAB* or *DNAJB11* will also explain a small proportion of patients with no mutation detected. Finally, phenocopies for ADPKD, such as autosomal dominant tubulointerstitial nephritis or nephronopthesis may be recognized by the experienced clinician, but could be misdiagnosed as ADPKD. A differential diagnosis of other diseases that include renal cysts is in Table 14.2.

14.8 CONCLUDING REMARKS

As sequencing costs continue to fall and carefully phenotyped cohorts grow, we will continue to gain greater insight into the genetic contribution to cystic kidney disease progression. Currently, genetic testing is clinically indicated when the diagnosis is in question, such as with a lack of family history or when an early diagnosis is required, but clinically useful information can be gained in scenarios with unusual clinical presentations. Genetic testing in combination with detailed clinical phenotyping, imaging, and biomarker evaluation currently forms a precision medicine approach to ADPKD.

ACKNOWLEDGMENTS

M.B.L. received a Ben Lipps Postdoctoral Fellowship from the American Society of Nephrology and is a fellow of the Krescent Program, a national kidney research training partnership of the Kidney Foundation of Canada, the Canadian Society of Nephrology, and the Canadian Institutes of Health Research. The Canadian Institutes of Health Research Strategy for Patient-Oriented Research in Chronic Kidney Disease program grant to Y.P. supported this work. M.B.L. and Y.P. received compensation for participating in advisory and consultancy boards with Otsuka Pharmaceuticals. There are no additional competing interests or disclosures.

REFERENCES

1. Lanktree, M. B. et al. Prevalence estimates of polycystic kidney and liver disease by population sequencing. *J. Am. Soc. Nephrol.* **29**, 2593–2600 (2018).
2. Chebib, F. T. and Torres, V. E. Autosomal dominant polycystic kidney disease: Core Curriculum 2016. *Am. J. Kidney Dis.* **67**, 792–810 (2016).
3. Lanktree, M. B. and Chapman, A. B. New treatment paradigms for ADPKD: Moving towards precision medicine. *Nat. Rev. Nephrol.* **13**, 750–768 (2017).
4. Hwang, Y.-H. et al. Refining genotype-phenotype correlation in autosomal dominant polycystic kidney disease. *J. Am. Soc. Nephrol.* **27**, 1861–1868 (2016).
5. Irazabal, M. V. et al. Imaging classification of autosomal dominant polycystic kidney disease: A simple model for selecting patients for clinical trials. *J. Am. Soc. Nephrol.* **26**, 160–172 (2015).
6. Pei, Y. et al. Imaging-based diagnosis of autosomal dominant polycystic kidney disease. *J. Am. Soc. Nephrol.* **26**, 746–753 (2015).
7. Iliuta, I.-A. et al. Polycystic kidney disease without an apparent family history. *J. Am. Soc. Nephrol.* **28**, 2768–2776 (2017).
8. Pei, Y. et al. Unified criteria for ultrasonographic diagnosis of ADPKD. *J. Am. Soc. Nephrol.* **20**, 205–212 (2009).
9. Lanktree, M. B., Iliuta, I.-A., Haghighi, A., Song, X. and Pei, Y. Evolving role of genetic testing for the clinical management of autosomal dominant polycystic kidney disease. *Nephrol. Dial. Transplant.* (2018). doi: 10.1093/ndt/gfy261.
10. Cornec-Le Gall, E. et al. The value of genetic testing in polycystic kidney diseases illustrated by a family with PKD2 and COL4A1 mutations. *Am. J. Kidney Dis.* **72**, 302–308 (2018).

11. Heyer, C. M. et al. Predicted mutation strength of nontruncating PKD1 mutations aids genotype-phenotype correlations in autosomal dominant polycystic kidney disease. *J. Am. Soc. Nephrol.* **27**, 2872–2884 (2016).
12. Song, X., Haghighi, A., Iliuta, I.-A. and Pei, Y. Molecular diagnosis of autosomal dominant polycystic kidney disease. *Expert Rev. Mol. Diagn.* **17**, 885–895 (2017).
13. Song, X. et al. Systems biology of autosomal dominant polycystic kidney disease (ADPKD): Computational identification of gene expression pathways and integrated regulatory networks. *Hum. Mol. Genet.* **18**, 2328–2343 (2009).
14. Porath, B. et al. Mutations in GANAB, encoding the Glucosidase IIα Subunit, cause autosomal-dominant polycystic kidney and liver disease. *Am. J. Hum. Genet.* **98**, 1193–1207 (2016).
15. Cornec-Le Gall, E. et al. Monoallelic mutations to DNAJB11 cause atypical autosomal-dominant polycystic kidney disease. *Am. J. Hum. Genet.* **102**, 832–844 (2018).
16. Cornec-Le Gall, E., Torres, V. E. and Harris, P. C. Genetic complexity of autosomal dominant polycystic kidney and liver diseases. *J. Am. Soc. Nephrol.* **29**, 13–23 (2018).
17. Cornec-Le Gall, E. et al. The PROPKD score: A new algorithm to predict renal survival in autosomal dominant polycystic kidney disease. *J. Am. Soc. Nephrol.* **27**, 942–951 (2016).
18. Perrichot, R. et al. Novel mutations in the duplicated region of PKD1 gene. *Eur. J. Hum. Genet.* **8**, 353–359 (2000).
19. Rossetti, S. et al. A complete mutation screen of the ADPKD genes by DHPLC. *Kidney Int.* **61**, 1588–1599 (2002).
20. Tan, Y.-C. et al. A novel long-range PCR sequencing method for genetic analysis of the entire PKD1 gene. *J. Mol. Diagn.* **14**, 305–313 (2012).
21. Audrézet, M.-P. et al. Autosomal dominant polycystic kidney disease: Comprehensive mutation analysis of PKD1 and PKD2 in 700 unrelated patients. *Hum. Mutat.* **33**, 1239–1250 (2012).
22. Rossetti, S. et al. Identification of gene mutations in autosomal dominant polycystic kidney disease through targeted resequencing. *J. Am. Soc. Nephrol.* **23**, 915–933 (2012).
23. Qi, X.-P. et al. Genetic diagnosis of autosomal dominant polycystic kidney disease by targeted capture and next-generation sequencing: Utility and limitations. *Gene* **516**, 93–100 (2013).
24. Trujillano, D. et al. Diagnosis of autosomal dominant polycystic kidney disease using efficient PKD1 and PKD2 targeted next-generation sequencing. *Mol Genet Genomic Med.* **2**, 412–421 (2014).
25. Liu, B. et al. Identification of novel PKD1 and PKD2 mutations in a Chinese population with autosomal dominant polycystic kidney disease. *Sci. Rep.* **5**, 17468 (2015).
26. Li, H. and Durbin, R. Fast and accurate short read alignment with Burrows-Wheeler transform. *Bioinformatics.* **25**, 1754–1760 (2009).
27. McKenna, A. et al. The Genome Analysis Toolkit: A MapReduce framework for analyzing next-generation DNA sequencing data. *Genome Res.* **20**, 1297–1303 (2010).
28. Ali, H. et al. PKD1 duplicated regions limit clinical utility of whole exome sequencing for genetic diagnosis of autosomal dominant polycystic kidney disease. *Sci. Rep.* **9**, 4141 (2019).
29. Eisenberger, T. et al. An efficient and comprehensive strategy for genetic diagnostics of polycystic kidney disease. *PLoS One* **10**, e0116680 (2015).
30. Bullich, G. et al. A kidney-disease gene panel allows a comprehensive genetic diagnosis of cystic and glomerular inherited kidney diseases. *Kidney Int.* **94**, 363–371 (2018).
31. Mertes, F. et al. Targeted enrichment of genomic DNA regions for next-generation sequencing. *Brief. Funct. Genomics* **10**, 374–386 (2011).

32. Xu, C., Nezami Ranjbar, M. R., Wu, Z., DiCarlo, J. and Wang, Y. Detecting very low allele fraction variants using targeted DNA sequencing and a novel molecular barcode-aware variant caller. *BMC Genomics.* **18**, 5 (2017).
33. Lionel, A. C. et al. Improved diagnostic yield compared with targeted gene sequencing panels suggests a role for whole-genome sequencing as a first-tier genetic test. *Genet. Med.* **20**, 435–443 (2017).
34. Mallawaarachchi, A. C. et al. Whole-genome sequencing overcomes pseudogene homology to diagnose autosomal dominant polycystic kidney disease. *Eur. J. Hum. Genet.* **24**, 1584–1590 (2016).
35. Trost, B. et al. A comprehensive workflow for read depth-based identification of copy-number variation from whole-genome sequence data. *Am. J. Hum. Genet.* **102**, 142–155 (2018).
36. Karczewski, K. J. et al. Variation across 141,456 human exomes and genomes reveals the spectrum of loss-of-function intolerance across human protein-coding genes. *bioRxiv* 531210 (2019). doi: 10.1101/531210.
37. Wang, K., Li, M. and Hakonarson, H. ANNOVAR: Functional annotation of genetic variants from high-throughput sequencing data. *Nucleic Acids Res.* **38**, e164 (2010).
38. Ghosh, R., Oak, N. and Plon, S. E. Evaluation of in silico algorithms for use with ACMG/AMP clinical variant interpretation guidelines. *Genome Biol.* **18**, 225 (2017).
39. Li, J. et al. Performance evaluation of pathogenicity-computation methods for missense variants. *Nucleic Acids Res.* **46**, 7793–7804 (2018).
40. Hofherr, A. et al. Efficient genome editing of differentiated renal epithelial cells. *Pflugers Arch.* **469**, 303–311 (2017).
41. Cruz, N. M. et al. Organoid cystogenesis reveals a critical role of microenvironment in human polycystic kidney disease. *Nat. Mater.* **16**, 1112–1119 (2017).
42. Nagao, S., Kugita, M., Yoshihara, D. and Yamaguchi, T. Animal models for human polycystic kidney disease. *Exp. Anim.* **61**, 477–488 (2012).
43. Tan, A. Y. et al. Autosomal dominant polycystic kidney disease caused by somatic and germline mosaicism. *Clin. Genet.* **87**, 373–377 (2015).

Index

F

G

H

N

Q

R

U

V